Lecture Notes in Electrical Engineering

Volume 387

About this Series

"Lecture Notes in Electrical Engineering (LNEE)" is a book series which reports the latest research and developments in Electrical Engineering, namely:

- Communication, Networks, and Information Theory
- Computer Engineering
- Signal, Image, Speech and Information Processing
- Circuits and Systems
- Bioengineering

LNEE publishes authored monographs and contributed volumes which present cutting edge research information as well as new perspectives on classical fields, while maintaining Springer's high standards of academic excellence. Also considered for publication are lecture materials, proceedings, and other related materials of exceptionally high quality and interest. The subject matter should be original and timely, reporting the latest research and developments in all areas of electrical engineering.

The audience for the books in LNEE consists of advanced level students, researchers, and industry professionals working at the forefront of their fields. Much like Springer's other Lecture Notes series, LNEE will be distributed through Springer's print and electronic publishing channels.

More information about this series at http://www.springer.com/series/7818

Ping Jack Soh · Wai Lok Woo
Hamzah Asyrani Sulaiman
Mohd Azlishah Othman · Mohd Shakir Saat
Editors

Advances in Machine Learning and Signal Processing

Proceedings of MALSIP 2015

Springer

Editors
Ping Jack Soh
Universiti Teknikal Malaysia Melaka
Durian Tunggal, Melaka
Malaysia

Mohd Azlishah Othman
Universiti Teknikal Malaysia Melaka
Durian Tunggal, Melaka
Malaysia

Wai Lok Woo
Newcastle University
Singapore
Singapore

Mohd Shakir Saat
Universiti Teknikal Malaysia Melaka
Durian Tunggal, Melaka
Malaysia

Hamzah Asyrani Sulaiman
Universiti Teknikal Malaysia Melaka
Durian Tunggal, Melaka
Malaysia

ISSN 1876-1100 ISSN 1876-1119 (electronic)
Lecture Notes in Electrical Engineering
ISBN 978-3-319-32212-4 ISBN 978-3-319-32213-1 (eBook)
DOI 10.1007/978-3-319-32213-1

Library of Congress Control Number: 2016939910

Printed on acid-free paper

This Springer imprint is published by SpringerNature
The registered company is Springer International Publishing AG Switzerland

Preface

The 2015 International Conference on Machine Learning and Signal Processing (MALSIP) was organized by the Malaysia Technical Scientist Association (MALTESAS) together with Newcastle University (Singapore campus). For this first installment, it was held in the historical city of Melaka, Malaysia, from June 12 to June 14, 2015. The conference chiefly focused on advances in and innovative technologies related to machine learning, signal processing, and their applications. The event, which drew participants from around the globe, was intended to bring together researchers and scientists working on signal processing and machine learning topics, offering them a forum to discuss recent advances in their research areas, exchange ideas, and lay the groundwork for future collaborations. Of the 95 manuscripts received, only 27 were ultimately accepted for publication in these conference proceedings.

The organizers of MALSIP 2015 were also delighted to have Professor Dr. M. Iqbal Saripan from Universiti Putra Malaysia (UPM) as a keynote speaker. He presented his work on "Digital Image Processing Applications in Nuclear Medical Imaging Systems," discussing both single modality and hybrid modality approaches in nuclear medical imaging (NMI), as well as the implementation of digital image processing to enable the use of NMI in medical diagnostics.

In total, 357 experts in signal processing and machine learning from various countries across the globe were selected to act as reviewers of manuscripts for MALSIP 2015, covering a wide range of topics relating to signal processing techniques and their applications.

This book highlights the latest technologies and trends in machine learning and signal processing techniques, offering a valuable resource for researchers and practitioners alike.

The editors wish to express their gratitude to the Malaysia Technical Scientist Association (MALTESAS) and Newcastle University (Singapore campus), as well as Event Management, Skaievent Technovation and Narujaya Consultant Enterprise, for organizing and supporting this very successful conference.

Ping Jack Soh
Wai Lok Woo
Hamzah Asyrani Sulaiman
Mohd Azlishah Othman
Mohd Shakir Saat

Contents

User Identification of Keystroke Biometric Patterns with the Cognitive RAM Weightless Neural Net

Weng Kin Lai, Beng Ghee Tan, Ming Siong Soo and Imran Khan

Abstract A user identification system which matches the keystroke dynamics of the users with the Cognitive RAM (CogRAM) weightless neural net is discussed in this paper. The keystroke patterns are made up of a common password for all users. While there are several common approaches to represent the users' keystroke patterns, the approach adopted here is based on the force applied to each key. Effectively, they will then constitute a fixed length passkey. In addition, the system was developed based on an 8-bit AVR enhanced, RISC microcontroller. From the experimental results obtained, it can be seen that the identity of the users can be successfully matched just from their keystroke biometric patterns alone.

Keywords Biometrics · Keystroke dynamics · Artificial neural networks

1 Introduction

User *identification* and *verification* are two common but different applications of biometric technologies. While verification relates to matching or verifying the patterns against a single user's stored identity, identification on the other hand, involves finding the one unique identity amongst the many stored identities. Essentially, identification seeks to determine the user's identity whereas verification attempts to prove the claimed identity. Although a variety of authentication devices may be used to verify a user's identity, passwords remain the most preferred method especially when the keyboard is the preferred data entry device, due to both the long history of the use of this mechanism to gain access as well as the fact that it is still relatively inexpensive compared to other more sophisticated solutions. However, like most modern technologies, unless it is used correctly, the level of security provided by

W.K. Lai (✉) · B.G. Tan · M.S. Soo
Tunku Abdul Rahman University College, Kuala Lumpur, Malaysia
e-mail: laiwk@acd.tarc.edu.my

I. Khan
IIUM, Kuala Lumpur, Malaysia

P.J. Soh et al. (eds.), *Advances in Machine Learning and Signal Processing*,
Lecture Notes in Electrical Engineering 387, DOI 10.1007/978-3-319-32213-1_1

passwords can be weak. Hence, multi-factor approaches are needed to extend and strengthen the security level that passwords provide. This reinforcement should be transparent and indiscernible to users and does not require any additional efforts while entering their account identity and password. However, in addition to different and personalized passwords for each user, the users are also known to have developed a unique typing style to enter their important account information. For example, a user may type the characters that constitute the password at different speeds. By leveraging on such differences, one can develop an approach that may be used to enhance the system's security with keystroke biometrics (or in some literature, typing biometrics) to reinforce password-authentication mechanisms. Since the earliest reported work on keystroke biometrics [1] in 1980, interest in this area has gradually gained momentum [2].

The remainder of this paper is organised as follows. The next section gives a short review of some of the significant work on keystroke biometrics. Section 3 discusses the hardware design of the keystroke biometrics user identification system which captures the force exerted by users on a numeric keypad. In Sect. 4, we describe the important details of weightless neural networks (WNN's) for pattern recognition. It also introduces the *Cognitive RAM* network (CogRAM) and details its architecture and learning rules. Section 5 discusses the major design issues as well as the important aspects of the experiments and the results obtained. Finally, in Sect. 6 we present some conclusions and potential areas for further work.

2 Related Work and Motivation

Previous research [3–6] has shown that it is possible to identify a user via his or her typing patterns in the form of keystroke biometrics which attempts to analyze a user's keystroke patterns. It is well known that an individual's keystroke biometrics pattern may be based on any combination of the following features [5, 7], viz. keystroke duration, latency between consecutive keystrokes or the force exerted on the keys. Some prior work has been done on the use of keystroke biometrics as a password hardening technique [8–11]. In addition to the duration between each pair of keystrokes, Obaidat and Sadoun [12] investigated the use of the holding time for each key pressed. The results reported of authenticating users based on just their keystroke have been encouraging. Nonetheless, these works were centered primarily on the common QWERTY computer keyboard. While the QWERTY layout has been used extensively in the past, the simple and inexpensive numeric keypad is gaining popularity. However, the typing style on the numeric keypad which is used in this work is significantly different than that on the QWERTY layout. Maxion and Killourhy [13] investigated keystroke biometrics on a numeric keypad based on 3 key features of hold time, diagram latency and diagram interval.

In this study, we will be investigating how we may identify an individual's keystroke biometric pattern on a numeric keypad based on the force (or amount of pressure) exerted on each key using a weightless neural network (WNN). WNNs do not use weighted connections between nodes but essentially uses a different kind of neuron model,—usually based on RAM memory devices [14]. One advantage of this approach is that training only involves a single pass through the data. In addition, these networks may easily be implemented in digital hardware, where they will be able to give better performance in terms of operating speed. And this is where we are motivated to investigate the use of this neural network model, as we would like to eventually develop a standalone biometrics system. The biometric sensors of the system are force sensitive resistors which were used to capture and translate the amount of force exerted on the keys into their equivalent electrical values, so as to give an accurate representation of the amount of force that each user applies while typing.

Two main authentication issues are emphasized during the overall design of the system, viz.

(a) the numeric password representing the normal passkey entered by the user, which consists of a combination of numeric keys of the appropriate length created by the user and saved in the system,
(b) the keystroke biometrics pattern associated with the user's password in the form of a "*typing template*". This is the second factor which will be analyzed by the weightless neural net.

3 System Design

3.1 Force Sensor

A major component of the system is the force sensor that measures the amount of force applied by the user. The force sensor used here are the ones made from a conductive polymer that changes its resistance linearly with respect to the amount of force applied to its surface [15]. Such a force sensitive resistor was then placed underneath the numeric keypad as shown in Fig. 1.

3.2 Microprocessor Design with Arduino

The force applied to the keys is transmitted to the force sensor and this is processed by the *Arduino Leonardo* micro controller (Fig. 2) into a form that is suitable for further processing. The *Arduino* has a total of 20 input-output pins, of which 7 can be used as PWM outputs and 12 as analogue inputs. Furthermore, it uses the *ATmega32u4* processor, a low-power CMOS 8-bit microcontroller based on the AVR enhanced RISC (*Reduced Instruction Set Computing*) architecture that has 32 8-bit general purpose working registers. The program and data in the AVR are

Fig. 1 The system showing the FSR in the *centre*, and the Arduino Leonardo on the *right*

Fig. 2 The Arduino Leonardo

stored in separate physical memory systems which would appear in different address spaces. Once the system has captured the force exerted by the user on each key, these keystroke biometric patterns will then be further processed by the weightless neural net to perform user identification.

4 Weightless Neural Networks

Pattern recognition tasks are often performed using artificial neural networks (ANNs). However, the multi-layer perceptron network (MLP) used are usually trained with the backpropagation learning rule [16] where the nodes within the different layers are modified. As the MLP is able to form complex decision regions rather than just hyperplanes, it is able to deal with nonlinear classification problems.

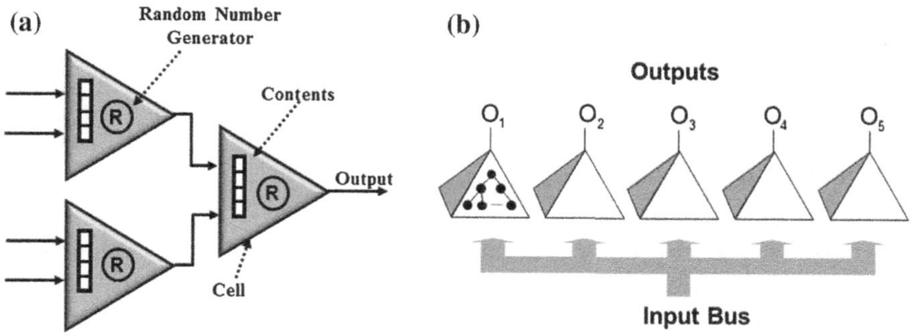

Fig. 3 Weightless neural network **a** The probabilistic logic neuron (PLN) **b** typical pyramid structure of the PLN

Each node in the first layer creates a decision region which when combined in the second layer, creates convex decision regions. These in turn will form additional concave regions when the previous convex regions are combined by the nodes in the third layer. Theoretically, any arbitrary region can be formed with just two hidden layers and sufficient hidden units.

Nevertheless, the backpropagation while it is popular has been regarded to be slow especially when dealing with problems with high input dimensionality and large data sets. A faster solution which requires one pass learning can be found in weightless neural networks (WNNs) that have been in existence since the 1960s. A comprehensive review on this can be found in Ludermir et al. [17]. The WNN used in the system described here is based on the work of Aleksander [18].

For ease of illustration, a simple example is shown in Fig. 3. Nevertheless, there can be many input layers with as many as 4 or even 8-bit input address lines in reality. The network is often triangular or pyramidal (for two-dimensional input data) in shape, because the network tends to map the usually large number of inputs to a smaller set of output nodes—as illustrated in Fig. 3b.

A sufficient number of pyramids, each with its own desired output value but a common input vector, have to be built if more than one output class is required. The binary input vector acts as an address generator during training, with all the cell's contents set to an undefined state at the beginning. As each undefined location is addressed, a random number of either a 1 or 0 will be generated with equal probability. Filho et al. [19] had proposed the goal seeking neural network which allows the output to generate an undefined value that is propagated towards the next layer whenever an undefined location is selected by the input address to a cell. It has an appropriate set of rules to deal with such occurrences. *Bowmaker and Coghill* further developed this idea into the *Deterministic Adaptive RAM Network* (DARN) [20] which has a better generalisation performance although the DARN's capabilities are still poorer than that of the MLP. Enhancements to DARN which resulted in the *Cognitive RAM Network* (CogRAM) however yielded much better results [21]. In CogRAM, all the registers in the network cells are first set to zero which is regarded as an undefined value but may become positive or negative during training. How the CogRAM learns and recalls are described in the next two sections.

4.1 Learning

The CogRAM learns by setting the desired output of every cell in the pyramid to be the same as the desired output of the pyramid. For example, if the desired output of the neurons at the input layer is 0, the contents of the addressed location are decremented by one. This is illustrated in Fig. 4 where the input pattern of '10_2' was addressed. The initial contents which had been *undefined* were then subsequently changed to become a negative value of −1; which effectively represents the desired output of '0'.

The outputs from each of the cells of this layer are then combined using the **Recall** function to form the address vector for the cells in the next layer. Details of this **Recall** function will be described in the next section. This is continued for every layer in a similar manner until the final output layer is reached. If there is more than one pyramid, then the same process is applied to each of the other pyramids. While this is repeated to each of the other pyramid, nevertheless, the entire training pattern set is only presented once.

4.2 Recall

In many ways, the **Recall** function is not unlike the earlier **Learning** function. Starting from the input layer, the content of each addressed location is interpreted as either U (undefined), 0 (zero), or 1 (one). When the contents of the cell is undefined (U), the *Goal Seeking Network* (GSN) approach will be used to compute the addressable set for the next layer. The same approach will also be used to identify the correct location for conflicting forward addresses as well. Just like the **Learning** function, **Recall** operates layer-by-layer, starting from the input layer until the output is reached. Figure 5a shows the CogRAM used to classify an arbitrary class

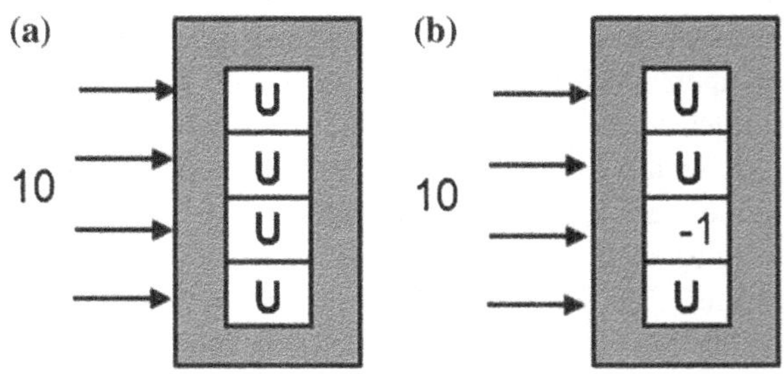

Fig. 4 An example of training an input pattern in CogRAM **a** Before training **b** desired output value '0' (or −1) stored in selected location ('10')

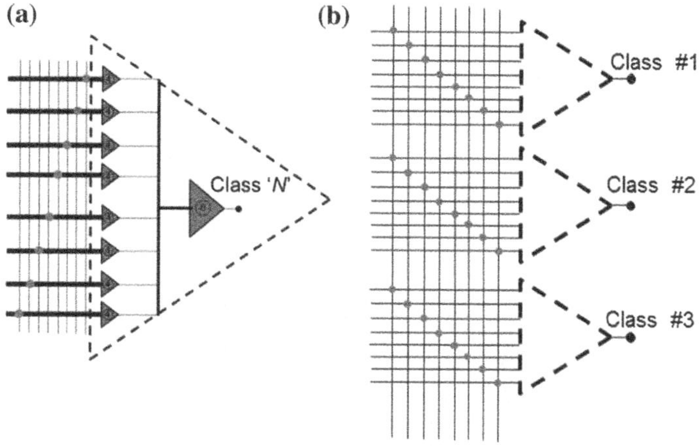

Fig. 5 Typical architecture of CogRAM **a** CogRAM for only one class **b** CogRAM for 3 classes

('*N*'). If there are 3 different classes instead, such a structure has to be repeated 3 times, as shown in Fig. 5b.

4.3 Improvements

We had observed that there are many cells that contain UNDEFINED values after the training phase. Basically this means that these cells were not addressed by the input patterns. Not surprisingly, this is more pronounced in data sets which have a smaller number of training patterns. Nevertheless, such locations would only contribute to a poor performance for the CogRAM, and hence such UNDEFINED cell locations needs to be reconciled to improve its performance. Yee and Coghill [21] did this by looking at each of the UNDEFINED locations in the output cell and computing the nearest address up to a certain Hamming distance to the defined location's address (either 0 or 1) for the whole training set. This Hamming distance should not be more than half of the cell size. They found that this resulted in a significant reduction of all the UNDEFINED values in the output cell of each class, contributing to a significant overall improvement in the classification results. However, because of the way we have represented the typing patterns, any two patterns with a Hamming Distance of 1 may not necessarily mean that the two patterns representing the typing force exerted onto the key are similar. For example, while the force for two patterns are only '*slightly*' dissimilar, their binary representation may be significantly different, and hence, not correct. Rather than modifying just one bit between these two patterns, we should instead look at the overall picture where the binary representation of the similar pattern should also be reflective of their actual physical value. Let's take the example for a force of 0.7 and 0.9 exerted on the same key. Their binary representation should share a high degree of similarity but if a simple linear transformation scheme had been

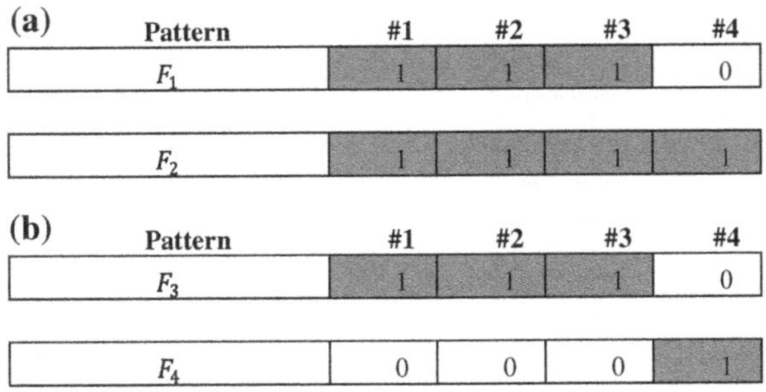

Fig. 6 Dissimilar and similar patterns **a** Dissimilar input patterns **b** similar input patterns

used, where 0.7 is represented as $1\,1\,1\,0_2$ in binary, and 0.9 would just be $1\,0\,0\,1_2$. The Hamming distance for a new pattern of $1\;1\;1\;1_2$ (representing 1.5) and $1\;1\;1\;0_2$ (representing a force of 0.7) would just be one bit, as illustrated in Fig. 6a. It may be clear that even though the Hamming distance may indicate they are quite similar, nevertheless, the actual force represented by these two values are not even remotely similar. Hence, a more accurate pattern representing similar forces would be one that would map back to the actual physical representation of the actual force. A similar pattern would then be one that is at the *next force level*(s) which, depending on the binary coding scheme, may not be close in the Hamming distance sense—as illustrated in Fig. 6b, which shows two similar patterns F_3 and F_4 and their binary coded representation.

5 Experimental Setup and Results

We have tested this new numerical keypad to authenticate the identity of the users based on their keystroke biometric signatures in our previous work [22]. Each user's keystroke biometric signature was based on their preferred individual password, each of which consists of 8 digits. In contrast, we are now using the keystroke biometric signatures generated from this numerical keypad based on the *SAME* 8 digit password for all users. Clearly, this will introduce a more challenging set of conditions as the individual identities are now based on nothing more than their keystroke biometric signatures. 10 samples were obtained from each of the 10 users, making a total of 100 sets of data. These users are familiar with the keyboard and have used the keyboard in their daily work but to minimise any inconsistent typing rhythm, they were given sufficient opportunities to familiarise themselves with this common password. Moreover the data were pre-processed using a standard sigmoid function (Eq. (1) to convert the biometric data to the range [0.0, 1.0] i.e.

$$f(x) = 1/1 + \exp(\lambda - x) \quad (1)$$

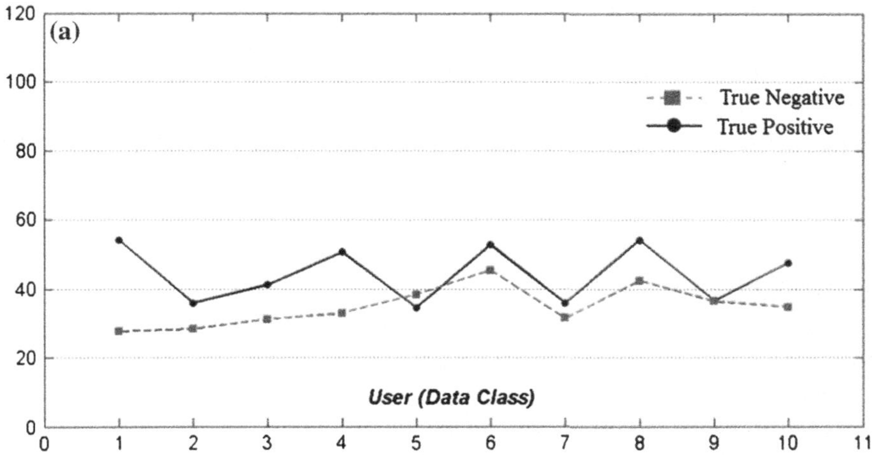

Fig. 7 Average True Positive and True Negative results **a** 3 samples used in training **b** 5 samples used in training

where λ, a constant was chosen arbitrarily from several values and the value of 10 was finally used in this investigation. One class of pressure patterns at a time was randomly selected for registration and to train the system. Correct identification of the test samples for the corresponding user contributes towards a *True Positive* (TP) measure. Furthermore, patterns from the remaining 9 classes were also used to test against this 1 class with the trained neural net so as to evaluate the CogRAM's ability to identify these invalid patterns. This would contribute to the *True Negative* (TN) measure. This validation strategy was repeated for a total of 100 times, each time with a randomly chosen set of training patterns. Figure 7 shows the average TP and TN values obtained with 3 and 5 samples taken per class for training over 100 runs. It can be seen that for most cases (except user 5) the True Positive identification rate is higher than the True Negative, meaning that correct identification takes place more often than incorrect identification. Furthermore, when 5 samples instead of 3 samples were used for training, the True Positive identification rates for some more users (such as users 2, 3 and 4) increases dramatically. This seems to suggest that these users have naturally more consistent typing styles which our system is able to identify.

6 Conclusions

Together with the electronic keyboard, passwords have been used for a long time as a simple and convenient means to gain access, especially for electronic systems. However, once the passwords are compromised, there cannot be any more protection from unauthorized entry. This is where keystroke biometrics may be used to strengthen this mode of access control. Nonetheless much of the work in this area has been done with the common QWERTY keyboard using either timing or pressure features. However, the typing style for most people on a numeric keypad is significantly different when compared with that on a QWERTY layout. In this paper we have discussed the development of a keystroke biometrics with a novel numeric keypad which can generate the keystroke biometric patterns based on the amount of pressure exerted on each key as the users entered the password. This keystroke biometric system was designed with a force sensitive resistor integrated into a common numeric keypad, and the pressure data from this force sensitive resistor was then acquired via the *Arduino* microcontroller. The keystroke biometric signatures generated from this numerical keypad were based on the *SAME* 8 digit password for all 10 subjects. The length of password was chosen as it resembles a telephone number and hence many of the subjects may find this very familiar and reassuring. The *Cognitive RAM* (CogRAM) weightless neural net was then used to process and authenticate the individuals based on these keystroke signatures. Preliminary results on this system had produced encouraging results for user authentication on a similar password. An advantage of using the CogRAM in such an application is that we can easily implement the neural net in hardware, and some work has been done in this direction by Nitish and his co-investigators [23]. This paper extends some of the earlier work for user identification with the CogRAM weightless neural net which had been used on different passwords [22]. Our future work would be directed into developing a standalone hardware implementation of the CogRAM to process the typing data as they are generated in situ. The experimental results from the keystroke biometrics data captured from the micro-controller system have shown that the system was able to correctly identify many of the users and this is especially true when we had used a higher number of keystroke patterns to train the CogRAM.

References

1. Gaines RS, Lisowski W, Press SJ, Shapiro N (1980) Authentication by keystroke timing: some preliminary results, Technical report R-2526-NSF, RAND Corporation Santa Monica, California, USA
2. Teh PS, Teoh ABJ, Yue S (2013) A Survey of Keystroke Dynamics Biometrics. Sci World J 2013:24
3. Joyce R, Gupta G (1990) Identity authentication based on keystroke latencies. Comm ACM 33:168–176
4. Monrose F, Rubin A (1997) Authentication via keystroke dynamics, Proceedings of the fourth ACM conference on computer and communications security, Zurich, Switzerland, p 48–56
5. De Ru W, Eloff J (1997) Enhanced password authentication through fuzzy logic. IEEE Expert 12:38–45
6. Obaidat MS, Sadoun B (1997) Verification of computer users using keystroke dynamics. IEEE Trans Syst Man Cybern 27:261–269
7. Tee ER, Selvanathan N (1996) Pin signature verification using wavelet transform. Malays J Comput Sci 9(2):71–78
8. Maisuria LK, Ong CS, Lai WK (1999) A comparison of artificial neural networks and cluster analysis for typing biometrics authentication, Proceedings of the international joint conference on neural networks (IJCNN), Washington, DC, 10–16 July 1999
9. Loy CC, Lim CP, Lai WK (2005) Pressure-based typing biometrics user authentication using the fuzzy ARTMAP neural network, Proceedings of the twelfth international conference on neural information processing (ICONIP 2005), Taipei, Taiwan ROC, p 647–652, 30 Oct–2 Nov 2005
10. Eltahir WE, Salami MJE, Ismail AF, Lai WK (2008) Design and evaluation of a pressure-based typing biometric authentication system. EURASIP J Inf Secur 2008 (345047):14. doi:10.1155/2008/345047
11. Yong S, Lai WK, Coghill GG (2004) Weightless neural networks for typing biometrics authentication, Proceedings of the 8th international conference on knowledge-based intelligent information & engineering systems, KES'2004, vol II, Wellington, New Zealand, p 284–293, 22–24 Sept 2004
12. Obaidat MS, Sadoun B (1996) Keystroke dynamics based authentication. In: Jain AK, Bolle R, Pankanti S (eds) Biometrics: personal identification in networked society. Chapter 10, p 213–229, Springer, U.S.
13. Maxion RA, Killourhy KS (2010) Keystroke biometrics with number-pad input, Proceedings of IEEE/IFIP international conference on dependable systems and networks p 201–209
14. Aleksander I, De Gregorio M, Franca FMG, Lima PMV, Morton H (2009) A brief introduction to weightless neural systems, proceedings of European symposium on artificial neural networks. Bruges, Belgium
15. Force sensing resistor. http://en.wikipedia.org/wiki/Force-sensing_resistor. Accessed on 18 Mar 2015
16. Rumelhart DE, Hinton GE, William RJ (1986) Learning internal representation by error propagation. In: Parallel distributed processing: explorations in the microstructures of cognition, vol 1, MIT Press, Cambridge, MA
17. Ludermir TB et al (1999) Weightless neural models: a review of current and past works. Neural Comput Surv 2:41–60
18. Kan WK, Aleksander I (1987) A probabilistic logic neuron network for associative learning, Proc. IEEE 1st Int Conf Neural Networks II:541–548, San Diego, June 1987
19. ECDBC F, Fairhurst MC, Bisset DL (1991) Adaptive pattern recognition using goal seeking neurons. Pattern Recogn Lett 12:131–138
20. Bowmaker RG, Coghill GG (1992) Improved recognition capabilities for goal seeking neuron. Electron Lett 28(3):220–221

21. Yee P, Coghill GG (2004) Weightless neural networks: a comparison between the discriminator and the deterministic adaptive RAM network, KES 2004:319–328
22. Lai WK, Tan BG, Soo MS, Khan I (2014) Two-factor user authentication with the CogRAM weightless neural net, World Congr Comput Intell (WCCI 2014), 6–11 July 2014, Beijing, China
23. Patel ND, Nguang SK, Coghill GG (2007) Neural network implementation using bit streams. IEEE Trans Neural Networks 18(5):1488–1504

Topology-Aware Mechanism to Improve Routing in Mobile Ad Hoc Networks

Baidaa Hamza Khudayer, Mohammad M. Kadhum and Tat-Chee Wan

Abstract Mobile Ad Hoc Networks (MANETs) consist of several nodes that are mobile in nature. The random and unpredictable node mobility in MANETs has negative impact on the network performance as it makes the network prone to frequent link breakages which increase packet loss and control overhead. Reliable MANET can be achieved through fast link failures detection that enables fast notification to be delivered to sources in order to make proper reaction. Therefore, the research work in this paper presents an effective topology-aware method, called Route Failure Prediction (RFP), which predicts the link breakage before it actually occurs in source routing networks. A node that utilizes RFP divides the area within transmission range into regions based on the signal strength received from neighbors. It calculates the changing rate of its location and update the upstream node if it is about to move out of the transmission range to avoid communication disruption. The performance of RFP was evaluated using ns-2 and compared to that of DSR. The experimental results gained from simulations were analyzed in terms of Packet Delivery Ratio (PDR), Control Overhead (COH), and

B.H. Khudayer (✉)
Department of Information Technology, Al Buraimi University College, Al Buraimi, Oman
e-mail: baidaa@buc.edu.om

B.H. Khudayer · M.M. Kadhum · T.-C. Wan
National Advanced IPv6 Centre (NAV6), Universiti Sains Malaysia (USM), 11800 Gelugor, Penang, Malaysia
e-mail: kadhum@usm.my; kadhum@cs.queensu.ca

T.-C. Wan
e-mail: tcwan@usm.my

M.M. Kadhum
Telecommunication Research Laboratory, School of Computing, Queen's University, Kingston, ON K7L 3N6, Canada

T.-C. Wan
School of Computer Sciences, Universiti Sains Malaysia (USM), 11800 Gelugor, Penang, Malaysia

P.J. Soh et al. (eds.), *Advances in Machine Learning and Signal Processing*,
Lecture Notes in Electrical Engineering 387, DOI 10.1007/978-3-319-32213-1_2

the End-to-End Delay with respect to the number of events of link breakage happened during the experiments. The numerical results showed that RFP outperforms DSR significantly and offers reliable data transfer between sources and destinations.

Keywords Routing protocols · DSR · Link failure detection · MANET

1 Introduction

Mobile Ad Hoc Networks (MANETs) are formed by group of mobile nodes that communicate with each other whether directly or over multihop intermediate nodes. MANETs provides instant networking between groups of people who move in different transmission ranges of each other [1]. Applications for MANET vary from university campus and conference to military and emergency operations. The number of users who use these applications ranges from few users in emergency cases to hundreds of users in campus and conference, to thousands of military applications users [2]. Routing process is one of the main issues that affect the performance of MANET seriously. The nodes mobility and frequent topology changes in MANET have negative impact on routing decision in MANET. In dynamic environment such as MANET, where the link failure is expected to happen frequently, it is important to have an alternate path to destination in order to avoid the communication disruption between source and destination [3]. Based on routing process, routing protocols developed for MANETs are categorized into three main categories [4]: (i) Proactive routing (table-driven) protocols (ii) Reactive routing (on-demand) protocols (iii) Hybrid routing protocols. Several studies on these protocols confirmed that, compared to proactive and hybrid routing protocols, on-demand routing protocols offer better performance in terms of control overhead and memory consumption reduction [5]. In addition, proactive and hybrid routing protocols do not offer satisfactory performance in dynamic environment with frequent topology changes due to slow detection of (and reaction to) route breakages as well as the unnecessary exchange of periodic updates [6].

2 Topology Changes Problem in MANET

The main challenge in MANET is that when link breakages happen, new routes should be built in order to keep the network functioning efficiently. In MANETs, nodes move randomly and unpredictably which makes the network subject to frequent dis-connectivity due link breakages. In order to have reliable MANET, it requires fast detection of link failures to enable fast notification. Reducing the overhead of finding route can be achieved if prior knowledge about the destination is available at the source, which in turn can help reducing the amount of broadcast traffic. Mobile nodes in MANETs play an important role in finding and maintaining routes from the source to the destination or from a node to another. This is the major issue to such networks. If link breakages occur, the network has to stay operational by building new routes. In reactive protocols, a route is searched when only needed and, therefore, eliminating the necessity to search for routes to all other active nodes in the network; which is an essential procedure in table-driven protocols. Furthermore, in reactive protocols, the route cache information is utilized efficiently by the intermediate nodes in order to reduce the control overhead. Advanced information about the status of a link before the breakage actually occurs can be of great benefits. Thus, using link breakage prediction allows reconstruction of a new route before the current active route becomes unavailable. This would help avoiding (or reducing) data packet loss and improve the network performance. The aim of the research work presented in this paper is to understand, detect and ultimately predict and avoid failure conditions. We are approaching link failure problem in MANETs by designing a route state prediction method which would make source routing protocols and network operations more robust to failure.

The rest of the paper is organized as follows; Sect. 3 presents the proposed route state prediction method. The performance evaluation of the proposed method, including simulation scenarios setup, is presented in Sect. 4 Simulation results are discussed in Sect. 5. Finally, conclusions and future work are presented in Sect. 6.

3 Route State Prediction Method

In this section, we present a reliable method for route failure prediction, called RFP that would improve the performance of MANETs. The proposed method allows detecting the signal sent by neighbors via beacons and compares it to a predetermined two thresholds in order to take proper link status update decisions.

3.1 Signal Strength Detection and Region Assignment

For a MANET with high mobility, the signal strength information can be of great benefits to detect the probability of link failures. The proposed link state prediction method utilizes the transmitted and received signal strengths between two blind mobile nodes to estimate whether the two nodes are going to be disconnected soon or not. Signal strength (S_{stg}) function in Signal Strength Detection module is responsible of collecting the current link signal status at the intermediate node M using the beacon received from the next hop node on the way to destination. The value of the collected S_{stg}, which is based on the Received Signal Strength Indicator (RSSI), is compared to double predefined thresholds ($R/2$ and $3R/4$) and classified into strong, normal, and weak categories. The classified S_{stg} then will be mapped to the corresponding region; *Range* 1, *Range* 2, and *Range* 3, as shown in Fig. 1. The next hop node is considered to be located in *Range* 1 if resulting value is $S_{stg} \leq R/2$, or in

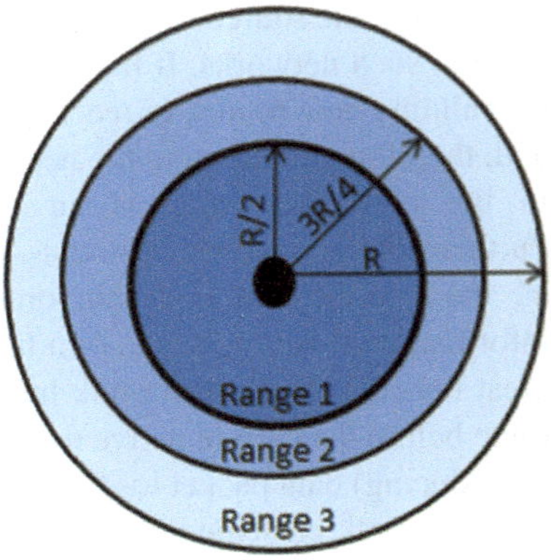

Fig. 1 Transmission range regions

Range 2 if $R/2 < S_{stg} < 3R/4$, or in *Range* 3 if $S_{stg} \geq 3R/4$. Node M also performs the same procedure for the upstream node to estimate its region with respect to the upstream node.

3.2 Rate of Change of Node Position

When a mobile node is moving from a point $P(x1, y1)$ to $P(x2, y2)$ as illustrated in Fig. 2, we concern the instantaneous (*inst*) rate of change of the mobile node position, which means how fast it is moving at any time instant. If we plot the node movement as a function of time, we can figure out the average speed of the node (the average rate of change of the node position). It represents the slop of line (dotted-dashed- line in Fig. 2) from zero (when the node is standstill) to a position at point, for example, (1, 20), which is the change in the distance $f(x)$ and time x.

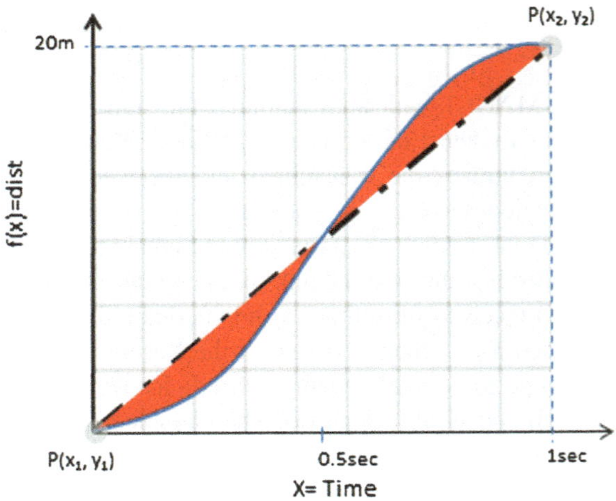

Fig. 2 The rate of change of node position wrt the slop of the line between positions

Therefore, the average speed avg_s of the node is calculated as follows:

$$avg_s = \frac{\Delta \text{dist}}{\Delta \text{t}} = \frac{\Delta \text{f(x)}}{\Delta x} \quad (1)$$

Thus, from the figure, the avg_s is:

$$avg_s = \frac{20-0}{1-0} = \frac{20}{1} = 20\,\text{m/s} \quad (2)$$

To know how the avg_s is different from *inst*, we can study the plot of the distance that the node crossed, relative to time. As the node cannot move in speed of 20 m/s immediately, it then has to accelerate. It start slower at the beginning of the movement; this means that the slop is a lot lower than the average slop between 0.1 and 0.3 s time intervals; and then accelerates as the time passes. So, the speed and the slop will get steeper and steeper. Then, at points near the second position, the node starts slowing down. So, the *dist* plotted against time in Fig. 2 is the curve representing the change in the speed when the node moves from the first position to the second position. It is noticeable from the graph that at any given time the slop is different; at the beginning, the node has a slower moving rate of change of distance, then it accelerates to level above its average at x = 0.5, and then it starts to slow down again.

Therefore, the instantaneous rate of change (*inst*) of node position at any given time (t) can be calculated as follows:

$$inst(x) = \lim_{\Delta x \to 0} \frac{f(x+\Delta x) - f(x)}{\Delta x} \quad (3)$$

inst give us the speed and the direction of the node movement. Note that the speed is the absolute value of *inst*. As the best linear approximation of the function at any point of interest can be described by the function derivative at that point, the derivative of the function $f(x)$ at time t can be defined as follows:

$$\frac{df}{dx}(t) = \lim_{\Delta x \to 0} \frac{f(t+\Delta x) - f(t)}{\Delta x} \quad (4)$$

That is, the derivative at point $f(t)$ is the slope of the tangent line to the graph at f (t). This lead to the conclusion that, if the position of node M_i at time t is $p(M_i, t)$, then the relative rate of change in the movement $inst(M_i, M_{i+1}, t)$ between nodes M_i and M_{i+1} at time t is:

$$inst(M_i, M_{i+1}, t) = \frac{d}{dt}(p(M_i, t) + p(M_{i+1}, t)) \quad (5)$$

Considering the movement *mov* as a function of the relative motion of nodes M_i and M_{i+1}, then *mov* between any two nodes (M_i, M_{i+1}) is defined as absolute average relative rate of change in their movement over the time period of T. Thus, *mov* can be obtained as follows:

$$mov_{M_i,M_{i+1}} = \frac{1}{T} \int_{t_0 \leq t \leq t_0+T} |inst(M_i, M_{i+1}, t)| dt \quad (6)$$

Averaging *mov* in (6) over all mobile nodes in the network can result in the following formula:

$$mov = \frac{1}{|M_i, M_{i+1}|} \sum\nolimits_{M_i,M_{i+1}} mov_{M_i,M_{i+1}} \quad (7)$$

$$mov = \frac{2}{n(n-1)} \sum\nolimits_{M_i=1} \sum\nolimits_{M_{i+1}=M_i+1} \quad (8)$$

where $|M_i, M_{i+1}|$ is the number of mobile node pairs (M_i, M_{i+1}), while n is the number of mobile nodes in the network, that being numbered from 1 to n.

Therefore, *mov* indicate the average relative rate of change in the movement between mobile nodes in the network. As a result, *mov* for a set of nodes moving parallelly at the same velocity, or standing still, is zero.

3.3 Node Behavior

When it comes to link failure events, a mobile node can be the detector of the event or the forwarder of the notification of the event. Node M send beacons when it about to move of the range of the parent node (i.e. the upstream[1] node that the data packets are received from). Node M checks the connectivity status to its parent, based on the signal strength, and updates (notifies) the parent if there is a possibility that node M going to be out of its parent transmission range in the near future; so that the upstream node can take action as the link to its child going to be broken soon. Node M uses two above mentioned thresholds to perform its update decisions:

- First threshold $R/2$ (called $inst_{thr}$ in this procedure) is the value after which node M needs to calculate its instantaneous velocity with respect to its parent.
- Second threshold $3R/4$ (called $beacon_{thr}$ in this procedure) is the value at which node M sends beacon to its upstream node, which includes the movement information (the calculated speed and direction of node M). Node M will not trigger the updates or notifications (send beacon) or calculates the instantaneous velocity unless these threshold values are reached, to reduce the control overhead.

[1]In this paper, the terms *parent* and *upstream* node are used interchangeably to refer to the node that data is received from over an active link.

With such information, the parent node will have an idea about the connectivity to its child in the near future condition. If the child node is going to be out of the transmission range soon, then it will update the upstream node; the one on the way back to source node S. Therefore, the beacon massage can carry two type of information:

- The instantaneous velocity calculation, based on which the upstream node has to calculate its instantaneous velocity with respect to its child who sent the massage.
- and/or the notification of that some links on the way to destination(s) going to be broken soon (as a forwarder of the notification on the reverse back to source node S).

Upon receiving the notification, the parent node will confirm its position with respect to its child and whether its child will move out of the range soon by checking the current distance ($dist_{cur}$) to its child and compare it to the last known distance ($dist_{dwn}$). The distance is calculated based on RSSI as in [1].

3.4 Operation Scenario

To make it clear of how an intermediate mobile node on an active route to destination acts, we captured the movement information of a node after the simulation time is collapsed. The movement of the intermediate node M to point $P(x_2, y_2)$ is found to follow the function line of $f(t) = 2t^5 + 6t^4 + 8t^2 - 3$, as shown in Fig. 3a. Figure 3b shows the intermediate node M with respect to its neighbors before starting moving. The figure shows that the connectivity is already established and the route is active. Figure 3c shows the new location of node M while maintaining the connectivity to its upstream and downstream nodes, and it is able to forward data packets towards their destination. The calculation of the new position confirms that node M is still within the transmission range of upstream node and even more closer to it. In this case, there is no prediction that the link going to be broken soon. Figure 3d illustrates the situation where node M moves out of the upstream transmission range and it should have notified the upstream node about its prediction of the link connectivity before it is about to be out of the transmission range of the upstream node.

4 Performance Evaluation

To verify the effectiveness of the proposed link state prediction method, a MANET of 150 nodes was simulated using ns-2. These nodes are communicating and moving randomly within an area of 1500 * 1500 m^2 with speed accelerating from

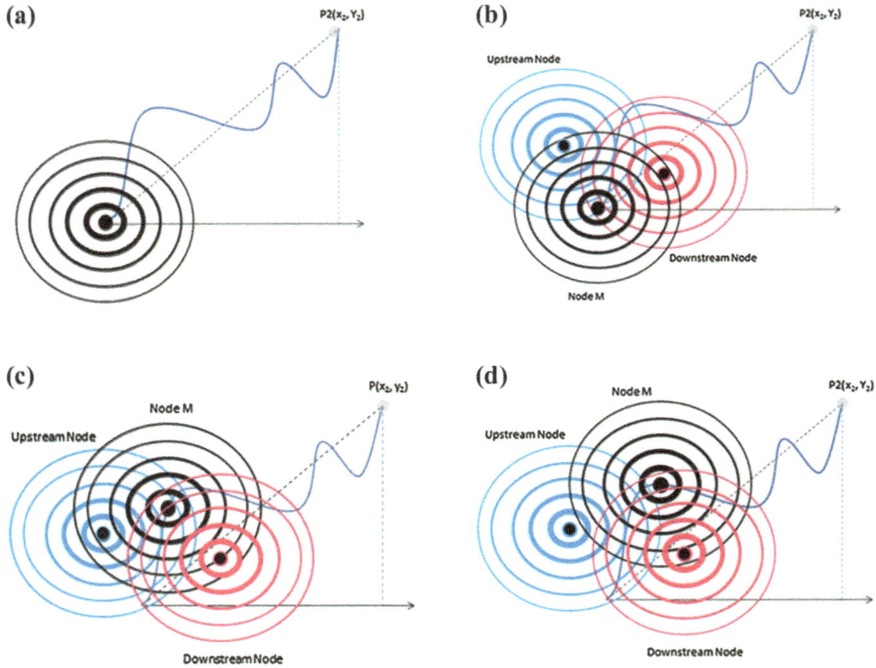

Fig. 3 **a** Movement line of Node M. **b** Node M position wrt upstream and downstream nodes. **c** Node M movement wrt upstream and downstream nodes, **d** The case when Node M is out of the range of the upstream node

1 m/s to maximum speed of 20 m/s with random pause time from 0 to 60 s. Node transmission range is 90 m. For generating random nodes movement, RWP has been used. The duration of simulation is 800 s. A random number of sources nodes (between 20 and 40 nodes) connections were configured to send CBR data packets, with a sending rate of 0.1 Mbps and packet size of 552 bytes, to their intended destinations. Similar to nodes in DSR, the source nodes in the proposed method use source routing procedure for route discoveries. The experiments were repeated 20 times to ensure that packet loss is due to the delay in the delivery of the notification of the link breakage event and not because the unavailability of routes to destinations.

5 Results and Discussion

The performance of the proposed route failure prediction (RFP) method has been analyzed based on the numerical results gained from simulations in terms of Packet Delivery Ratio (PDR), Control Overhead (COH), and the End-to-End Delay with respect to the number of events of link breakage happened during the experiments. The measured performance of proposed method is then compared to that of DSR routing protocol [7].

5.1 Packet Delivery Ratio (PDR)

Figure 4a shows the PDR gained over the proposed method and DSR. It is clear that DSR suffers significant degradation in PDR as the route changes increase. This is due the procedure that DSR follow in the presence of link failure event. DSR was expected to perform better as it utilizes the link layer feedback that is supposed to detect the link breakages fast, and hence, has lower packet loss. The poor performance is justified through DSR link failure detection procedure that is triggered only after the link breakage is already taken place. In contrast, utilizing the proposed link state prediction method offers early detection and notification that would help the source node in taking a proper action (early enough to avoid dis-connectivity with the destination). When a link breakage is about to happen, intermediate nodes will immediately send notification to their upstream nodes

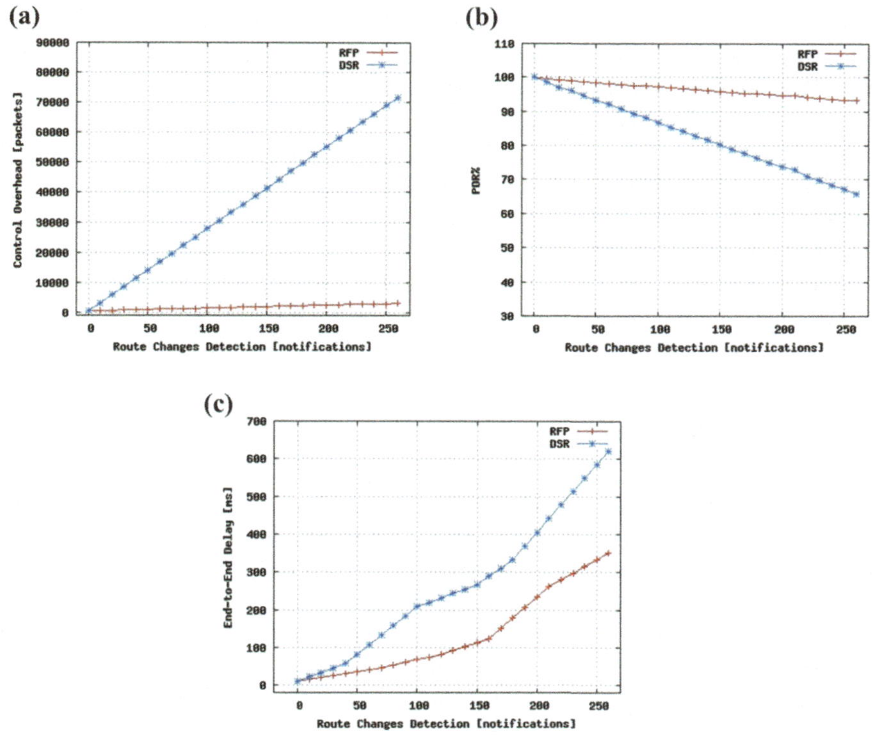

Fig. 4 **a** PDR gained using DSR and the proposed RFP method. **b** Control overhead of DSR and the proposed RFP method. **c** End-to-End delay of DSR and the proposed RFP method

informing about the anticipated situation. That is why the proposed method present excellent PDR compared to that of DSR. From the figure, it is noticeable that link failures have serious negative impact on the performance of DSR. The PDR offered by DSR drops almost linearly as the link breakage increases, while the proposed method is capable of delivering 30–35 % more data packets than DSR. It obvious that both DSR and the proposed method are delivered 100 % of their data packets when no link breakage has occurred.

5.2 Control Overhead

From Fig. 4b, it's clear that utilizing the proposed method leads to considerable reduction in routing overhead as the proposed method executes significantly less route discoveries. This important improvement in the performance is due the use of link failure prediction which confirms our hypothesis that early notification of the link status would help avoid unnecessary packet drops and eliminate the need for flooding the network with routing packets to find alternative routes. According to failure detection theory [8], communication systems should respond quickly to sudden changes in the connectivity. DSR shows slow reaction when the network topology changes frequently. When more link failures occur, DSR needs to perform route switching more frequently, which include three processes: route breakage detection, new route discover, and routing data through the new route to destination. This confirms and justifies the high control overhead presented by DSR in Fig. 4b compared to that of the proposed method. Same as in DSR, salvaging mechanism used by the proposed method helps in reducing the control overhead and improves the PDR as when the link breakage happens, the queues in the intermediate nodes are almost full and, therefore, a lot of packets are salvaged. Recall that notification messages are only sent from intermediate nodes on active routes and only when the route breakage about to happen, thus, no much overhead is contributed by the proposed method.

5.3 End-to-End Delay

Figure 4 shows the end-to-end delay for packets reached their destinations using the proposed RFP method and DSR. Noticeably, DSR introduce more delay compared to that of the proposed method. DSR takes fairly long time to know about the link breakage event and the queue at the intermediate nodes are full with packets that might eventually drop as there is no route to destination(s). Hence, salvaging used by DSR will increase the delay and worsening the contention in the network; while with RFP, source node will be already informed on the event earlier so that the source will find alternative router, whether from its cache or through new route discovery. Thus, the connection will not be interrupted and its throughput at the destination will be high. This is confirmed by the graph of RFP in Fig. 4c.

6 Conclusion and Potential Future Work

In this paper, we have proposed a reliable method for route failure prediction, namely RFP, to improve the performance and connection stability in MANETs. We evaluated the proposed method in simulated wireless multihop mobile ad hoc network. Our experimental evaluation showed that RFP performs well in medium-sized highly dynamic environment. It was verified through the numerical results that, by using RFP, nodes can adapt themselves to topology change faster due to the notification triggered based on link state prediction. It was shown that RFP offers better PDR and lower control overhead and delay compared to DSR. These metrics were measure with respect to the rate of link breakages occur in active routes according to the notifications delivered to source nodes. RFP needs to be investigated further in large scale MANETs with random sending rates.

Acknowledgments This work was supported by National Advanced IPv6 Centre of Excellence (NAv6), Universiti Sains Malaysia (USM).

References

1. Abusubaih M, Rathke B, Wolisz A (2007) A dual distance measurement scheme for indoor ieee 802.11 wireless local area networks. In: 9th IFIP international conference on mobile wireless communications networks, 2007, pp 121–125, Sept 2007
2. Wan Z (2011) Commercial issues in hybrid cellular-ad hoc networks. In: Fifth international conference on Management of e-Commerce and e-Government (ICMeCG), pp 226–228, Nov 2011
3. Khudayer BH, Kadhum MM, Chee WT (2015) Multi-disjoint routes mechanism for source routing in mobile ad-hoc networks. In: Submitted to ICISA2015: international conference on information science and applications, Pattaya, Thailand. Springer Lecture Notes in Electrical Engineering series (LNEE). 24th–26th Feb 2015
4. Boukerche A (2004) Performance evaluation of routing protocols for ad hoc wireless networks. Mob Netw Appl 9(4):333–342
5. Rohankar R, Bhatia R, Shrivastava V, Sharma DK (2012) Performance analysis of various routing protocols (proactive and reactive) for random mobility models of adhoc networks. In: 2012 1st international conference on Recent Advances in Information Technology (RAIT), pp 331–335, Mar 2012
6. Khudayer BH, Kadhum MM (2014) Reliability of dynamic source routing in heterogeneous scalable mobile ad hoc networks. In: IEEE international conference on communication, networks and satellite—COMNESAT, pp 71–79, Jakarta, Indonesia. Institute of Electrical and Electronics Engineers (IEEE), 4–5 Nov 2014

7. Johnson DB, Maltz DA, Broch J (2000) DSR: the dynamic source routing protocol for multihop wireless ad hoc networks. In: Ad hoc networking. Addison-Wesley Longman Publishing Co., Inc., Boston, pp 139–172, Dec 2000
8. Willsky, A (1976) A survey of design methods for failure detection in dynamic systems. Automatica 12:601–611

An Adaptive Learning Radial Basis Function Neural Network for Online Time Series Forecasting

Mazlina Mamat, Rosalyn R. Porle, Norfarariyanti Parimon and Md. Nazrul Islam

Abstract Most of the neural network based forecaster operated in offline mode, in which the neural network is trained by using the same training data repeatedly. After the neural network reaches its optimized condition, the training process stop and the neural network is ready for real forecasting. Different from this, an online time series forecasting by using an adaptive learning Radial Basis Function neural network is presented in this paper. The parameters of the Radial Basis Function neural network are updated continuously with the latest data while conducting the desired forecasting. The adaptive learning was achieved using the Exponential Weighted Recursive Least Square and Adaptive Fuzzy C-Means Clustering algorithms. The results show that the online Radial Basis Function forecaster was able to produce reliable forecasting results up to several steps ahead with high accuracy to compare with the offline Radial Basis Function forecaster.

1 Introduction

Forecasting has become an important research area and is applied in many fields such as in sciences, economy, meteorology, politic and to any system if there, exist uncertainty on that system in the future. Before the emergence of mathematical and

M. Mamat (✉)
Artificial Intelligence Research Unit (AIRU), Universiti Malaysia Sabah, 88450 Kota Kinabalu, Sabah, Malaysia
e-mail: mazlina@ums.edu.my

R.R. Porle · N. Parimon · Md.N. Islam
Faculty of Engineering, Universiti Malaysia Sabah, 88450 Kota Kinabalu Sabah, Malaysia
e-mail: rlyn39@ums.edu.my

N. Parimon
e-mail: fara2012@ums.edu.my

Md.N. Islam
e-mail: nazrul@ums.edu.my

P.J. Soh et al. (eds.), *Advances in Machine Learning and Signal Processing*, Lecture Notes in Electrical Engineering 387, DOI 10.1007/978-3-319-32213-1_3

computer models so called machine learning algorithms, forecasting was carried out by human experts. In this approach, all parameters that are possibly give effect to the system to be forecasted are considered and judged by the experts before producing the forecasting output. Unfortunately, forecasting using human experts is very vague and sometimes arguable since it is totally depends on the expert's knowledge, experiences an interest. Other than using human experts, there exists other prediction approach: Statistical Prediction Rules which is more reliable and robust [1]. One of the popular methods in Statistical Prediction Rules is Time Series Prediction where it uses a model to forecast future events based on known past events: to forecast future data points before they are measured.

A time series consists of sequence of numbers which explained the status of an activity versus time. In more detail, a time series is a sequence of data points, measured typically at successive times, spaced at (often uniform) time intervals. A time series has features that are easily understood. For instance a stock price time series has a long term trend, seasonal and random variations while a cereal crops price time series contains only seasonal components [2]. There exist many approaches to perform time series forecasting. Among the approaches are Box-Jenkins Approach (ARMA/ARIMA) [3], Regression analysis [4], Artificial Neural Networks (ANN) [5], Fuzzy Logic (FL) [6] and Genetic Algorithms [GA] [7]. However among them, the computational intelligence technique such as ANN, FL and GA are getting more attention in time series forecasting because they are non-linear in nature and able to approximate easily complex dynamically system [8–11].

In its typical implementation, ANN will be trained by using the existing set of previous data. After the ANN reaches its optimized performance, the training process is stopped. Then the optimized ANN will be used to estimate the forthcoming output based on the current received inputs. This form of implementation is called as offline mode and is implemented in real-world, especially in power generator station [12, 13]. However, studies shown forecasting in the offline mode has several disadvantages. The major disadvantage is that the ANN parameters must be updated from time to time to suite with the various changing of the incoming data. This requires the ANN to be trained again by including the latest available data for the training. If the updating process is neglected, the ANN will generate incorrect forecasting whenever they receive unseen input data beyond the training data set. In situation where the data are non-stationary, the offline forecaster will be under performance, unless it is continuously being updated to track non-stationarities. This calls for online forecasting technique, where the parameters of the ANN will be adaptive to the latest available data.

2 Materials and Methods

2.1 *Radial Basis Function*

The Radial basis function (RBF) neural networks typically have three layers: an input layer, a hidden layer with a non-linear RBF activation function and a linear output layer. Each layer consists of one or more nodes depending on the design. The nodes in input layer are connected to the nodes in hidden layer while the nodes in hidden layer are connected to the nodes in the output layer via linear weights [14]. The output from each node in the hidden layer is given as:

$$Z_j = \Phi(||v(t) - c_j(t)||) \quad j = 1, 2, 3.\ldots\ldots.n_h. \tag{1}$$

where $c_j(t)$ and n_h representing the RBF centre and number of hidden nodes, $v(t)$ is the input vector to the RBF and $\Phi(\bullet)$ representing the temporary activation function while $||\bullet||$ represents the Euclidean distance between the input vector and RBF centre. The initial value of the RBF centre, $c_j(t)$ is given by taking the first data of the series as the RBF centre. The activation function $\Phi(\bullet)$ used is Thin Plate Spline given by $\Phi(a) = a^2 \log(a)$, where $a = a = ||v(t) - c_j(t)||$ is Euclidean distance. The Euclidean distance for each hidden node is given by:

Euclidean distance,

$$a_j = \sqrt{\sum_{i=1}^{n} \left(c_{ij}(t) - v_j(t)\right)^2}. \tag{2}$$

where $c_{ij}(t)$ = RBF centre for the j-th hidden node and i-th input, and $v_i(t)$ = the i-th input. The RBF output is given by:

$$y_k(t) = w_{k0} + \sum_{j=1}^{n_h} w_{kj}\Phi(||v(t) - c_j(t)||); \quad k = 1, 2, 3, \ldots\ldots m. \tag{3}$$

where w_{kj}, is the weight between hidden node and output node and m is the number of output node.

2.2 *Adaptive Learning Algorithms*

Two parameters, namely, the RBF centre in hidden nodes and weights between the hidden nodes and the output nodes were updated using the *Adaptive Fuzzy C-Means Clustering* (*AFCMC*) and the *Exponential Weighted Recursive Least Square* (*e-WRLS*) algorithms respectively [15, 16].

Adaptive Fuzzy C-Means Clustering. Give initial value to $c_j(0)$, $\mu(0)$ and q, where $0 \leq \mu(0) \leq 1.0$ and $0 \leq q \leq 1.0$ (the typical value is between 0.30 to 0.95 respectively). Then compute the Euclidean distance $d_j(t)$ between input $v(t)$ and centre $c_j(t)$. Obtain the shortest distance, $d_s(t)$ longest distance $d_l(t)$, nearest centret $c_s(t)$ and distant centre $c_l(t)$.

For $j = 1$ to $j = n_c$, where n_c = number of RBF centre,

(a) Updates the square distance between centre and input $v(t)$ using:

$$\gamma(t) = \frac{1}{n_c}\sum_{k=1}^{n_c} [||v(t) - c_k(t)||]^2. \tag{4}$$

(b) if $j \neq s$ that is if that centre is not c_s centre, updates the centre by referring to:

$$\Delta c_j = \mu(t)\vartheta(t)\left[v(t) - c_j(t-1)\right]. \tag{5}$$

where

$$\vartheta(t) = \begin{cases} D_l(t)D_j(t)\exp[-D_a(t)] & \text{if } d_j > 0 \\ D_l(t)\exp[-D_a(t)] & \text{if } d_j = 0 \end{cases}. \tag{6}$$

and

$D_l(t) = \frac{\gamma(t)}{d_l^2}, D_j(t) = \frac{d_a^2(t)}{d_j^2(t)}$ and $D_a(t) = \frac{d_a^2(t)}{\gamma(t)}$ with $a = s$ if $d_s(t) > 0$ and $a = z$ if $d_s(t) = 0$, ($d_z(t)$ = smallest nonzero distance between $v(t)$ and $c_j(t)$).

(c) updates $c_s(t)$ using

$$\Delta c_s(t) = \mu(t)\varphi(t)[v(t) - c_s(t-1)]. \tag{7}$$

Where

$$\varphi(t) = \begin{cases} \exp\left(-\frac{d_s^2(t)}{\gamma(t)}\right) & \text{if } d_s(t) > 0. \\ 0 & \text{if } d_s(t) = 0. \end{cases} \tag{8}$$

Measure the distance between $c_s(t)$ with all centres, $h_k(t) = (||c_s(t) - c_k(t)||)$, $k = 1, 2, 3, \ldots n_c$ and $k \neq s$. If the shortest distance $h_c(t)$, is $h_c(t) < \mu(t)d_a(t)$, move the nearest distance, $c_c(t)$ to new location based on:

$$\Delta c_c(t) = -\frac{\mu(t)d_a^2(t)}{d_l^2(t)}(c_c(t) - c_s(t)). \tag{9}$$

Set $t = t + 1$, and repeat the above for each data sample. The diffusion coefficient $\mu(t)$, is updates by using:

$$\mu(t) = \mu(0)\exp\left(-\frac{qt^2}{n_c^2}\right) + \frac{\exp(-q\mu(t-1))}{n_c}. \quad (10)$$

Exponential Weighted Recursive Least Square. Set $\hat{\beta}_0^j = 0$ and construct matrix $P_0 = \alpha I$. The typical value for α is $10^4 \leq \alpha \leq 10^6$ and I is Identity matrix of n_h (number of hidden nodes). Read the output from the hidden nodes, X_k^T, and calculate K_k and P_{k+I} using:

$$K_k = \frac{P_k X_{k+1}}{\lambda + X_{k+1}^T P_k X_{k+1}}, \quad (11)$$

and

$$P_{k+1} = \frac{P_k}{\lambda}\{I - K_k X_{k+1}^T\}. \quad (12)$$

where λ is a forgetting factor with its typical value of $0.95 \leq \lambda \leq 0.99$. The λ can also be computed by:

$$\lambda(t) = \lambda_0 \lambda(t-1) + (1-\lambda_0). \quad (13)$$

where $\lambda_0 = 0.99$. Estimates $\hat{\beta}_{k+1}^j$ by using:

$$\hat{\beta}_{k+1}^j = \hat{\beta}_k^j + K_k\left[y_{k+1} - X_{k+1}^T \hat{\beta}_k^j\right]. \quad (14)$$

Set $k = k + 1$, $k = 1, 2, 3, \ldots N$ where N is the number of data. Repeat steps 2–4 till converges. Because $\xi(k)$ cannot be computed, $\hat{\varepsilon}(k)$ is used to replace $\xi(k)$ where $\hat{\varepsilon}(k)$, is measurement error and can be computed by $\hat{\varepsilon}(k) = y(k) - \hat{y}(k)$ where $\hat{y} = X_k^T \hat{\beta}_{k-1}$.

3 Results and Discussion

3.1 Data

The forecasting performance is evaluated by using two simulated data and one real data. The simulated data are the Mackey-Glass nonlinear time series and Set A from Santa Fe Competition (SantaFe-A), while the real data is the IBM Stock Price data. The forecasting based on time series produced by the Mackey-Glass equation is regarded as a criterion for comparing the ability of different predicting method and is used in many time series forecasting researches [17, 18]. The SantaFe-A data were recorded from a Far-Infrared-Laser in a chaotic state. These data were chosen because they are a good example of the complicated behavior that can be seen in a clean, stationary, low-dimensional non-trivial physical system for which the underlying governing equations dynamics are well understood. The IBM data are the daily closing price of IBM stock from January 1, 1980 to October 8, 1992 [19].

3.2 *Forecaster Optimization*

The selection of input lag and the number of hidden node in the RBF have strong impact on the performance of a neural network based forecaster [3]. In parallel to this, the analysis on input selection (input lag) and number of hidden node that produce the optimized forecaster is conducted. The analysis starts by setting the RBF input with the data at elapsed time $(t-1)(t-2)$ … $(t-8)(t-9)(t-10)$ and increasing the hidden nodes one by one until it reaches 50. For each number of hidden nodes, the R^2 value for one step ahead forecasting were recorded. The same process is repeated for the other input lag as tabulated in Table 1. The R^2 values obtained from the analysis were plotted and the number of hidden node which gives the highest R^2 values for all five input lags was used in the analysis to obtain the correct number of input lag. This analysis was conducted by setting the hidden node to the value obtained and varies the input lag from $(t-1)$ to $(t-1)(t-2)$ … $(t-98)(t-99)(t-100)$. Table 2 shows the input lag and the number of hidden node which produce the best multiple steps ahead forecasting performance for the three data.

Table 1 List of input lags used to determine the correct number of hidden node

Name	Input lag
Input lag 1	$(t-1)(t-2)(t-3)$ … $(t-9)(t-10)$
Input lag 2	$(t-1)(t-2)(t-3)$ … $(t-19)(t-20)$
Input lag 3	$(t-1)(t-2)(t-3)$ … $(t-29)(t-30)$
Input lag 4	$(t-1)(t-2)(t-3)$ … $(t-39)(t-40)$
Input lag 5	$(t-1)(t-2)(t-3)$ … $(t-49)(t-50)$

Table 2 The best input lag and number of hidden node for the three data

Data	Input lag	Hidden node
Mackey-Glass	$(t-1)(t-2)(t-3)$ … $(t-36)(t-37)$	32
SantaFe-A	$(t-1)(t-2)(t-3)$ … $(t-29)(t-30)$	25
IBM	$(t-1)(t-2)(t-3)$ … $(t-40)(t-41)$	39

3.3 Forecasting Performance

The findings obtained from the analyses in Sect. 3.2 were used to construct a universal online RBF forecaster for Mackey Glass data, SantaFe-A data and IBM Stock Price data. The performance of the forecaster to forecast the three data were tested. Figure 1 presented one to four steps ahead forecasting for the last 500 Mackey Glass data. For SantaFe-A data, due to the nature of data which is too fluctuating, only the last 100 actual and forecasted data are displayed in Fig. 2. Figure 3 presented the forecasting on 500 IBM Stock Price data. From plots in Figs. 1, 2, and 3, it can be observed that the online RBF forecaster is able to produce reliable and close forecasted values in most of the time. For both data, the input lags and number of hidden nodes which produce the best forecasting performance and the R^2 values for one to four steps ahead forecasting are presented in Table 3.

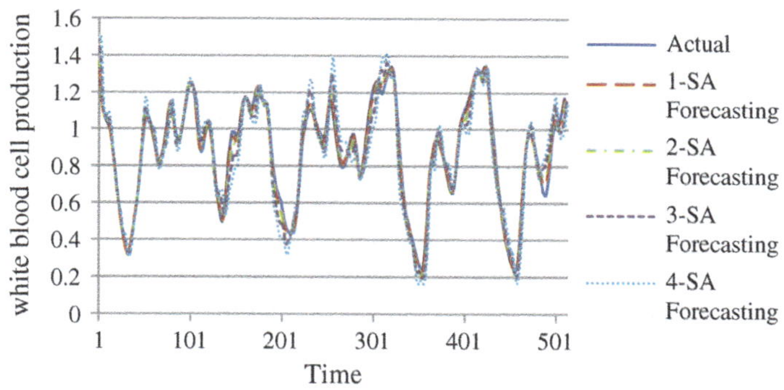

Fig. 1 One to four steps ahead forecasting for the last 500 Mackey Glass data

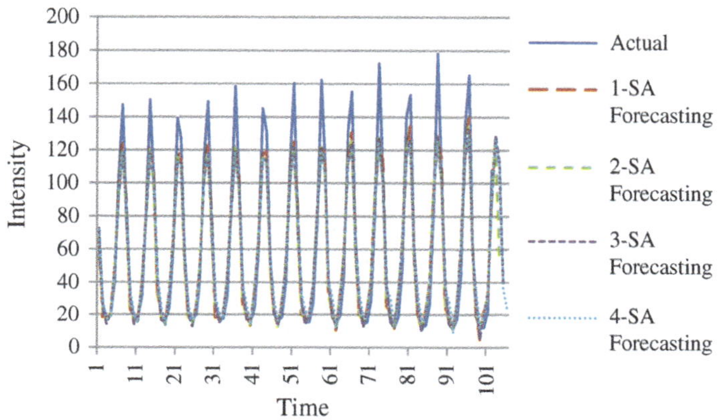

Fig. 2 One to four steps ahead forecasting for the last 100 SantaFe-A data

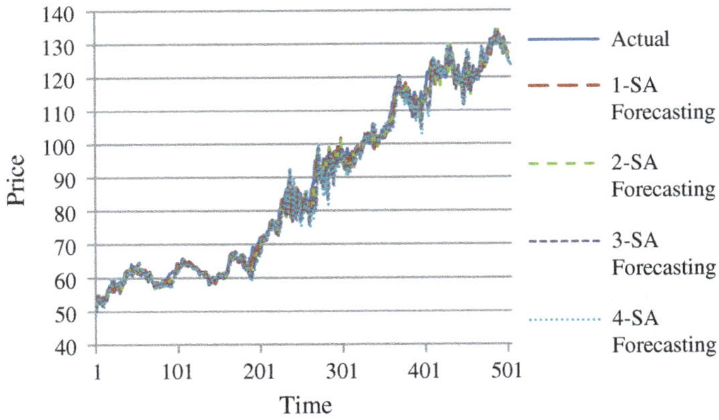

Fig. 3 One to four steps ahead forecasting for the last 500 IBM data

Table 3 RMSE and R^2 values for one to four steps ahead forecasting obtained by the online RBF forecaster

Data	Mackey-Glass		SantaFe-A		IBM	
Input lag	(t−1)(t−2)(t−3) … (t−36)(t−37)		(t−1)(t−2)(t−3) … (t−29)(t−30)		(t−1)(t−2)(t−3) … (t−40)(t−41)	
Hidden node	32		25		39	
Indicator	RMSE	R^2	RMSE	R^2	RMSE	R^2
1-step ahead	0.019	0.995	25.477	0.653	1.872	0.994
2-steps ahead	0.064	0.947	27.436	0.597	6.211	0.939
3-steps ahead	0.099	0.872	28.527	0.565	8.634	0.882
4-steps ahead	0.137	0.758	28.510	0.565	10.512	0.825

Table 4 RMSE and R^2 values for one to four steps ahead forecasting obtained by the offline RBF forecaster

Data	Mackey-Glass		SantaFe-A		IBM	
Input Lag	(t−1)(t−2)(t−3) … (t−36)(t−37)		(t−1)(t−2)(t−3) … (t−29)(t−30)		(t−1)(t−2)(t−3) … (t −40)(t−41)	
Hidden node	32		25		39	
Indicator	RMSE	R^2	RMSE	R^2	RMSE	R^2
1-step ahead	0.030	0.988	23.733	0.699	7.215	0.918
2-steps ahead	0.083	0.911	25.317	0.657	16.424	0.573
3-steps ahead	0.140	0.746	26.617	0.621	28.026	−0.243
4-steps ahead	0.214	0.412	26.687	0.619	42.152	−1.812

The forecasting performance of the online RBF forecaster versus the offline RBF forecaster was also evaluated. Table 4 presents the RMSE and R^2 values for four steps ahead forecasting achieved by the offline RBF forecaster.

For Mackey Glass data, the performance of the offline RBF for one step ahead forecasting are slightly lower to compare with the online RBF. However for multiple steps ahead forecasting, the performance of offline RBF was absolutely poorer where significant deviations were recorded as the forecasting distance increases. The analysis on IBM stock price data also favors online RBF over offline RBF. It can be noted that the offline RBF is able to produce good one step ahead forecasting with 91 % accuracy. However it is considered low to compare with the accuracy obtained by online RBF which is 99 %. For other forecasting distance, namely two to four steps ahead, the offline RBF was failed to generate the acceptable forecasting performance. The forecasting performance obtained by the two, three and four steps ahead was lower than 0.6 and can be regarded as poor to compare with online RBF.

The superiority of online RBF over offline RBF on the Mackey Glass and IBM stock price data can be explained by the nature of the data themselves. It can be observed that both data display different patterns over times especially in the first half and second half of the data. Therefore by using the first half of the data for training are insufficient for the offline RBF to cover all patterns that were exhibited by the data in the next second half. This finding shows that the offline RBF is unable to generate good forecasting for any system which shows chaotic and non-stationary patterns over time.

However, different observation was obtained from the analysis on SantaFe-A data. For this data, the offline RBF produces higher performance in one to four steps ahead forecasting to compare with online RBF. This is again can be explained by the data itself where it can be noted that SantaFe-A data exhibit almost consistent patterns throughout the time. In brief, it can be said that the data is repeating themselves over times. Therefore by training the offline RBF using the first 500 data repeatedly was enough to cover the next 500 testing data. While for online RBF, continuous learning contributes to over fitting which degraded its forecasting ability.

4 Conclusion

More and more fields including science, financial, economy and meteorological adapt time series forecasting to solve uncertainty situation or outcomes in their respective fields. Due to the nature that problems to be solved are affected much by other parameters which change over time, the requirement of the online forecasting model is practical in real-world applications. This paper presented a tool to perform online multiple steps ahead time series forecasting using Radial Basis Function, which shows reliable and accurate forecasting capability.

References

1. Bishop MA, Trout JD (2004) Epistemology and the psychology of Human judgment. Oxford Uni. Press, USA
2. Vemuri VR, Rogers RD (1994) Artificial neural networks forecasting time series. IEEE Computer Society Press, California
3. Montanes E, Quevedo JR, Prieto MM, Menendez CO (2002) Forecasting time series combining machine learning and box-jenkins time series. Advances in artificial intelligence—IBERAMIA 2002, vol 2527
4. Lin K, Lin Q, Zhou C, Yao J (2007) Time series prediction on linear regression and SVR. Third international conference on natural computation, Hainan, China, pp 688–691
5. Boznar M, Lesjak M, Mlakar P (1993) A neural network-based method for short-term predictions of ambient SO2 concentrations in highly polluted industrial areas of complex terrain. Atmos Environ 27(2):221–230
6. Chen SM, Chung NY (2006) Forecasting enrollments using high-order fuzzy time series and genetic algorithms. Int J Intell Syst 21:485–501
7. Nunnari G, Bertucco L (2001) Modelling air pollution time-series by using wavelet function and genetic algorithms. International conference on artificial neural network and genetic algorithms, Prague, pp 489–492
8. Ferrari S, Robert F (2005) Smooth function approximation using neural networks. IEEE Trans Neural Netw 16(1):24–38
9. Crippa P, Turchetti C, Pirani M (2004) A stochastic model of neural computing. In: Lecture notes in artificial intelligence (Subseries of lecture notes in computer science), Part II, vol 3214. Springer Verlag, Berlin, Germany, pp 683–690
10. Zhang GP (2012) Neural networks for time-series forecasting. In: Rozenberg G, Kok JN, Back T (eds) Handbook of natural computing. Springer, Berlin, Germany, pp 461–477
11. Leu Y, Lee C, Hung C (2010) A fuzzy time series-based neural network approach to option price forecasting. In: Nguyen NT, Le MT, Swiatek J (eds) Lecture notes in computer science, vol 5990. Springer, Berlin, Germany, pp 360–369
12. Khotanzad A, Hwang RC, Abaye A, Maratukulam DJ (1995) An adaptive modular artificial neural network hourly load forecaster and it's implementation at electric utilities. IEEE PES winter meeting, vol 95. New York, pp 290–297
13. Mohammed O, Park D, Merchant R (1995) Practical experiences with an adaptive neural network short-term load forecasting system. IEEE Trans PWRS 10(1):254–265
14. Haykin S (1994) Neural networks a comprehensive foundation. Prentice Hall, USA
15. Mashor MY (2001) Adaptive fuzzy c-means clustering algorithm for a radial basis function network. Int J Syst Sci 32(1):53–63
16. Karl JA, Bjorm W (1997) Computer controlled systems: theory and design, 3rd edn. Prentice Hall, New Jersey
17. Gao CF, Chen TL, Nan TS (2007) Discussion of some problems about nonlinear time-series prediction using v-support vector machine. Commun Theor Phys 48:117–124
18. Muller KR, Smola AJ, Ratsch G, Scholkopf B, Kohlmorgen J (1999) Using support vector machines for time series prediction. In: Advances in kernel methods: support vector learning. MIT Press, Cambridge, USA, 1999)
19. Hyndman RJ (2010) Time series data library. http://robjhyndman.com/TSDL, 10 Feb 2010

Incremental-Eclat Model: An Implementation via Benchmark Case Study

Wan Aezwani Bt Wan Abu Bakar, Zailani B. Abdullah, Md. Yazid B. Md Saman, Masita@Masila Bt Abd Jalil, Mustafa B. Man, Tutut Herawan and Abdul Razak Hamdan

Abstract Association Rule Mining (ARM) is one of the most prominent areas in detecting pattern analysis especially for crucial business decision making. With the aims to extract interesting correlations, frequent patterns, association or casual structures among set of items in the transaction databases or other data repositories, the end product of association rule mining is the analysis of pattern that could be a major contributor especially in managerial decision making. Most of previous frequent mining techniques are dealing with horizontal format of their data repositories. However, the current and emerging trend exists where some of the research works are focusing on dealing with vertical data format and the rule mining results are quite promising. One example of vertical rule mining technique is called Eclat which is the abbreviation of Equivalence Class Transformation.

W.A.B.W.A. Bakar (✉) · Z.B. Abdullah · Md.Y.B. Md Saman · Masita@Masila B.A. Jalil · M.B. Man
Department of Computer Science, School of Informatics and Applied Mathematics, Universiti Malaysia Terengganu, 21030 Kuala Terengganu, Terengganu, Malaysia
e-mail: beny2194@yahoo.com

Z.B. Abdullah
e-mail: zailania@umt.edu.my

Md.Y.B. Md Saman
e-mail: yazid@umt.edu.my

Masita@Masila B.A. Jalil
e-mail: masita@umt.edu.my

M.B. Man
e-mail: mustafaman@umt.edu.my

T. Herawan
Department of Information Systems, Faculty of Computer Science and Information Technology, University of Malaya, Lembah Pantai, 50603 Kuala Lumpur, Malaysia
e-mail: tutut@um.edu.my

A.R. Hamdan
Data Mining and Optimization Research Group, Fakulti Teknologi & Sains Maklumat, Universiti Kebangsaan Malaysia, 43650 Bangi, Selangor, Malaysia
e-mail: arh@ukm.edu.my

P.J. Soh et al. (eds.), *Advances in Machine Learning and Signal Processing*, Lecture Notes in Electrical Engineering 387, DOI 10.1007/978-3-319-32213-1_4

In response to the promising results of the vertical format and mining in a higher volume of data, in this study we propose a new model called an Incremental-Eclat adopting via relational database management system, MySQL (My Structured Query Language) that serves as our association rule mining database engine in testing benchmark Frequent Itemset Mining (FIMI) datasets from online repository. The experimental results of our proposed model outperform the traditional Eclat with certain order of magnitude.

Keywords Association rule mining · Relational database · Mysql · Frequent itemset · Eclat algorithm

1 Introduction

Association rules mining (ARM) is first defined by [1] remains as one of the prominent and advance techniques in data mining. With the objectives to find the correlations, associations or casual structures among sets of items, association rules are the if-then statements that uncover relationships between unrelated data in transactional database, relational database or other types of data repositories. There are two (2) major aspects of ARM i.e. mining frequent itemset and generate interesting rules or also called positive association rule (PAR) and also mining infrequent itemset and generate interesting rules or also called negative association rule (NAR) [2, 3].

Most of the previous efforts on ARM have utilized the traditional horizontal transactional database layout format such as in [2, 4]. However, recently a number of vertical association rules mining algorithms have been proposed in works done by [5–8]. A general survey and comparison for association rule mining algorithms has been comprehensively initiated by [9] where a group of researchers have systemized the approaches of ARM algorithms into detailed schematic diagram. Since the introduction of frequent itemset mining, it has received a major attention among researchers [3, 10–14] and various efficient and sophisticated algorithms have been proposed to do frequent itemset mining. Among the three basic frequent itemset mining and best-known algorithms are Apriori [1, 2], Eclat [5, 15] and FP-Growth [4, 16, 17]. The state of the art in association rule mining algorithm is dealing with the extraction of frequent itemsets that occur with high frequency of support, s% in transactional database. The prerequisite feature that must taking into major account is the database format or sometimes called as database layout. The database format (either in horizontal or vertical) might be a major determinant of how far and how fast association rule is mined prior to generation of frequent itemsets from a database. Among existing works on vertical data association rules mining [4–7, 9–11, 15], Eclat algorithm is known for its 'fast' intersection of its tidlist whereby the resulting number of tids is actually the support (frequency) of each itemsets [5, 9]. That is, we should break off each intersection as soon as the resulting number of tids is below minimum support threshold that we have set.

Studies on Eclat algorithm has attracted many development including the works of [6, 8, 18]. Motivated to its 'fast intersection', this paper proposed a new Incremental-Eclat model by taking Eclat as well as Eclat-variants as the based models. This new model proposes a new incremental mechanism in Eclat dimension model.

The rest of the paper is organized as follows. Section 2 describes the related works of vertical algorithm in ARM. Section 3 explains the theoretical background. Section 4 outlines on Traditional Eclat and Eclat-variants versus Incremental-Eclat concept. This is followed by the experimentation of Eclat and Incremental-Eclat in Sect. 5. Finally, conclusion and future direction is reported in Sect. 6.

2 Related Works

The first aspect in association rule mining is looking on the frequent itemset mining as applied in FP-Growth and Eclat algorithms. The FP-Growth is defined in [4, 17] employs a divide-and-conquer strategy and a FP-tree data structure to achieve a condensed representation of the transaction database. It has become the fastest algorithms for frequent pattern mining. In large databases, it's not possible to hold the FP-tree in the main memory. A strategy to cope with this problem is to firstly partitioned the database into a set of smaller databases (called projected databases), and then construct an FP-tree from each of these smaller databases. The generation of FP-tree is done by counting occurrences and depth first search (refer to Fig. 2) in searching the nodes [9].

The Eclat is first initiated by [5] stands for *Equivalence Class Transformation* [15, 16] and as an acronym of *Equivalence Class Clustering and bottom up Lattice Traversal* [18]. It also takes a depth-first search in searching nodes and intersecting, in which each item is represented by a set of transaction IDs (called a tidset) whose transactions contain the item. The tidset of an itemset is generated by intersecting tidsets of its items. Because of the depth-first search, it is difficult to utilize the downward closure property like in Apriori [1, 2] that based on breadth-first searching. However, using tidsets has an advantage that there is no need for counting support, the support of an itemset is the size of the tidset representing it. The main operation of Eclat is intersecting tidsets, thus the size of tidsets is one of main factors affecting the running time and memory usage of Eclat. The bigger tidsets are, the more time and memory are needed.

Continuing the work by [5], a new vertical data representation, called Diffset is proposed in [6], and introduced dEclat, an Eclat-based algorithm using diffset. Instead of using tidsets, they use the difference of tidsets (called diffsets). Using diffsets has drastically reduced the set size representing itemsets and thus operations on sets are much faster. The dEclat has shown to achieve significant improvements in performance as well as memory usage over Eclat, especially on dense databases. However, when the dataset is sparse, diffset loses its advantage over tidset. Therefore, the researchers suggested using tidset format at the start for sparse databases and then switching to diffset format later when a switching condition is met.

The VIPER (Vertical Itemset Partitioning for Efficient Rule Extraction) is established by [7] uses compressed vertical bitmaps for association rule mining has shown to outperform an optimal horizontal algorithm that has complete apriori knowledge of the identities of all frequent itemsets and only need to find their frequency.

Following the efforts in [5, 6], a novel approach for vertical representation wherein the authors used the combination of tidset and diffset and sorted the diffset in descending order to represent databases [8] which is called sortdiffset. The technique is claimed to eliminate the need of checking the switching condition and converting tidset to diffset format regardless of database condition either sparse or dense. Besides, the combination can fully exploit the advantages of both tidset and diffset format where the prelim results have shown a reduction in average diffset size and speed of database processing.

Motivating on the support measure in frequent item mining in [6], an improvement work by [11] is done whereby a conjecture of support count and improvement of traditional Eclat are proposed. The new Bi-Eclat algorithm sorted on support is introduced such that items are in descending order according to frequencies in transaction cache while itemsets use ascending order of support during support count. As compared to traditional Eclat, it has gained better performance when tested on several public selected datasets.

3 Theoretical Background

Following is the formal definition of the problem defined in [11]. Let $I = \{i_1, i_2, \ldots i_m\}$ for $|m| > 0$ be the set of items. D is a database of transactions where each transaction has a unique identifier called tid. Each transaction T is a set of items such that $T \subseteq I$. An association rule is an implication of the form $X \subseteq Y$ where X represent the antecedent part of the rule and Y represents the consequent part of the rule where $X \subseteq I$, $Y \subseteq I$ and $X \cap Y = \emptyset$. A set $X \subseteq I$ is called an *itemset*. An itemset with k-items is called a k—*itemset*. The itemset that satisfies minimum support is called frequent itemset. The rule $X \Rightarrow Y$ holds in the transaction set D with confidence c if c% of transactions in D that contain X also contain Y. The rule $X \Rightarrow Y$ has support s in the transaction set D if s% of transaction in D contains $X \cup Y$. A rule is *frequent* if its support is greater than minimum support (min_supp) threshold. The rules which satisfy minimum confidence (min_conf) threshold is called *strong rule* and both min_supp and min_conf are user specified values [15]. An association rule is considered *interesting* if it satisfies both min_supp and min_conf thresholds [18].

4 Traditional Eclat and Eclat-Variants Versus Incremental-Eclat Concept

A. Traditional Eclat

An Eclat algorithm is first proposed by [13, 18, 19] for discovering frequent itemsets from a vertical database layout of a transaction database. It uses prefix-based equivalence relation, θ_1 along with bottom up search. It enumerates all frequent itemsets. There are two main steps: candidate generation and pruning.

1. Candidate Generation

In candidate generation, each *k-itemset* candidate is generated from two frequent (k-1)-itemsets and its support is counted, if its support is lower than the threshold, then it will be discarded, otherwise it is frequent itemsets and used to generate (k + 1)-itemsets. Since Eclat uses the vertical layout, counting support is trivial. Depth-first searching strategy is done where it starts with frequent items in the item base and then 2-itemsets from 1-itemsets, 3-itemsets from 2-itemsets and so on.

The first scan of the database builds the transaction id (tids) of each single items. Starting with single item (k = 1), then the frequent (k + 1)-itemset will grow from the previous k-itemset will be generated with a depth first computation order similar to FP-Growth [15]. The computation is done by intersecting tids of the frequent k-itemsets to compute the tidsets of the corresponding (k + 1)-itemsets. The process is repeated until no more frequent candidate itemsets can be found.

2. Equivalence Class Clustering

An equivalence class $E = \{(i_1, t(i_1 \cup P)), \ldots, (i_k, t(i_k \cup P))|P\}$, considering the set $\{i_1, \ldots, i_k\}$ as an item base, it will have a tree of itemsets over this item base and if the prefix P is appended to all itemsets in this new tree, it will have a set of all itemsets sharing the prefix P. in the search tree over the item base B. In other word, from this equivalence class, a set of all itemsets sharing the prefix P could be generated and this set forms a sub tree of the initial search tree.

Eclat starts with prefix {} and the search tree is actually the initial search tree. To divide the initial search tree, it picks the prefix {a}, generate the corresponding equivalence class and does frequent itemset mining in the sub tree of all itemsets containing {a}, in this sub tree it divides further into two sub trees by picking the prefix {ab}: the first sub tree consists of all itemset containing {ab}, the other consists of all itemsets containing {a} but not {b}, and this process is recursive until all itemsets in the initial search tree are visited. The search tree of an item base {a,b, c,d,e} is represented by the tree as shown in Fig. 1.

In the vertical layout, each item i_k in the item base B is represented as i_k : $\{i_k, t(i_k)\}$ and the initial transaction database consists of all items in the item base.

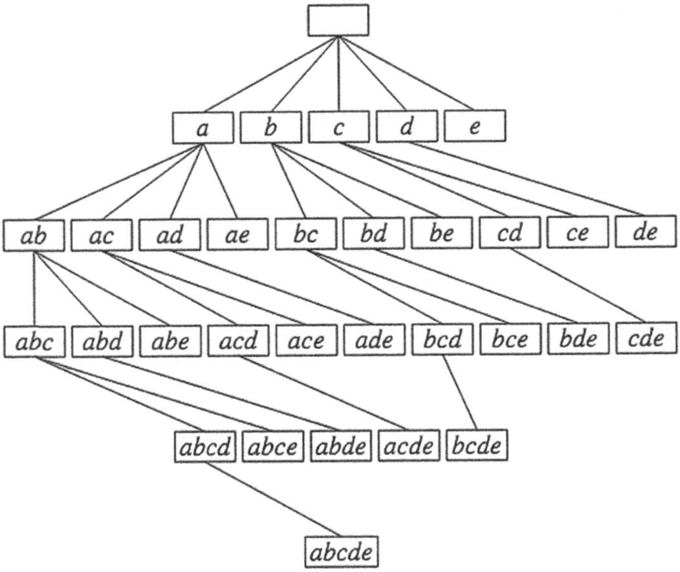

Fig. 1 Prefix Tree for 5 items {a,b,c,d,e} with null set

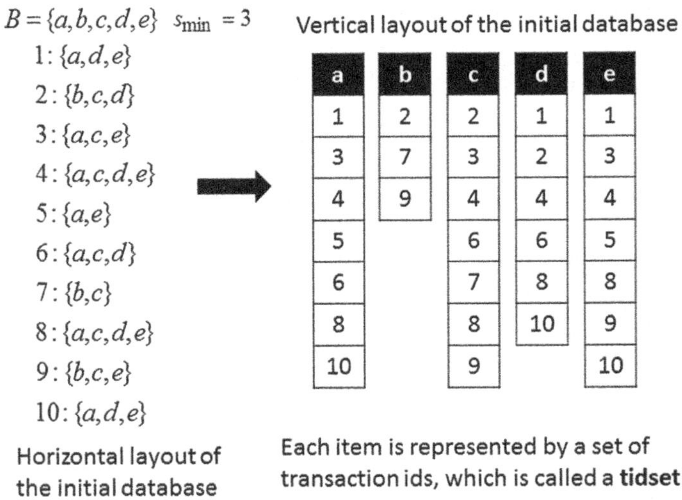

Fig. 2 Transformation from horizontal to vertical layout

For both layouts, it is possible to use the bit format to encode tids and also a combination of both layouts can be used [20, 21]. Figure 2 illustrates how data in horizontal layout is transformed by a set of transaction ids or tidset in vertical layout [20].

In Fig. 2, the items in B consist of {a,b,c,d,e} and each itemsets are allocated with unique identifiers (tids) for each transactions. This is clearly visualized in horizontal format. To switch to vertical format, every items {a,b,c,d,e} are then organized where all items are allocated with their corresponding tids. When this is done, it is clearly visualized the support of each items through the counting number of every item's tids.

B. Traditional Eclat (Tidset) and Eclat-Variants (Diffset and SortDiffset)

A detail steps taken in Eclat-tidset algorithm when assuming that the initial transaction database is in vertical layout and represented by an equivalence class E with prefix { } is shown in Fig. 3 (refer to steps *1,2,3,4,5,6,7,8,9,10*). The itemset in the database table is first sorted in ascending order into each column and then, the looping value is getting based on the number of columns occupied by the itemset. Starting with the first column and second column, the intersection of items in both columns is done and save the intersecting tidlist into database. This process is repeated until all frequent itemsets have been enumerated. The looping number is determined by the number of attributes of the dataset read.

The Eclat-diffset or named as dEclat (different set or diffset) is proposed by [6] where the authors represent an itemset by tids that appear in the tidset of its prefix but do not appear in its tidsets. In other words, diffset is the difference between two (2) tidsets (i.e. tidset of the itemsets and its prefix). Using diffset, the cardinality of sets representing itemsets is reduced significantly and this results in faster intersection and less memory usage. The dEclat is shown to achieve significant improvements in performance and memory usage over traditional Eclat especially in dense database. However when database is sparse, it loses its advantages over tidsets. Then in [6] the authors suggested to use tidset format at starting for sparse database and later switch to diffset format when switching condition is met. The pseudocode of diffset is given in Fig. 3 (refer to steps *1,2,3,4,5,6.1,7,8,9,10*).

```
1.  start
2.  Sort data by itemset
2.1 Sort data by itemset with descending order of dataset value.
3.  looping=numberofcolumn;
    //process tidset
4.  for(i=0;i<looping;i++)
5.  {
6.  get intersect data for column[i] with  column[i+1];
6.1 get diffset data for column[i] with  column[i+1];
7.  save to db;
8.  add next transaction data;
9.  }
10. end
```

Fig. 3 Eclat-tidset (in steps 1,2,3,4,5,6,7,8,9,10), Eclat-diffset (in steps 1,2,3,4,5,6.1,7,8,9,10), and Eclat-sortdiffset (in steps 1,2.1,3,4,5,6,7,8,9,10)

The Eclat-SortDiffset is established in [8] that applied the sorting of diffset in descending order. It claims to achieve a significant reduces in running time and memory usage. Since $p(PXY) = sup(PX) - |d(PXY)| = sup(PY) - |d(PYX)|$, both $d(PXY)$ and $d(PYX)$ could be used to calculate $sup(PXY)$. Therefore, the smaller one of the two should be used to calculate $sup(PXY)$ to reduce the memory usage and processing time. Because $d(PXY) = d(PY) - d(PX)$ and $d(PYX) = d(PX) - d(PY)$, if $d(PX)$ is smaller than $d(PY)$ then $d(PYX)$ is smaller than $d(PXY)$. In general, diffsets in an equivalence class should be sorted in descending order according to size to generate new itemsets represented by diffsets with smaller sizes. The portion of SortDiffset algorithm is given in Fig. 3 (refer to steps *1,2.1,3,4,5,6.1,7,8,9,10*) where the difference only in step 2 as compared to diffset algorithm.

C. Incremental Eclat Algorithm

The initial objective of Incremental Eclat is to handle the issues of big and dynamic data. In real application, data is becoming bigger prior to non-stop transaction being done in many of real world application domain. With respect to association rule mining, items may incur either in two (2) different ratios i.e. increment of itemset or increment of records in typical database. To mine frequent items, it may require higher specification of memory and spaces of the computer hardware. Incurring itemsets result in bigger cardinality of data in equivalence class clustering whereas incurring records consumes higher volume of data. Thus, Incremental Eclat attends to reduce a memory and spaces requirement by implementing flushing of memory prior to each itemset being visited before intersecting the next itemsets. The current or last transaction data that is in-memory will be flushed before proceeding into next transaction data. Adopting in a structured and relational MySQL database, the incremental of either itemset or records of transaction is easier and more efficient in structuring the data. The pseudocode of the proposed algorithm is denoted in Fig. 4. The only difference in Incremental-Eclat engine is depicted in step 9.

```
1.   start
2.   Sort data by itemset
2.1  Sort data by itemset with descending order of dataset value.
3.   looping=numberofcolumn;
4.   for(i=0;i<looping;i++)
5.   {
6.   get intersect data for column[i] with  column[i+1];
6.1  get diffset data for column[i] with  column[i+1];
7.   save to db;
8.   add next transaction data;
9.   flush value for current/last transaction data;
10.  }
11.  end
```

Fig. 4 Incremental-Eclat approach

5 Experimentation

A. Database Platform

All experiments are performed on a Dell N5050, Intel ® Pentium ® CPU B960 @ 2.20 GHz with 2 GB RAM in a Win 7 64-bit platform. The software specification used is MySQL version 5.5.27—MySQL community server (GPL) for our database server, Apache/2.4.3 (Win32) OpenSSL/1.0.1c PHP/5.4.7 for our web server and phpMyAdmin with version 3.5.2.2. For the kick-off experimentation, we start with simple synthetic dataset. In addition, we have retrieved benchmark datasets from http://fimi.ua.ac.be/data/ in a *.dat file format. For the ease of use in MySQL, we convert datasets into comma separated value (csv) format. The characteristics of benchmark datasets with the average size include chess and mushroom is depicted in Table 1. For the faster results of a depth first with intersection searching strategy in database mining, we have split chess, connect and mushroom datasets into benchmark trained datasets. There are three (3) sub divisions of each i.e. chess1000 × 12, chess2000 × 12, chess3000 × 12. The same sub division is done in connect.

B. Empirical Results

The experimentation is done with regards to Eclat algorithm (tidset) in [5], dEclat algorithm (diffset) in [6] and sortdiffset algorithm in [8]. Figure 5, 6, 7, 8 show the graph of performance result in execution time between Eclat and Eclat-variants versus Incremental-Eclat within chess and connect datasets prior to running with Eclat engine versus Incremental-Eclat engine. The graphs indicate the result of sortdiffset algorithm with an order of magnitude outperforms diffset and tidset algorithm in Eclat engine. The execution time shows a slight decreased in connect dataset for about 0.31 % in Incremental-Eclat as compared to Eclat engine. However, in chess, the execution of Incremental-Eclat decreases tremendously for 21.03 % as compared to Eclat engine. As overall, Incremental-Eclat performs better than Eclat in certain order of magnitude. The incremental process either in itemsets

Table 1 Database characteristics

Database	Size (KB)	Average length	Records
Chess (original)	335	37	3196
Chess1000 × 12	32	12	1000
Chess2000 × 12	63	12	2000
Chess3000 × 12	95	12	3000
Connect (original)	9039	43	67,557
Connect1000 × 12	34	12	1000
Connect2000 × 12	67	12	2000
Connect3000 × 12	100	12	3000

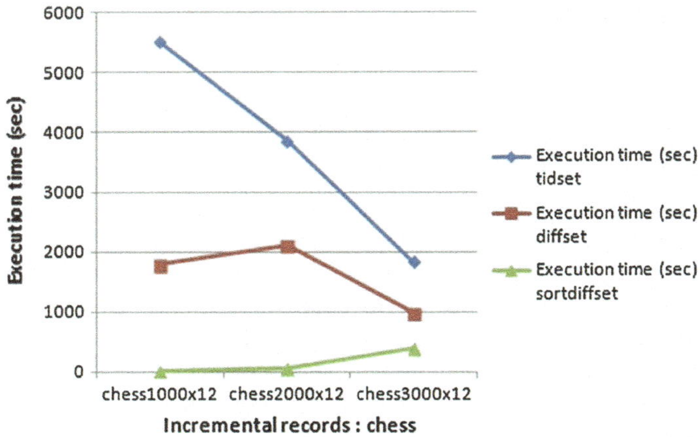

Fig. 5 Chess with eclat

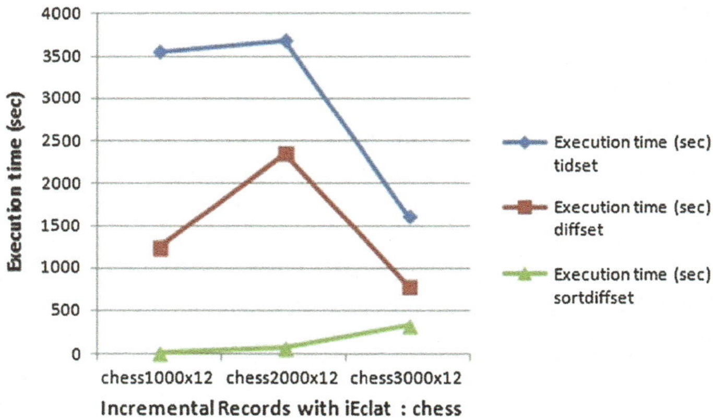

Fig. 6 Chess with incremental-Eclat

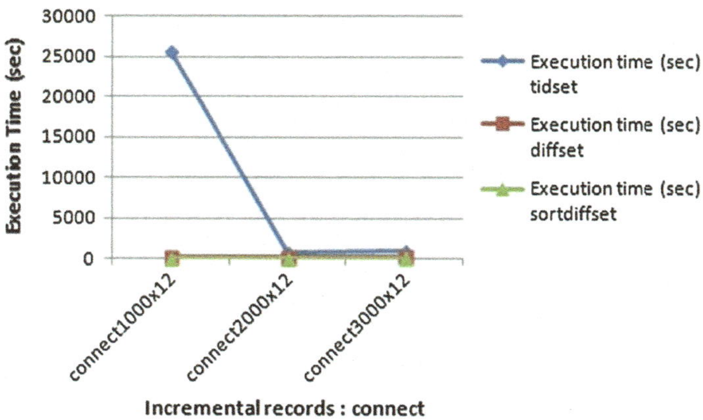

Fig. 7 Connect with eclat

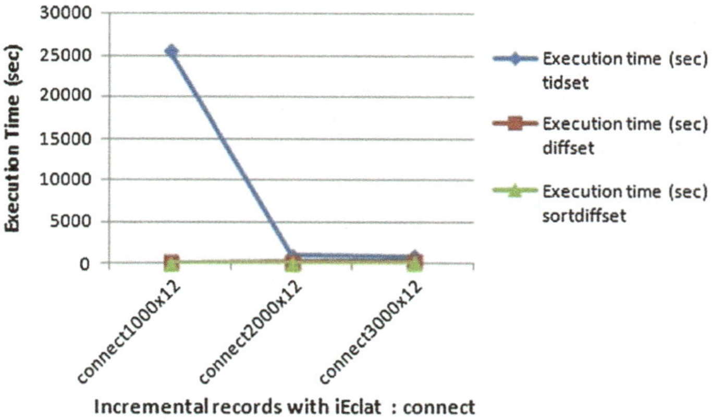

Fig. 8 Connect with incremental-eclat

or in transaction records are efficiently conducted with the adoption of MySQL database where the adhock query either addition or deletion of data can be efficiently manipulated through SQL query in phpMyAdmin software.

6 Conclusion and Future Direction

Experimenting ourselves in association rule database mining with the selected benchmark datasets conforms to what the other previous researchers have proven. In this paper, we have successfully adopted a depth first search (DFS) with intersection strategy through Eclat and our proposed algorithms within a benchmark transaction database in mining association rules. The important advantages in database mining that we disclose here are firstly, the ease of indexing mechanism. Secondly the ad hoc query support mechanism and thirdly, is the interoperability and flexibility of data storage to facilitate the altering (either adding or deleting of row/column) in a data table.Our proposed algorithm seems to benefit with dynamic database where data is always incur in volume from time to time. In conjunction with big data explosion and when the database integration method as in [21] needs to be adopted, then the use of this incremental method will give benefits to end users.

Acknowledgment We express our gratitude to MyPhD scholarship under MyBrain15 of Kementerian Pendidikan Malaysia (KPM) and also to UM research grant and UKM research grant from Research Acceleration Center Excellence (RACE) for the financial foundation of this work.

References

1. Agrawal R, Srikant R (1994) Fast algorithms for mining association rules. In: Proceedings of 20th international conference on very large data bases (VLDB), vol 1215, pp 487–499
2. Agrawal R, Imielinski T, Swami A (1993) Mining association rules between sets of items in large databases. ACM SIGMOD Record 22(2):207–216
3. Abdullah Z, Herawan T, Deris MM (2010) Scalable model for mining critical least association rules. In: Information computing and applications. Springer Berlin Heidelberg, pp 509–516
4. Han J, Pei J, Yin Y (2000) Mining frequent patterns without candidate generation. ACM SIGMOD Record 29(2):1–12
5. Zaki MJ, Parthasarathy S, Ogihara M, Li W et al (1997) New algorithms for fast discovery of association rules. In: Proceedings of the ACM SIGKDD international conference on knowledge discovery and data mining (KDD'97), pp 283–286
6. Zaki MJ, Gouda K (2003) Fast vertical mining using diffsets. In: In Proceedings of the ninth ACM SIGKDD international conference on knowledge discovery and data mining, pp 326–335
7. Shenoy P, Haritsa JR, Sudarshan S, Bhalotia G, Bawa M, Shah D (2000) Turbo-charging vertical mining of large databases. ACM SIGMOD Record 29(2):22–33
8. Trieu TA, Kunieda Y (2012) An improvement for declat algorithm. In: Proceedings of the 6th international conference on ubiquitous information management and communication (ICUIMC'12), vol 54, pp 1–6
9. Hipp J, Güntzer U, Nakhaeizadeh G (2000) Algorithms for association rule mining: a general survey and comparison. ACM SIGKDD Explor Newslett 2(1):58–64
10. Borgelt C (2003) Efficient implementations of apriori and eclat. In: Proceedings of the IEEE ICDM workshop on frequent itemset mining implementations (FIMI03)
11. Schmidt-Thieme L (2004) Algorithmic features of eclat. In: Proceedings of the IEEE ICDM workshop on frequent itemset mining implementations (FIMI04)
12. Goethals B (2010) Frequent set mining. In: Data mining and knowledge discovery handbook. Springer, pp 321–338
13. Borgelt C, Kruse R (2002) Induction of association rules: apriori implementation. In: Compstat. Springer, pp 395–400
14. Bakar WAWA, Saman MYM, Jalil MA (2014) Mining educational data: a review on student's pattern of behaviours and performances. Int J Adv Comput Sci Appl 4:247–252
15. Zaki MJ (2000) Scalable algorithms for association mining. IEEE Trans Knowl Data Eng 12(3):372–390
16. Han J, Cheng H, Xin D, Yan X (2007) Frequent pattern mining: current status and future directions. Data Min Knowl Disc 15(1):55–86
17. Han J, Pei J, Yin Y, Mao R (2004) Mining frequent patterns without candidate generation: a frequent-pattern tree approach. Data Min Knowl Disc 8(1):53–87
18. Yu X, Wang H (2014) Improvement of eclat algorithm based on support in frequent itemset mining. J Comput 9(9):2116–2123
19. Toivonen H (1996) Sampling large databases for association rules. In: Proceeding of the 22nd international conference on very large data bases (VLDB '96), pp 134–145
20. Slimani T, Lazzez A (2014) Efficient analysis of pattern and association rule mining approaches. Int J Inf Technol Comput Sci 6(3):70–81
21. Man M, Rahim MSM, Zakaria MZ, Bakar WAWA (2011) Spatial information databases integration model. In: Manaf AA et al (eds) ICIEIS 2011. Springer, Informatics Engineering and Information Science, pp 77–90
22. Savasere A, Omiecinski ER, Navathe SB (1995) An efficient algorithm for mining association rules in large databases. In: Proceeding of the 21th international conference on very large data bases (VLDB '95), pp 432–444

Fall Detection Using Visual Cortex Bio-inspired Model for Home-Based Physiotherapy System

Nor Surayahani Suriani

Abstract A home-based physiotherapy system aim to provide support for rehabilitation patients in home environment. Home-based physiotherapy system is useful for those who do not like to travel and lives in rural areas. The therapist can just monitor the physiotherapy exercise via online. This paper focuses on walking exercise monitoring and analyses the motion of the patient to detect any abnormality movement. The patient's movement was captured by the low cost camera and this paper proposed visual cortex bio-inspired methodology to analyses the motion patterns. Visual cortex bio-inspired models mimics human vision system, mainly the primary visual cortex V1 and MT layer. The motions patterns are encoded using 3D Gabor spatio-temporal filter in V1 layer. Then generated spiking neuron model in MT layer formed active motion map based on Gaussian distribution. The extracted active motions plan was formed according to the direction, orientation and speed of the object movement. Then, the SVM classifier will identify the patient state either as a normal or as an anomaly movement. The robustness and accuracy of the system has been extensively tested with rigorous dataset that contained various types of fall. It can be scalable to become an online monitoring system that links with the rehabilitation centre for better support.

1 Introduction

Research in home-based rehabilitation is important to provide support for the rehabilitation patients and reducing financial burden for laboratory setup and human resources. Typical rehabilitation patients are those who suffer from stroke or

N.S. Suriani (✉)
Embedded Computing System (EmbCoS) Research Focus Group,
Department of Computer Engineering, Faculty of Electrical
and Electronics Engineering, Universiti Tun Hussein Onn Malaysia,
86400 Batu Pahat, Johor, Malaysia
e-mail: nsuraya@uthm.edu.my

P.J. Soh et al. (eds.), *Advances in Machine Learning and Signal Processing*,
Lecture Notes in Electrical Engineering 387, DOI 10.1007/978-3-319-32213-1_5

physical operation, surgery and brain injury. This research area keeps on interest due to the majority cases of long-term disability of rehabilitation patient increasing as the elderly population grows. Hence, with the help of computer vision approach, home-based physiotherapy system used to monitor patients while they doing their exercise to improve physical functional abilities. This physiotherapy system could benefit patients in terms of timing and also safety for speedy recovery. A web based rehabilitation monitoring system [1] only dealing with gathering and storing information of patient's movement data. Healthcare professionals will have accessed to these patients information's to monitor the rehabilitation progress. In addition, sensor-based rehabilitation system [2] able to support early detection of abnormal conditions from continuous and real-time monitoring. However, there are some limitations occur for wearable device monitoring system such as the wires or electrodes attached to the patient's body may limit the patients activity and they might not comfortable which may resulting false alarm. Moreover, most researchers focused to improve motor ability for upper limb by utilizing the advancement in game industry [3]. The gaming consoles used to monitor patient gait, postural control and movement ability for stroke patient.

Therefore, this paper aims to bridge the gap in existing research methodology through computer vision approach. The utilization of camera and video processing could benefit patients for continuous rehabilitation monitoring system. Recently, research in image processing techniques such as 3D Haar like feature [4], segmentation [5], resampling and template generation [6] for patient rehabilitation gain a lot of interest to many researchers. Hence, machine learning based classifier such as Support Vector Machine (SVM), K-Nearest Neighbours (K-NN), Principal Component Analysis (PCA) used to classify the movement either normal or abnormal. In this paper, bio-inspired methodology which mimics the visual perception mechanism was applied to extract unique and novel features to differentiate and classify the movement of objects. The primary visual cortex can achieve a similar observation in recognizing different actions and can be model in video sequences of human actions using spatiotemporal filter to determine the motion orientation in time-space. A bio-inspired model based on active motion map (BIM-AMM) framework is proposed for rehabilitation monitoring system. Our contributions are as follows: (1) A novel spatio-temporal active motion detector based on BIM-AMM, which simulates the dorsal stream of primate visual system (V1), that performs well in scaling, orientations and speed using Gabor 3D, (2) Develop active motion map based Gaussian distribution in MT layer. This paper is organized as follows: the details explanation on proposed BIM-AMM framework, which mainly involves V1 and MT area in visual cortex. Results present simulation, evaluation and analysis of the proposed framework conducted under various types of dataset or normal and abnormal human movements. Finally, conclusion of this paper.

2 BIM-AMM Framework

Research in bio-inspired visual cortex was first implemented by Serre et al. [7] for object recognition. Then, Jhuang et al. [8] and Escobar et al. [9] utilized the bio-inspired visual cortex mechanism for human action recognition based on previous study of biological movement and action by Giesse and Paggio [10]. Jhuang et al. [8] constructed biological model in the visual cortex to formed processing pathway at V1 level with simple and complex cells. Escobar et al. [9] extended the modeling of dorsal stream to areas V1 and MT and utilized spiking neuron model to integrate stimulation between low-level and high-level nerve. Previously, the shape information from V1 and MT cells were pooled according to the maximum speed and direction of moving visual stimuli. Figure 1 presents new mechanism of pooling information as proposed in this paper. Visual stimuli was captured from two pathways which is V1 and MT layer. Then, active motion map represents motion information according to Gaussian probability of maximum direction, orientation and speed of visual stimuli.

2.1 V1 Layer

This work focuses on extraction of motion features in the primary visual cortex V1. The cortex V1 in vision system is corresponding to the first stage processing visual in brain. This paper divide V1 model into two stages: (1) V1 simple and complex cell to obtain local motion; (2) spiking neuron layer which transfer motion information into spike train. Set of spatio-temporal Gabor filter were used to determine preferred direction, orientation and speed tuning property of receptive field of cells in V1 area.

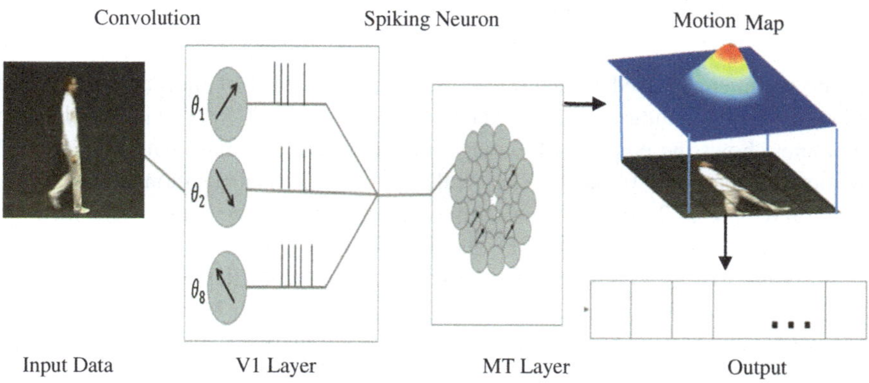

Fig. 1 The proposed BIM-AMM framework which encoded motion patterns using 3D Gabor spatio-temporal filter in V1 layer, generates spiking neuron model in MT layer and formed active motion map (AMM) based on Gaussian distribution of preferred direction, orientation angle and speed

(a) V1 Simple Cell

Gabor filter 2D space function represents only preferred direction of movement along the *x* axis of time, *t*. However, preferred spatial orientation which give the distance ratio of *x* and *y* did not shown in spatial 2D Gabor filter. Therefore, combination between both Gabor 2D function *G*(*x*, *y*) and temporal Gabor *G*(*x*, *t*) produced spatio-temporal Gabor 3D with preferred direction, orientation and speed [11]. This BIM-AMM framework used 3D Gabor spatio-temporal G(*x*, *y*, *t*) filter to model the receptive field profiles of cells in V1 area. The 3D Gabor function considers speed of movement to enhance the detection accuracy of normal and abnormal activity. The speed element is significant feature to distinguished abnormal activity like sudden fall or slip during walking exercise for rehab patient. The 3D Gabor spatio-temporal model is given in the following equation.

$$G(x,y,t) = \frac{\gamma}{2\pi\sigma^2} exp\left[-\frac{\left((\bar{x} + v_f t)^2 = y^2 \bar{y}^2 \right)}{2\sigma^2} \right] \\ \times cos\left(\frac{2\pi}{\lambda} (\bar{x} + vt) \right) + \phi \times \frac{1}{\sqrt{2\pi t}} \exp\left(-\frac{(t-\mu)^2}{2\tau^2} \right) U(t) \tag{1}$$

$$U(t) = \begin{cases} 1 & if\, t \geq 0 \\ 0 & if\, t < 0 \end{cases} \tag{2}$$

$$\bar{x} = x \cos\theta + y \sin\theta, \bar{y} = -x \sin\theta + y \cos\theta \tag{3}$$

γ aspect ratio,
μ minimum temporal Gaussian,
v_f spatial Gaussian envelope,
v_t temporal Gaussian envelope,
υ phase speed of cosine factor

The parameter v used to determine the preferred speed of motion. Petkov and Subramaniam [11] indicate that the speed is within range [0–8]. The phase offset, ϕ parameter shows the position of Gabor function in space domain with symmetrical position is when $\phi = 0$ and $\phi = \pi$, while anti-symmetrical position is when $\phi = -\pi/2$ and $\phi = \pi/2$. In this work, the receptive field size was determined by spatio-temporal oriented Gabor 3D function according to the orientation angle, θ within range $[0, 2\pi]$ with spatial frequency, f and standard deviation, $\sigma = 0.5622/f$ as proposed by Watson and Ahumada [12]. According to Mante and Carandini [13], spatial frequency is in range 0.05–0.2 cycle per pixel. While temporal frequency is in range 2-8 cycle per second. The maximum response of convolution determines that the preferred direction of motion according to the preferred direction and speed of the filter.

(b) Complex Cells

Complex cells have the same receptive field as simple cell. Complex cells respond to changes in the orientation of the object movement and also a directional selectivity cells which demonstrated through a combination of V1 response cell. Thus, complex cells produced by the convolution process between the input stimulus which is video image, $L(x, y, t)$ with the spatio-temporal filter $G(x, y, t)$.

$$C(x, y, t) = G(x, y, t) * L(x, y, t) \tag{4}$$

V1 complex cells modeled by the above equation were reorganized into V1 layers with V1 complex cells that have similar tuning spatio-temporal frequency for different orientation, θ. As proposed by Daugman [14], centers of V1 cells are disposed along a radial log-polar distribution grid.

2.2 Spiking Neuron Model

Considering a VI complex cell whose center is located in (x_0, y_0) in the visual space, the cell generates spike trains according to a conductance-driven Integrate-and-Fire equation as follows:

$$\frac{dV(t)}{dt} = G_\theta^{exc}(x, y, t)(E^{exc} - V(t)) + G_\theta^{inh}(x, y, t)\left(E^{inh} - V(t)\right) - g^L V(t) \tag{5}$$

where

$G_i^{exc}(t)$	excitatory conductance
$G_i^{inh}(t)$	inhibitory conductance
E^L	resting potential
E^{exc}	membrane cell potential
g^L	inert leaks in the cells's membrane

Spike when V = 1, otherwise reset. The excitatory conductance $G_i^{exc}(t)$ for (x, y) in the receptive field V1 cells are given as

$$G_i^{exc}(x, y, t) = k_c C_\theta(x, y, t) \tag{6}$$

where k_c is amplification factor. When cell fires a spike, the inhibitory conductance, $G_i^{inh}(t)$ is generated in the neighborhood cell which has similar direction, orientation angle and speed. The inhibitory conductance, G_i^{inh} given as

$$G^{inh} = G_{max}^{inh} e^{-\frac{d^4}{2R_{inh}^2}} \tag{7}$$

where d is the Euclidean distance between each cell with the neighborhood cell and R_{inh} is the size of receptive field. The firing rate is calculated by the summation of spikes that have been fired by neuron, i as shown in the firing rate equation as follow:

$$\gamma_i(t, \Delta t) = \frac{\eta_i(t - \Delta t, t)}{\Delta_t} \tag{8}$$

2.3 MT Layer

Escobar et al. [9] derived MT-like model by summing the output directional selective of spatial-temporal structure of histogram. The BIM-AMM model presented active motion vector according to the Gaussian distribution of MT cells preferred direction. The use of representative vectors for active motion map (AMM) based MT cells in Gaussian distribution is to avoid the noise on the information layer of MT cells. Shadlen et al. [15] mentioned that noise in the MT may reduce the relevant information and at the same time lead to errors in decision-making process. The AMM information is coded by calculating the Gaussian distribution $f(x, y)$ with output of MT layer, $j(x, y)$ as follows:

$$J(x, y) = j(x, y) * f(x, y) \tag{9}$$

Each image pixel represents a neuron cell, N. The most active neuron cell, N will have the highest firing rate. Therefore, input image is sampled to an optimum kernel size, N_v to speed up the process in MT layer. The kernel sizes should be determined properly so that all the important features of the image can be extracted. Hence, maximum number of active cell can be collected. The number of active cells for each channel are summed for each neighborhood. Then, the histogram is built based on the number of active cells in preferred direction of motion for each kernel area. The histogram representation based on Gaussian distribution form for AMM model is given as follows:

$$\widetilde{H}_G = \left(n_1^1, \ldots, n_1^{N_v}, n_2^1, \ldots, n_2^{N_v}, n_{N_{ch}}^1, \ldots, n_{N_{ch}}^{N_v}\right) \tag{10}$$

where N_v and N_{ch} number of kernel size and number of channel in retina model, respectively. The AMM will be the input feature vector to the SVM classifier [16], which have yielded excellent results in various classification problems and become most robust classifier in machine learning. In this case SVM classifier deals with two-class problems which is normal and abnormal (fall or slip) walking exercise.

Fig. 2 Sample image from **a** Wiezman **b** MILE and **c** SESRG dataset

Given a training samples set $\{(x_i, y_i), i = 1, \ldots, n, x_i \in R^d, y_i \in \{+1, -1\}\}$, where x_i is the feature vector anddeveloped is the label. SVM is developed for finding the optimal classification plane.

3 Results & Discussion

3.1 Database & Settings

As shown in Fig. 2, this physiotherapy system has been simulated and tested under various dataset such as MILE,[1] Weizman and SESRG dataset for walk and fall action only. These dataset are expected as walking exercise for the rehab patient. In this paper, our own dataset named SESRG was captured using IP camera for various types of possible anomaly movement such as bending, forward fall, backward fall and slip fall while doing walking exercise. The simulation was implemented using Matlab on Intel® Core 2.93 GHz machine. The videosequencese runs at 30fps.

3.2 Parameter Settings

The parameter setting of V1 layer is set as follow: eight different directions are selected for simple cells, namely, (0°, 45°, 90°, 135°, 180°, 225°, 270° and 315°). The motion speed v is set to [1–3], spatial frequency, f_s = [0.05, 0.15, 0.2] and temporal frequency, t_f is [2, 4, 8]. The settings for V1 layer have a total of 24 layers for each spatial and temporal frequencies, formed by 3 speeds and 8 orientations. The spiking neuron models initial voltage, threshold voltage and reset voltage of spiking neural model are set to 80, −10 and 0 mv respectively. The Weizmann

[1]http://www.milegroup.net/.

dataset was taken from publicly binary image in the database. While preprocessing mechanism was applied to extract the object of interest and remove the background information for MILE and SESRG dataset.

The simulation was carried out by cross-validation testing procedure. The cross-validation for group of 5 subject with walk and fall actions was randomly chosen for training and the remaining subject for testing using SVM classifier. This setup uses to examine the ability of proposed method (BIM-AMM) to recognize sudden fall in the videos.

3.3 Recognition Performance

The recognition rate for all test datasets was different at each of the preferred speeds. The results in Table 1 shows the sensitivity and specificity of our proposed systems at preferred speed v is set to constant value 3 to detect anomaly during walking rehabilitation activities. The results in Table 1 shows that on average, the percentage achieved for both sensitivity and specificity using BIM-AMM model are more than 95 % for various types of dataset.

Figure 3 shows sample of walking patterns from SESRG dataset and the output of AMM that represents the motion patterns of MT cells which estimated from

Table 1 Recognition rate of BIM-AMM model

Dataset	Sensitivity (%)	Specificity (%)
Weizmann	97.2	96.5
MILE	93.5	96.3
SESRG	99.3	98.4

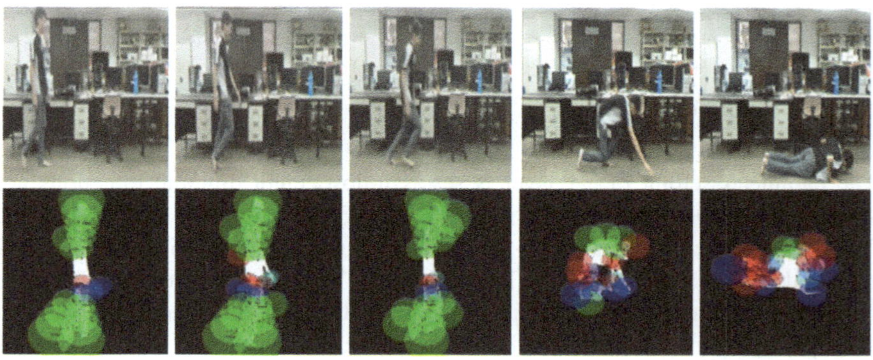

Fig. 3 Result shows the original video dataset of SESRG (*top*) and the AMM estimated from Gaussian distribution of MT cells (*below*). The different color coded indicate different orientation angle of the motion patterns

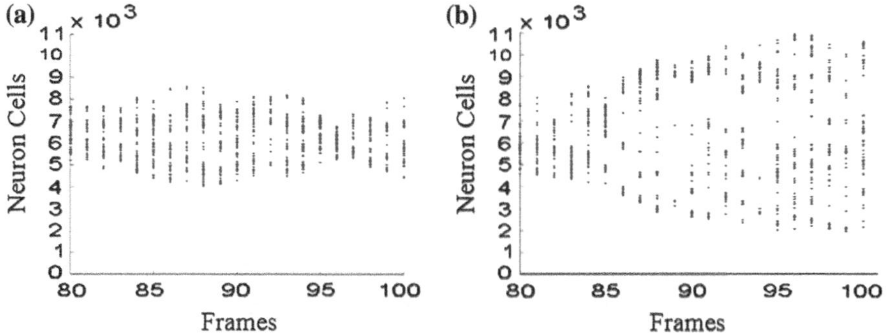

Fig. 4 Raster plot of all MT neurons corresponding to **a** walk and **b** fall

Table 2 Comparison with previous work

Previous work	Overall (%)
JHuang et al. [8]	92.7
Escobar et al. [9]	97.8
BIM-AMM	98.5

Gaussian distribution. This results proved the assumption that center position of the body has the highest Gaussian distribution to represent AMM of MT cells. The AMM model of MT cells was distributed in log-polar form and presented in color coded which indicate the eight angle of motion directions. Figure 4 shows the raster plot which represents spike trains of all MT neurons corresponding to a given orientation and speed for walking exercise and sudden fall event in Fig. 3.

To present with fair and meaningful results, the performance comparison of different bio-inspired motion processing approaches is made on the same database with similar number of frames. Therefore, the simulations were carried out for previous bio-inspired model applied by JHuang [8] and MT-like model proposed by Escobar [9] using the dataset in Table 2. This extended simulation work used to verify the effectiveness of our proposed BIM-AMM model for detection of fall during walking exercise in home-based physiotherapy system. Table 2 gives a comparison among JHuang model, Escobar model and our proposed BIM-AMM framework.

Our proposed model gives average recognition rate higher than dense C2 features of JHuang model. BIM-AMM model based on 3D Gabor can simulate complete information in spatial and temporal domain compared to Escobar's which might have omitted some information using 2D Gabor filter. The Gaussian distribution estimation in BIM-AMM framework gave better estimation of preferred motion direction compared to mean motion map in Escobar's. It can be seen that the spatio-temporal 3D Gabor filter and the estimation of AMM through Gaussian distribution in BIM-AMM model improved the average recognition of Escobar V1/MT model. The Gaussian distribution of active motion map able to reduce noise in MT layer and improved the decision-making process using SVM classifier. Comparing with different datasets, it shows that our approach able to overcome the effect of scaling and lighting variation. Therefore, it can be concluded that our model is highly effective.

4 Conclusion

In conclusion, this paper proposed a bio-inspired visual cortex system for home-based rehabilitation system. An accurate physiotherapy system to support home-based rehabilitation programs for the elderly and their caregivers is constructed using bio-inspired model. Our main contribution is encoded motion patterns from incoming videos by 3D Gabor filter and estimate active motion map through Gaussian distribution of MT preferred cells directions. This paper presents some evaluation results compared with other previous methods. In our future work, object tracking method can be incorporated to further improve the localization of the object, extended with multimodal framework like shape by utilizing ventral pathway like V1, V2, V3, V4 and IT area. The analysis of shape through ventral stream may be also important for recognition of motion in human activity.

References

1. Zheng H, Davies RJ, Black ND (2005) Web-based monitoring system for home-based rehabilitation with stroke patients. In: Proceedings of symposium on computer-based medical system
2. Jovanov E, Milenkovic A, Otto C, de Groen PC (2005) A wireless body area network of intelligent motion sensors for computer assisted physical rehabilitation. J Neuroeng Rehabil 2:6
3. Martins T (2013) Application for physiotherapy and tracking of patients with neurological diseases—preliminary studies. In: IEEE 2nd international conference on serious games and applications for health, pp 1–8
4. Ar I, Akgul YS (2014) A computerized recognition system for the home-based physiotherapy exercises using an RGBD camera. In: IEEE Trans Neural Syst Rehabil Eng 22(6):1160–1170
5. Loconsole C, Bann F, Frisoli A, Bergamasco M (2012) A new Kinect-based guidance mode for upper limb robot-aided neurorehabilitation. In: IEEE international conference on intelligent robots and systems, pp 1037–1042
6. Kalra S, Kalra B, Kumar N (2007) Prevention and management of diabetes: The role of the physiotherapist. Diab Voice 52(3):12–14
7. Serre T, Wolf L, Bileschi S, Riesenhuber M, Poggio T (2007) Robust object recognition with cortex-like mechanisms. IEEE Trans Pattern Anal Mach Intell 29(3):411–426
8. Jhuang H, Serre T, Wolf L, Poggio T (2007) A biologically inspired system for action recognition. In: 2007 IEEE 11th international conference on computer vision. doi:10.1109/ICCV.2007.4408988
9. Escobar M-J, Kornprobst P (2012) Action recognition via bio-inspired features: the richness of center–surround interaction. Comput Vis Image Underst 116(5):593–605. doi:10.1016/j.cviu.2012.01.002

10. Giese M, Poggio T (2003) Neural mechanisms for the recognition of biological movements and actions. Nat Rev Neurosci 4:179–192
11. Petkov N, Subramanian E (2007) Motion detection, noise reduction, texture suppression and contour enhancement by spatiotemporal Gabor filters with surround inhibition. Biol Cybern 97 (5–6):423–439
12. Watson A, Ahumada A (1985) Model of human visual-motion sensing. J Opt Soc Am 2 (2):322–342
13. Mante V, Carandini M (2005) Mapping of stimulus energy in primary visual cortex. J Neurophysiol 94:788–798
14. Daugmann J (1988) Complete discrete 2D-Gabor transforms by neural networks for image analysis and compression. IEEE Trans Acoust Speech Sig Process 36:1169–1179
15. Shadlen M, Britten K, Newsome W, Movshon J (1996) A computational analysis of the relationship between neuronal and behavioral responses to visual motion. J Neurosci 16 (4):1486–1510
16. Vapnik V (1995) The nature of statistical learning theory. Springer, New York

Online Clustering of Narrowband Position Estimates with Application to Multi-speaker Detection and Tracking

Maja Taseska, Gleni Lamani and Emanuël A.P. Habets

Abstract Speaker detection, localization and tracking are required in systems that involve e.g. hands-free speech acquisition, or blind source separation. Localization can be done in the (TF) domain, where location features extracted using microphone arrays are used to cluster the TF bins corresponding to the same source. The TF clustering approaches provide an alternative to the Bayesian tracking approaches that are based on Kalman and particle filters. In this work, we propose a maximum-likelihood approach where detection, localization, and tracking are achieved by online clustering of narrowband position estimates, while incorporating the speech presence probability at each TF bin in a unified manner.

Keywords Multi-speaker tracking · Maximum likelihood · Number of source estimation · Distributed arrays

1 Introduction

To provide high-quality capture of speech in communication and entertainment systems without requiring close-talking microphones, the spatial diversity of microphone arrays is exploited to extract sources of interest. Such systems need to localize and track a desired source, and use the location information to compute a spatial filter (beamformer) that extracts the source signal and reduces interferers. A multitude state-of-the-art tracking approaches estimate the evolution of the source positions using Kalman or particle filters (see [1, 2] and references therein). Although these approaches are elegantly formulated within a Bayesian framework and provide excellent tracking performance, the estimated source positions can only be used to steer data-independent beamformers to the estimated source locations.

A joint institution of the University Erlangen-Nuremberg and Fraunhofer IIS.

M. Taseska (✉) · G. Lamani · E.A.P. Habets
International Audio Laboratories Erlangen, Erlangen, Germany
e-mail: maja.taseska@audiolabs-erlangen.de

P.J. Soh et al. (eds.), *Advances in Machine Learning and Signal Processing*,
Lecture Notes in Electrical Engineering 387, DOI 10.1007/978-3-319-32213-1_6

From the vast literature on spatial filters, it is known that such filters often provide insufficient performance in reverberant environments.

The localization can alternatively be done by exploiting the speech sparsity in the TF domain [3]. Such approaches often involve clustering of location features, such as phase differences between sensors [4, 5], (DOAs) [6, 7], or narrowband positions [8]. As a by-product of the localization, each TF bin is classified to the dominant source providing means to track the second-order statistics of the sources [7–9]. The latter can be used to compute data-dependent beamformers which offer better performance in reverberant and noisy environments than data-independent beamformers. Note that after clustering of DOAs or phase differences, an additional step is required to obtain the Cartesian coordinates of the sources, in case they are required. Due to the non-linear and non-injective relation between the DOA and the position, this step is non-trivial. Approaches that estimate the positions have been proposed in [5], by deriving the position from clustered phase differences, and in [8], by clustering narrowband position estimates. However, these approaches assume that the number of sources is fixed, which is restrictive in practice. DOA-based clustering that detects the number of sources online has been proposed in [7, 10].

For the application of source localization, source tracking, and clustering of the TF bins, we propose a (ML) framework based on narrowband position estimates obtained from distributed arrays. Narrowband positions were used in our previous work for localization [8] and tracking of a known number of sources [11]. In this paper, we address dynamic scenarios where the number of sources is unknown and time-varying. The proposed framework consists of (i) estimating the parameters of a mixture model by an online (EM) algorithm (Sects. 2 and 3) and (ii) a data-driven mechanism to detect appearing/disappearing sources and add/remove the corresponding clusters (Sect. 4). A further contribution is the unified treatment of speech presence uncertainty in the model, which increases the robustness to noise and reverberation without requiring voice activity detection in a pre-processing stage as in [11, 12]. The cluster information can be used for multi-speaker tracking, as well as for computation of data-dependent spatial filters for (BSS). The evaluation in Sect. 5 focuses on the tracking application, whereas the evaluation of a BSS system is an ongoing work.

2 Probabilistic Model of Narrowband Position Estimates

The signals from S sources in a noisy and reverberant enclosure are captured by A microphone arrays. The signal at microphone m from the a-th array, at time index n and frequency index k in the (STFT) domain is given by $Y_m^{(a)}(n,k) = \sum_{s=1}^{S} X_{m,s}^{(a)}(n,k) + V_m^{(a)}(n,k)$, where $X_{m,s}^{(a)}$ and $V_m^{(a)}$ denote the signal of the s-th source and the noise, respectively. The different signals represent realizations of mutually uncorrelated random processes. Assuming far field conditions and selecting an arbitrary reference microphone m', a DOA θ_a at array a can be obtained at each TF bin by a phase difference-based estimator [13] (TF indices omitted for brevity)

$$n_a = \begin{bmatrix} \cos(\theta_a) \\ \sin(\theta_a) \end{bmatrix} = \frac{c}{2\pi f}[\boldsymbol{D}_{m'}^{(a)}]^{\dagger} \arg \frac{\boldsymbol{y}^{(a)}}{Y_{m'}^{(a)}}, \tag{1}$$

where $\boldsymbol{D}_{m'}^{(a)} = [\boldsymbol{d}_1^{(a)} - \boldsymbol{d}_{m'}^{(a)}, \ldots, \boldsymbol{d}_{M_a}^{(a)} - \boldsymbol{d}_{m'}^{(a)}]^{\mathrm{T}}$, $\boldsymbol{d}_m^{(a)}$ for $m = 1, \ldots M_a$ denote the positions of the microphones from array a, $\boldsymbol{y}^{(a)}$ contains the signals from array a stacked in a vector, c and f are the speed of sound and the frequency in Hz, † denotes the Moore-Penrose pseudoinverse, and $\arg(\cdot)$ is taken element-wise. Although the DOAs consider only azimuth θ_a, an extension to elevation is possible. By triangulating two vectors $\boldsymbol{n}_{a_1}$ and $\mathbf{n}_{a_2}$ from different arrays, a position $\boldsymbol{\Theta}_{nk}$ is obtained for each bin (n, k). As the signals at each TF bin represent (RV), $\boldsymbol{\Theta}_{nk}$ is also an RV. The position estimates are used to cluster the TF bins based on their respective dominant source. If at TF bin (n, k), the energy of a given source is dominant over the other sources and the noise, the position $\boldsymbol{\Theta}_{nk}$ represents a good estimate of the source location. Note that the DOA estimates can be obtained with any narrowband estimator and the choice of an estimator influences the accuracy of the clustering.

Due to the speech sparsity in the STFT domain [3], it can be assumed that there is at most one dominant source at each TF bin. Let z_{nk} be a discrete RV that takes values from 0 to S, indicating the dominant source as follows

$$z_{nk} = 0 \quad \text{if} \quad \mathbf{y}(n,k) = \mathbf{v}(n,k) \text{ (i.e. only noise present)}, \tag{2a}$$

$$z_{nk} = s \quad \text{if} \quad \mathbf{y}(n,k) \approx \mathbf{x}_s(n,k) + \mathbf{v}(n,k), \tag{2b}$$

where the entries of $\mathbf{y}$, $\mathbf{x}_s$, and $\mathbf{v}$ contain the respective signals from all microphones. Marginalizing over the unobservable RV z, the distribution of the observable RV $\boldsymbol{\Theta}$ is given by $p(\boldsymbol{\Theta}) = \sum_z p(z)p(\boldsymbol{\Theta}|z)$ (subscript nk omitted for brevity). We propose the following parametric model for the likelihood $p(\boldsymbol{\Theta}|z)$

$$p(\boldsymbol{\Theta} \mid z) = \mathcal{N}(\boldsymbol{\Theta}; \boldsymbol{\mu}_z, \boldsymbol{\Sigma}_z), \quad \text{for} \quad z \neq 0, \tag{3a}$$

$$p(\boldsymbol{\Theta} \mid z) = \mathcal{U}(\boldsymbol{\Theta}), \quad \text{for} \quad z = 0, \tag{3b}$$

where $\mathcal{U}$ is a uniform distribution (noise is localized across the room with equal probability) and $\mathcal{N}(\boldsymbol{\Theta}; \boldsymbol{\mu}_z, \boldsymbol{\Sigma}_z)$ is a two-dimensional Gaussian distribution with mean $\boldsymbol{\mu}_z$ and covariance matrix $\boldsymbol{\Sigma}_z$. The mean $\boldsymbol{\mu}_z$ represents the true location of the z-th source in the xy plane. By writing the marginal distribution of z as

$$p(z) = p(z \mid z \neq 0)p(z \neq 0) + p(z \mid z = 0)\, p(z = 0), \tag{4}$$

and introducing $\pi_z = p(z \mid z \neq 0)$ and $\pi_0 = p(z \neq 0)$, where π_0 is the (SPP), the following holds

$$p(z) = \begin{cases} \pi_z \pi_0, & \text{if } z \neq 0 \\ (1 - \pi_0) & \text{if } z = 0. \end{cases} \tag{5}$$

The parametrized distribution of the observable RV $\boldsymbol{\Theta}$ can now be written as

$$p(\boldsymbol{\Theta}; \mathcal{P}, \pi_0) = \sum_{z \neq 0} \pi_z \pi_0 \mathcal{N}(\boldsymbol{\Theta}; \boldsymbol{\mu}_z, \boldsymbol{\Sigma}_z) + (1 - \pi_0)\,\mathcal{U}(\boldsymbol{\Theta}), \tag{6}$$

where $\mathcal{P} = \{\pi_z, \boldsymbol{\mu}_z, \boldsymbol{\Sigma}_z\}_{z \in [1,S]}$ are unknown parameters. The SPP π_0 is treated as a known parameter as it can be estimated at each bin, independently of the clustering framework. The interested reader is referred to [14] and references therein for details on the computation of the SPP π_0.

The goal in this work consists of inferring and tracking the time-varying parameters $\mathcal{P}_n$ and the number of speakers S_n online as blocks of narrowband position estimates become available. The framework should be capable of removing and adding clusters for sources that disappear and appear, respectively.

3 Maximum Likelihood-Based Online Clustering

In this section, we derive the update of the parameters $\widehat{\mathcal{P}}_{n-1}$ at frame $n - 1$ to the new estimates $\widehat{\mathcal{P}}_n$ in light of the input data at frame n. For ease of exposition, we assume that the number of speakers is equal in the two successive frames. A mechanism that determines the number of speakers is detailed in Sect. 4.

The observable data at frame n is the set of position estimates and the corresponding SPP from the most recent L frames, across all K frequency bins

$$D_n = \{(\boldsymbol{\Theta}_{ik}, \pi_{0,ik}) | i \in [n - L + 1, n], k \in [1, K]\}. \tag{7}$$

Note that depending on the accuracy of the DOA estimates at different frequencies and on the spatial aliasing limit for the microphone array, data only from a subset of frequencies can be used as done e.g. in [7]. The parameters $\widehat{\mathcal{P}}_n$ are obtained by maximizing the following log likelihood, with respect to $\mathcal{P}_n$

$$L(D_n; \mathcal{P}_n) = \sum_{i,k} \ln p(\boldsymbol{\Theta}_{ik}; \mathcal{P}_n, \pi_0). \tag{8}$$

We omit the TF-bin index from π_0 for brevity, although π_0 is computed bin-wise. The mixture model given by (6) leads to a summation inside the logarithm in (8), which does not allow for a closed form maximization. Instead, maximizing the expectation of the complete data likelihood under the posterior distribution of the unobservable z, represents a significantly easier problem [15]. This expectation, also known as the Q-function, is given for our model by

$$\sum_{i,k} \Big\{ p(z_{ik} = 0 \,|\, \boldsymbol{\Theta}_{ik}; \widehat{\mathcal{P}}_{n-1}) \left[\ln\,(1-\pi_0) + \ln\, \mathcal{U}(\boldsymbol{\Theta}_{ik}) \right] \\ + \sum_{z \neq 0} p(z_{ik} = z \,|\, \boldsymbol{\Theta}_{ik}; \widehat{\mathcal{P}}_{n-1}) \left[\ln \pi_z \pi_0 \; p(\boldsymbol{\Theta}_{ik} \,|\, z_{ik} = z; \mathcal{P}_n) \right] \Big\}, \quad (9)$$

where the posterior distribution of z_{ik} is computed with respect to the old parameters $\widehat{\mathcal{P}}_{n-1}$. Evaluating (9), and maximizing it with respect to $\mathcal{P}_n$ represents an iteration of the EM algorithm, guaranteeing that $L(D_n; \mathcal{P}_n) > L(D_n; \mathcal{P}_{n-1})$ [15]. Setting the derivatives of (9) with respect to $\mathcal{P}_n$ to zero, the standard M-step for a (GMM) is obtained, as the terms due to $z = 0$ do not depend on $\mathcal{P}_n$. Introducing $P_z(n) = \sum_{i,k} p(z_{ik} = z \,|\, \boldsymbol{\Theta}_{ik})$, the new parameters are computed as

$$\pi_z = \frac{P_z(n)}{LK}, \quad \boldsymbol{\mu}_z = \frac{\sum_{i,k} p(z_{ik} = z \,|\, \boldsymbol{\Theta}_{ik})\, \boldsymbol{\Theta}_{ik}}{P_z(n)}, \\ \boldsymbol{\Sigma}_z = \frac{\sum_{i,k} p(z_{ik} = z \,|\, \boldsymbol{\Theta}_{ik})\, (\boldsymbol{\Theta}_{ik} - \boldsymbol{\mu}_z)(\boldsymbol{\Theta}_{ik} - \boldsymbol{\mu}_z)^{\mathrm{T}}}{P_z(n)}. \quad (10)$$

To compute the posteriors $p(z_{ik} = z \,|\, \boldsymbol{\Theta}_{ik})$, we express them as follows

$$p(z_{ik} = z \,|\, \boldsymbol{\Theta}_{ik}) = p(z_{ik} = z \,|\, \boldsymbol{\Theta}_{ik}, z \neq 0)\, \pi_0 \; + \; p(z_{ik} = z \,|\, \boldsymbol{\Theta}_{ik}, z = 0)\, (1 - \pi_0). \quad (11)$$

Next, noting that for $z_{ik} \neq 0$ the second term equals zero, the posterior for $z_{ik} \neq 0$ can be written by applying the Bayes theorem to the first term, i.e.,

$$p(z_{ik} = z \,|\, \boldsymbol{\Theta}_{ik}) = \pi_0 \cdot \frac{p(\boldsymbol{\Theta}_{ik} \,|\, z_{ik} = z)\, \pi_z\, \pi_0}{\sum_{z' \neq 0} p(\boldsymbol{\Theta}_{ik} \,|\, z_{ik} = z')\, \pi_{z'}\, \pi_0}. \quad (12)$$

Hence, by virtue of the proposed model in Sect. 2, the speech presence uncertainty is inherently considered when computing the model parameters. The EM iteration given by (9)–(12) can be efficiently implemented using sufficient statistics for GMMs. The reader is referred to our work in [11] where the EM algorithm was implemented in this manner.

Given the estimated parameter set $\mathcal{P}_n$ at each frame n, the sequence $\left[\boldsymbol{\mu}_{z,1}, \boldsymbol{\mu}_{z,2}, \ldots, \boldsymbol{\mu}_{z,n}\right]$ represents the estimated track for the s-th speaker, whereas the posterior probabilities can be used to design TF masks or spatial filters for BSS [7, 9, 11].

4 Robust Counting and Tracking of Sources

The EM iteration, described in Sect. 3, is coupled with an outlier control and a source counting mechanism. The outlier control is based on the likelihood of the incoming data under the current parameter estimates and implicitly imposes smooth

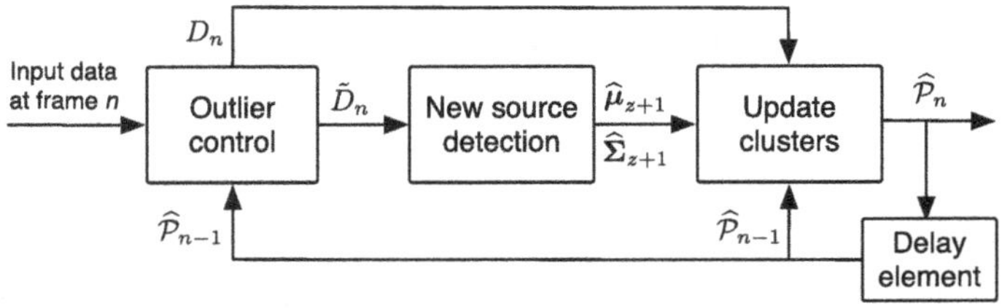

Fig. 1 Diagram of the proposed system. The parameters $\widehat{\mathcal{P}}_{n-1}$ from the previous frame are used to remove outliers and to run an EM iteration at the current frame n

speaker tracks, as described next. A block diagram of the proposed online clustering framework is illustrated in Fig. 1.

4.1 Outlier Control and Smoothness of Estimated Tracks

Noise and reverberation result in outliers which often exceed the number of reliable data points. If there are erroneous SPP estimates in the incoming data, the ML-based criterion in Sect. 3 forces the parameter estimates to fit noisy data that do not accurately represent the speaker locations. Without a motion model or track smoothness constraint, even moderate amount of outliers lead to significant tracking errors. We propose a data-driven approach for trimming the set D_n to contain only positions $\boldsymbol{\Theta}_{ik}$ that belong to a specified confidence region of at least one of the clusters. Given the confidence probability p, the data used in the EM step at frame n needs to satisfy the following, for at least one $z \in [1, S]$

$$(\boldsymbol{\Theta}_{ik} - \boldsymbol{\mu}_{z,n-1})^{\mathrm{T}} \boldsymbol{\Sigma}_{z,n-1}^{-1} (\boldsymbol{\Theta}_{ik} - \boldsymbol{\mu}_{z,n-1}) \leq \Psi_p. \tag{13}$$

As the quadratic form on the left hand side in (13) follows a Chi-squared distribution with two degrees of freedom, Ψ_p and p are related as $p = 1 - e^{-\frac{\Psi_p}{2}}$ [16]. The data trimming step implicitly imposes smoothness of the speaker tracks and assists the detection of new speakers, as described in Sect. 4.2.

Note that the points with low SPP, and hence a low impact on $\widehat{\mathcal{P}}_n$, can be removed from D_n without evaluating (13). In this manner, the storage and computation complexity are notably reduced, as due to speech sparsity the number of low SPP points is significant. Thresholding based on voice activity detection is often a required step in acoustic source clustering and tracking [6, 11, 12]. In this work, due to the incorporated SPP and outlier control, this step is optional and is done only to reduce the computational cost.

4.2 Removing and Adding Speakers

Removing speakers. Inspired by the sparse EM variant [17], where at frame n only the Gaussian components with large responsibility for observing the current data D_n are updated, we propose to first update the mixture coefficients π_z using the data D_n, update the mean and covariance according to (10) only for the components z for which $\pi_z > \pi_{\text{thr}}$, and freeze the means and covariances otherwise. The frozen parameters indicate inactive sources that are removed if their parameters are frozen longer than a pre-defined number of frames L_{frz}. The value L_{frz} is often referred to as *time-to-live* in tracking literature. In this work, L_{frz} was fixed to 60 frames, corresponding to 1.9 s. Alternatively, L_{frz} can be computed online based on the predicted travelled distance within a silent period. If a speaker travels a large distance without acoustic activity, the track can not be recovered when acoustic activity is resumed. This behavior is due to the smoothness requirement implicitly imposed by (13). Instead, the speaker is removed after L_{frz} frames, and promptly re-detected as a new speaker when activity is resumed.

Adding speakers. Let us denote by $\widetilde{D}_n$ the set of all the points that were removed from D_n by (13), and for which the SPP satisfies $\pi_0 > p_{sp}$. Assuming that there is no new speaker at frame n, the cardinality $|\widetilde{D}_n|$ is low as all points with high SPP are modeled by the current GMM and remain in the set D_n. On the contrary, if a new speaker appears, a cluster of points is present in $\widetilde{D}_n$. To verify the existence of a new cluster we first take the maximum $\theta_{a,\max}$ of the DOA histogram for each array a, computed using the TF bins in $\widetilde{D}_n$. Consider a set $\widetilde{D}'_n \subset \widetilde{D}_n$ such that for all TF bins (i, k) corresponding to the points in $\widetilde{D}'_n$, the following holds

$$\widetilde{D}'_n = \{\boldsymbol{\Theta}_{ik} \,|\, |\theta_{a,\max} - \theta_{a,ik}| < \Delta_\theta,\ \forall a\}. \tag{14}$$

The threshold Δ_θ is chosen such that the intersection region satisfying (14) is sufficiently small, so that a large cardinality $|\widetilde{D}'_n|$ indicates, with high probability, a new source activity in that region. Denote by $\boldsymbol{\mu}_{z+1}$ the mean of the data points in $\widetilde{D}'_n$. To ensure that the newly detected source does not model an already existing one we impose a limit on the minimum distance between two detected sources, leading to the last condition for adding a source

$$||\boldsymbol{\mu}_{z+1} - \boldsymbol{\mu}_s||_2 > d_{\min}, \forall s \wedge |\widetilde{D}'_n| \geq \xi \tag{15}$$

where ξ is the minimum number of points required to declare a new speaker. If (15) is satisfied, a new Gaussian is initialized with mean $\boldsymbol{\mu}_{z+1}$ and a scaled identity covariance, and the standard EM iteration is executed with the new model. Note that although only one new speaker per frame can be detected, there is no constraint on the number of new speakers that appear in the same frame. Due to the outlier-based speaker detection, both speakers will be detected, however with at least one frame delay between the two detections.

In addition, the following check is continuously exectuted $||\boldsymbol{\mu}_i - \boldsymbol{\mu}_j||_2 < d_{\mathrm{mrg}}$ so that clusters of speaker i and j are merged in the case of crossing tracks. After the speakers move away from each other, the two separate clusters and the corresponding trajectories are recovered, as shown in the results. Although in certain applications, it might be desirable to maintain separate speaker tracks regardless of crossing, the proposed system in this work was developed in view of the signal extraction applications using spatial filters, where speakers with small inter-speaker distances cannot be separated by the spatial filter as separate sound sources. Therefore, we choose to merge the the tracks of crossing trajectories.

5 Performance Evaluation

The proposed algorithm was evaluated in a simulated $6 \times 5 \times 3$ m room. Clean speech signals were convolved with room impulse responses for moving sources using the software in [19]. Diffuse babble noise [18] and uncorrelated sensor noise were added to the speech signals. The STFT frame length was 64 ms, with 50 % overlap, at a sampling rate of 16 kHz. Three uniform circular arrays of diameter 2.9 cm and three omnidirectional microphones per array were employed. All relevant processing parameters are summarized in Table 1. Scenarios with different reverberation times T_{60}, noise levels, number of sources, and motion patterns were examined. The system is tested in dynamic situations with appearance of new sources, speech pauses, and sources with crossing trajectories.

Experiment 1. In this experiment, we tested the tracker for a fixed number of moving speakers. We started the algorithm with an unknown number of speakers and once all were detected, the RMSE between the true and the estimated tracks was computed. The results in Table 2 are averaged over time frames, over speakers, and over three scenarios with different motion patterns. Two, three, and four simultaneously active speakers were tracked for signal-to-noise ratios of 11 and 21 dB and speeds of 0.23 and 0.34 m/s. It can be observed that while the system is robust to noise and reverberation, accuracy decreases for faster speaker movement. The sensitivity to speed can be controlled by the number of frames L that constitute

Table 1 Parameters used in the implementation

L	Δ_θ	p_{sp}	p	π_{thr}	L_{frz}	ξ	d_{min}	d_{mrg}
30	10°	0.85	0.95	0.07	60	30	0.5	0.3

Table 2 Average root mean squared error in meters between the true speaker location and the mean of the respective Gaussian

SNR (dB)	speed (m/s)	$T_{60} = 200$ ms			$T_{60} = 400$ ms		
		two	three	four	two	three	four
21	0.23	0.16	0.14	0.12	0.18	0.15	0.22
21	0.34	0.23	0.17	0.17	0.24	0.22	0.25
11	0.23	0.16	0.12	0.19	0.20	0.16	0.24
11	0.34	0.25	0.23	0.24	0.26	0.26	0.26

the short-term data. Smaller L allows for faster adaptation suited for higher speeds, however leads to quickly varying parameter estimates and less smooth tracks. Experiments showed that $L \in [25, 35]$ offers a good tradeoff between responsiveness and stability.

Experiment 2. We tested the tracker in several challenging scenarios which often occur in practice. In Fig. 2, we illustrate a four-speaker example with $T_{60} = 200$ ms and velocity 0.34 m/s for each speaker. Snapshot II shows a detection of a new speaker, where even in multitalk scenarios, a new speaker is detected in less than 0.5 s after the first activity. Snapshots I and III show stabilized clusters of three and four speakers, respectively. Finally, snapshot IV is taken at the frame where the third speaker is discarded, and only two Gaussians track the remaining two speakers.

Experiment 3. In the last experiment, the tracker was tested in a triple-talk scenario when the trajectories of two speakers cross. The tracking result is visualized in Fig. 3, where the x and y coordinates of the true and estimated tracks are plotted across time. The crossing happens around second 5 on the time axis, where it is visible that for a short period the two speakers are represented by a single Gaussian component. When speakers split, around second 7, the tracker promptly detects a new speaker, assigns a new Gaussian distribution to it, and continues to track the three speakers.

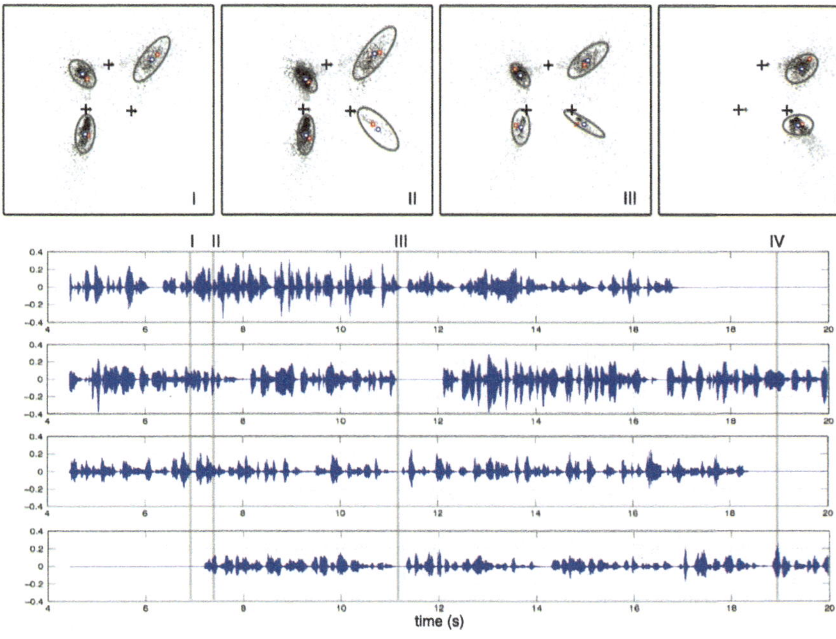

Fig. 2 Snapshots of the tracking at different times and the signals of each speaker. The number at the *right bottom corner* of each snapshot relates to the signal segment as indicated by the markers. SNR 21 dB, T_{60} = 200 ms, velocity 0.34 m/s. The *black dots* denote the points in D_n that are used in the current EM iteration, whereas the *gray dots* denote the discarded points $\widetilde{D}_n$. The *plus signs* denote the microphone arrays

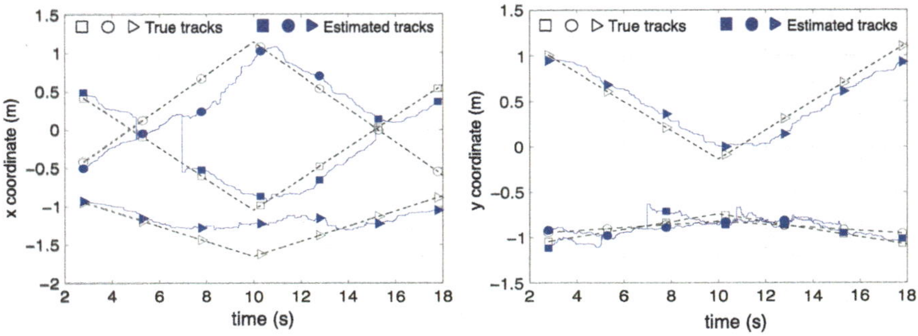

Fig. 3 Visualizing the speaker tracks for a scenario with crossing trajectories. SNR = 21 dB, T_{60} = 200 ms, velocity 0.34 m/s

6 Conclusions

A maximum likelihood framework for multi-speaker detection and tracking was proposed, based on clustering of narrowband position estimates. The position estimates are obtained by using at least two distributed arrays. Speech presence uncertainty and outlier control are incorporated in a unified manner, resulting in a system that is robust to noise and reverberation, estimates the number of speakers online, and allows for track recovery even in situations where the sources have crossing trajectories. As a by-product of the clustering-based tracking, each TF bin is classified to the dominant source providing means to design data-dependent spatial filters for blind source separation.

References

1. Fallon FC, Godsill JS (2012) Acoustic source localization and tracking of a time-varying number of speakers. IEEE Trans Audio Speech Lang Process 20:1409–1415
2. Gehrig T, Klee U, McDonough J, Ikbal S, Wölfel M, Fügen C (2006) Tracking and beamforming for multiple simultaneous speakers with probabilistic data association filters. Interspeech
3. Yilmaz O, Rickard S (2004) Blind separation of speech mixture via time-frequency masking. IEEE Trans Signal Process 52:1830–1847
4. Mandel M, Ellis D, Jebara M (2006) An EM algorithm for localizing multiple sound sources in reveberant environtments Proceedings of Neural Information Processing System
5. Schwartz O, Gannot S (2014) Speaker tracking using recursive EM algorithms. IEEE Trans Audio Speech Lang Process 22:392–402
6. Loesch B, Yang B (2008) Source number estimation and clustering for underdetermined blind source separation. Proceedings of international workshop on acoustic signal enhancement
7. Madhu N, Martin R (2011) A versatile framework for speaker separation using a model-based speaker localization approach. IEEE Trans Audio Speech Lang Process 19:1900–1912
8. Taseska M, Habets EAP (2014) Informed spatial filtering with distributed arrays. IEEE Trans Audio Speech Lang Process 22:1195–1207

9. Souden M, Kinoshita K, Delcroix M, Nakatani T (2014) Location feature integration for clustering-based speech separation in distributed microphone arrays. IEEE Trans Audio Speech Lang Process 22:354–367
10. Plinge A, Fink GA (2014) Multi-speaker tracking using multiple distributed microphone arrays. Proceedings of IEEE international conference on acoustics, speech and signal processing
11. Taseska M, Habets EAP (2013) An online EM algorithm for source extraction using distributed microphone arrays. Proceedings of European signal processing conference
12. Lehmann EA, Johansson AM (2007) Particle filter with integrated voice activity detection for acoustic source tracking. EURASIP J Appl Signal Process
13. Araki S, Sawada H, Mukai SMR (2006) DOA estimation for multiple sparse sources with normalized observation vector clustering. Proceedings of IEEE international conference on acoustics, speech and signal processing
14. Taseska M, Habets EAP (2012) MMSE-based blind source extraction in diffuse noise fields using a complex coherence-based a priori SAP estimator. Proceedings of international workshop acoustic signal enhancement
15. Demptster AP, Laird NM, Rubin DB (1977) Maximum likelihood from incomplete data via the EM algorithm. J R Statist Soc 39:1–38
16. Bar-Shalom Y (2001) Estimation with applications to tracking and navigation. Wiley & Sons
17. Neal RM, Hinton GE (1998) A view of the EM algorithm that justifies incremental, sparse, and other variants. Learning in Graphical Models, Kluwer Academic Publishers, p 355–368
18. Habets EAP, Gannot S (2007) MATLAB implementation for: generating sensor signals in isotropic noise fields. [Online]. Available: https://www.audiolabs-erlangen.de/fau/professor/habets/software/noise-generators
19. Habets EAP Available: http://www.audiolabs-erlangen.de/fau/professor/habets/software/signal-generator

Taguchi-based Parameter Setting of Cuckoo Search Algorithm for Capacitated Vehicle Routing Problem

Mansour Alssager and Zulaiha Ali Othman

Abstract The cuckoo search algorithm is a novel metaheuristic based on the obligate brood parasitic behavior of some cuckoo species in combination with the Lévy flight behavior of some birds and fruit flies. Its shows good performance when applied to various domains in continuous and discrete search space. This study presents a cuckoo search algorithm for capacitated vehicle routing problem; its parameter optimized using Taguchi method. The comparison results on 16 benchmark instances shows potential performance of the Taguchi based method which outperforms the basic one.

Keywords Taguchi method · Cuckoo search · Parameter setting · Capacitated vehicle routing problem

1 Introduction

Various metaheuristic algorithms have been successfully applied to solve the Vehicle Routing Problem (VRP). Most of the common algorithms introduced for VRPs can be found in the review by Gendreau et al. [1]. However, as yet none of these algorithms have been able to reach the optimal solution across all the benchmark datasets and the community is still looking for a more stable and robust algorithm. The main challenge of the VRP is to design least-cost (distance, time) routes for a fleet of vehicles to serve geographically scattered clients. Capacitated VRP (CVRP) is one of the VRP extensions, in which the total demand for any vehicle cannot exceed a present capacity. As the CVRP is classified as a NP-hard problem based on the theory of computational complexity, various

M. Alssager (✉) · Z.A. Othman
Data Mining and Optimization Group, Faculty of Information Sciences and Technology, Centre of Artificial Intelligence, University Kebangsaan Malaysia, Selangor, Malaysia
e-mail: mansour.alssager@gmail.com

Z.A. Othman
e-mail: zao@ftsm.ukm.com

P.J. Soh et al. (eds.), *Advances in Machine Learning and Signal Processing*, Lecture Notes in Electrical Engineering 387, DOI 10.1007/978-3-319-32213-1_7

approaches have been presented to solve the CVRP, and these can be divided into two types: exact and metaheuristic algorithms. To the best of our knowledge no work has attempted to apply Cuckoo Search (CS) to solving the CVRP. Consequently, and based on CS recent advantage particularly for routing problem such as in Travel Salesman Problem [2]. This research proposes a CS to solve the CVRP.

In most of the cases the parameters of the metaheuristics have been selected based on two ways, on the experience of the authors or other available approaches in the literature. Trial and error assortment of the parameters make the method less productive. Since selection of proper parameters of the metaheuristic algorithms is problem dependent and an important issue for any optimization approach, therefore a systematic and robust strategy of the parameter designing is required to make the method with more accuracy and efficiency.

However, the impact of those of CS parameters on the algorithm performance is not extensively investigated up-to-date since the suggested value made by Yang and Deb [3]. Yang and Deb [3] Provide some guiding suggestions for reasonable choices of those parameters but the opening question is: Is that parameter suggestion applicable for all optimization problems? In fact, related parameters setting of the most intelligent algorithms are confirmed based on the specific problems being studied [4]. Moreover, Finding a universal parameter setting for intelligence algorithms is rather and more difficult. Therefore, a method such as Taguchi method is a powerful tool for parameter design [5]. It is begin used to evaluate the problem related parameter. Furthermore, this method identifies the best settings of parameters and improves the performance characteristic with orthogonal array technique and signal-to-noise (SN) ratios. However, it's efficiency proven by many study [6–9]. This study uses the Taguchi method to analyze the effect of these parameters and identify the major and minor factors in influencing CS, which provides a basis for choosing the best settings of these parameters of CS. Moreover, the results are compared with the basic CS method.

This paper is organized as following. Section 2 explains the proposed method in detail. The results of the evaluation and analyze of our results are reported in Sect. 3 and finally the conclusion of the work is presented in Sect. 4.

2 The Taguchi parameters based Cuckoo Search Algorithm

The Cuckoo Search algorithm is one of the latest nature-inspired metaheuristic algorithms that belong to the swarm intelligence category [3]. This algorithm is inspired by the cuckoo's reproduction behaviour which consists of laying eggs in the nests of other birds. In the Cuckoo Search algorithm, cuckoos are abstracted as entity having associated a solution (i.e. the egg) of the optimization problem that they try to place in a solution container (i.e. the host bird's nest).

In the CS there are only two parameters to be adjusted i.e. the fraction of eggs to be abandoned each generation pa and the host nest size n. However, Yang and Deb [3] state that the convergence rate was not sensitive to the parameters values using

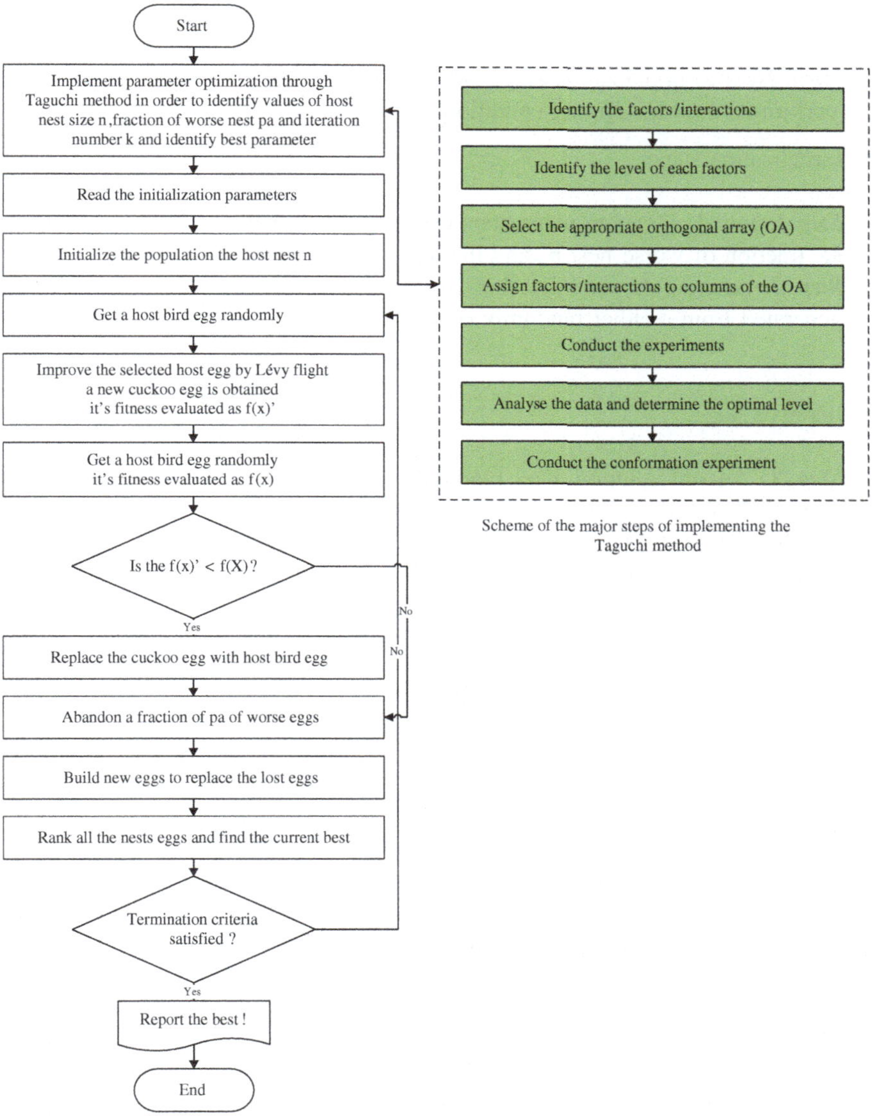

Fig. 1 The flow chart of the proposed method

well-known function benchmark problem which are a continuous search space problem. They came to the conclusion that the CS algorithm finds the optimum solution for the values of the parameter n from 15 to 25 and of the parameter pa from 0.15 to 0.25. Consequently, the open question is that parameters range is suitable for all problem even the discrete one such as CVRP.

The Cuckoo Search algorithm has an initialization stage and an iterative stage. In the initialization stage, a Taguchi method is employed for optimizing the parameters of CS named as (T-CS). Figure 1 shows the flowchart of the proposed T-CS, and then each container is initialized with a solution of the optimization problem. Next to initialization stage is the iterative stage; the improvements steps are performed until a stopping condition is satisfied:

- Initialization stage:

- ***Step 1***: Initial algorithm parameters using Taguchi method. They are: nest size *M*, fraction of worse nest p_a and the maximum number of iterations *K*.
- ***Step 2***: generate *M* candidate host nests. The initial candidate host nests are generated from a either randomly or by a heuristic.

- Iterative stage:

- ***Step 3***: An old solution x^t is randomly selected and a new solution x^{t+1} is generated by performing a random walk using Lévy flight which preserve a step length whether it's large or small with characteristic of balancing between the exploration and exploitation:

$$x^{t+1} = x^t + \alpha * \mathrm{L}(\mathrm{s}, \lambda) \quad (1)$$

Where α is a step-size-scaling factor, s is the step size, and *L* refers to a random walk defined as:

$$\mathrm{L}(\mathrm{s}, \lambda) = \frac{\lambda\,\Gamma(\lambda)\sin(\pi\lambda/2)}{\pi} * \frac{1}{s^{1+\lambda}} \quad (2)$$

There are large neighbourhoods structures have been used as a local search mechanism. The aim of using large neighbourhood structures is to provide significant improvements in the quality of the solution

- ***Step 4***: A container where the cuckoo may lay its solution is randomly chosen.
- ***Step 5***: The new solution will replace the container's solution, if the solution associated to the container has a lower (in case of maximization problems) or higher (in case of minimization problems) fitness than the fitness of the new solution.
- ***Step 6***: Some of the containers containing the worst candidate solutions are destroyed by the host bird (similar to the abandon of the nest by the host bird) with probability of $p_a \in (0, 1)$ and replaced with new ones containing new solutions.
- ***Step 7***: The containers with the best solutions are kept.
- ***Step 8***: The solutions are ranked and the current best solution is found.

3 Result and Discussion

Experiments are conducted to test the performance of the CS on benchmark datasets [10]. There are 16 CVRP instances, the total number of clients varies from 30 to 135 clients, and the total number of vehicles varies from 3 to 10 vehicles. The instances can be downloaded from the website www.branchandcut.org/VRP/data/.

Table 1 Level values obtained for factors

Factor	Level 1	Level 2	Level 3	Level 4	Level 5
p_a	0.1	0.2	0.25	0.4	0.5
n	10	20	50	70	100
S	100	200	300	400	500

Table 2 Parameter values obtained by the Taguchi method in line with parameter range values and results of the experiments

Standard order	p_a	n	S	Average of the result of the observation		S/N ratio
				Objective	Time to best (s)	
1.	0.1	10	100	747.8182	1.007273	−54.4656
2.	0.1	20	200	710	4.764545	−54.0151
3.	0.1	30	300	705.7273	7.235455	−53.9629
4.	0.1	40	400	700.2727	8.915455	−53.8957
5.	0.1	50	500	695.8182	10.92	−53.8407
6.	0.2	10	200	703.1818	3.376364	−53.9312
7.	0.2	20	300	690.7273	4.773636	−53.776
8.	0.2	30	400	688.4545	6.076364	−53.7475
9.	0.2	40	500	684.8182	6.696364	−53.7016
10.	0.2	50	100	708	8.615455	−53.991
11.	0.25	10	300	696.7273	4.065455	−53.8511
12.	0.25	20	400	685.7273	3.128182	−53.7128
13.	0.25	30	500	686.6364	7.382727	−53.7247
14.	0.25	40	100	709	7.915455	−54.0032
15.	0.25	50	200	689.9091	10.14727	−53.7665
16.	0.4	10	400	685.3636	2.678182	−53.7082
17.	0.4	20	500	681.8182	3.009091	−53.6632
18.	0.4	30	100	689.7273	4.082727	−53.7634
19.	0.4	40	200	681.8182	6.684545	−53.6635
20.	0.4	50	300	680.8182	9.337273	−53.6511
21.	0.5	10	500	680.6364	2.070909	−53.648
22.	0.5	20	100	687.0909	3.818182	−53.7301
23.	0.5	30	200	677.7273	6.372727	−53.6112
24.	0.5	40	300	676	7.531818	−53.5892
25.	0.5	50	400	676.9091	6.882727	−53.6008

To generate the Taguchi result we used Minitab software and each example was run for each factor at each level. The 'A-k30-n5.vrp' problem is used as the test problem, consisting of 30 customers and 5 vehicles, this choice is made based on previous study by Peker, Sen [6]. Table 1 presents amount of factors at each level for CS in Taguchi method. The parameters in Table 1, whose lowest and highest levels have been determined, are classified into 5 levels. The table displays the level values obtained for the factors. In the case of an experiment where classic full factorials are used, 35 = 243 experiments would be needed for each observation value. This method ensures the identification of effective parameters and levels with fewer experiments by providing balance among the orthogonal index, parameters, and levels.

Table 2, presents the results obtained from the experiments. However, the table lists the values of pa, n and S, created in the L25 octagonal order. The table also presents the solution travel cost values and the time to best solution as a respond value obtained by implementing the parameters to the problem. The parameters were runs for 11 independent run to increase the reliability of the experiments and the average of all different observation values was obtained. The table also shows the S/N ratio obtained from the experiment.

The results obtained by the Taguchi experimental design was evaluated by transforming the results into S/N ratios (see Table 3). This table shows us the order of importance of the variables and displays the level at which variables should be used in order to achieve the best result. Delta is the difference between the maximum and minimum values of the variable. Rank shows the level according to which variables should be used in order to receive the best solution. The level with the highest S/N value is the most suitable level. The S/N values were calculated using the 'least best' formula, since the shortest total routes length was desired for the CVRP.

Table 3 S/N ratios obtained in the Taguchi experimental design

Level	p_a	n	S
1.	−54.04	−53.93	−53.99
2.	−53.85	−53.79	−53.81
3.	−53.83	−53.78	−53.8
4.	−53.72	−53.81	−53.79
5.	−53.71	−53.84	−53.75
Delta	0.33	0.15	0.24
Rank	1	3	2

Table 4 Parameter set obtained through optimization

Parameter	Rank	Actual value
p_a	1	0.1
n	3	50
S	2	200

Table 5 Computational result for 16 benchmark problems for CS versus T-CS

Instances	BKS	Objective function (distance)						Computational time (s)		p-value
		CS			T-CS			CS	T-CS	
		Best	Avg.	Std.	Best	Avg.	Std.			
A-n33-k5	661	688	692.00	16.67	678	684	4.56	11.49	3.391	0.0
A-n46-k7	914	973	994.76	8.05	966	975.45	6.05	12.15	8.782	0.0
A-n60-k9	1354	1414	1413.49	11.16	1401	1412.63	9.24	17.89	12.72	0.008
B-n35-k5	955	962	966.37	24.87	***955***	961.54	4.84	22.42	5.31	0.001
B-n45-k5	751	770	794.88	10.45	769	785.90	8.16	12.91	5.62	0.0
B-n68-k9	1272	1320	1326.73	15.39	1303	1310.63	4.31	39.94	24.44	0.0
B-n78-k10	1221	1284	1307.94	26.79	1281	1301.6	13.00	38.86	27.18	0.0
E-n30-k3	534	565	559.07	24.41	545	555.18	4.40	20.19	6.67	0.029
E-n51-k5	521	590	618.28	7.98	586	599.54	6.94	31.33	14.38	0.0
E-n76-k7	682	746	763.69	22.25	742	755.54	9.24	59.25	56.62	0.0
F-n72-k4	237	255	265.27	19.35	245	249.18	3.84	81.12	80.41	0.0
F-n135-k7	1162	1301	1319.58	11.32	1289	1310.9	10.83	1178.85	1165.86	0.001
M-n101-k10	820	844	843.69	12.84	836	841.36	3.32	41.46	24.51	0.0
M-n121-k7	1034	1088	1110.22	15.11	1086	1094.81	6.58	140.85	126.45	0.0
P-n76-k4	593	679	688.27	8.72	676	686.63	5.95	67.78	66.65	0.02
P-n101-k4	681	758	760.75	14.16	744	753.45	5.16	175.46	167.47	0.11

However, based on Table 3, the best parameter set was identified as pa1-n3-S2 according to the S/N values. The real values of these levels are presented in Table 4.

After identified the best parameter suit for the problem of CVRP, we conducted a study to compare the performance of our proposed T-CS algorithm with the basic CS on all 16 benchmark problems of [10] in terms of the objective function found and the average computational time over 31 replications. Note that the smaller the objective function value the better the solution. The results of the basic CS and the T-CS are presented in Table 5.

The best results in Table 5 are shown in bold. In this table we can see the ability of the proposed method to get near to optimal solution for all instances. Moreover, the T-CS reached to the best results for one instances (B-n35-k5). However, the promising results of the method shows ability of the method and sensitivity of the parameter setting of the search process which can effects the ability of the algorithm for problem solving process.

A Wilcoxon test has been conducted to show a significant difference between the two methods with confidence level of 95 %. The p-value indicates significant differences in 15 out of 16 instances with percentage difference of 93 %. We can conclude that the Taguchi method has ability to find the more accurate design of parameters for the proposed algorithm.

4 Conclusion

This work has presented a CS applied for CVRP and its performance. Also this paper presented the performance of integrated CS with Taguchi method for CVRP over CS. The promising results of this method call for further improvements of the algorithm and motivate us to continue our study with an enhanced method using more potent algorithm in order to improve the quality of the CVRP problem.

References

1. Gendreau M et al (2008) Metaheuristics for the vehicle routing problem and its extensions: a categorized bibliography, in the vehicle routing problem. In: Golden B, Raghavan S, Wasil E (eds) Latest advances and new challenges, Springer, US. p 143–169
2. Ouaarab A, Ahiod B, Yang X-S (2013) Discrete cuckoo search algorithm for the travelling salesman problem. Neural Comput Appl 1–11
3. Yang X-S, Deb S (2009) Cuckoo search via Levy flights. In: Nature and biologically inspired computing, NaBIC 2009, World Congress on 2009
4. Cui L et al (2013) A new improved quantum evolution algorithm with local search procedure for capacitated vehicle routing problem. Math Probl Eng 2013:17
5. Dehnad K (2012) Quality control, robust design, and the Taguchi method, Springer Science & Business Media

6. Peker M, Sen B, Kumru PY, An efficient solving of the traveling salesman problem: the ant colony system having parameters optimized by the Taguchi method. Turk J Electr Eng Comput Sci 21(1):2015–2036
7. Sibalija TV, Majstorovic VD (2012) An integrated approach to optimise parameter design of multi-response processes based on Taguchi method and artificial intelligence. J Intell Manuf 23(5):1511–1528
8. Tsai P-W et al (2012) Enhanced parallel cat swarm optimization based on the Taguchi method. Expert Syst Appl 39(7):6309–6319
9. Hamta N et al (2013) A hybrid PSO algorithm for a multi-objective assembly line balancing problem with flexible operation times, sequence-dependent setup times and learning effect. Int J Prod Econ 141(1):99–111
10. Chen A-L, Yang G-K, Wu Z-M (2006) Hybrid discrete particle swarm optimization algorithm for capacitated vehicle routing problem. J Zhejiang Univ Sci A 7(4):607–614

IncSPADE: An Incremental Sequential Pattern Mining Algorithm Based on SPADE Property

Omer Adam, Zailani Abdullah, Amir Ngah, Kasypi Mokhtar, Wan Muhamad Amir Wan Ahmad, Tutut Herawan, Noraziah Ahmad, Mustafa Mat Deris, Abdul Razak Hamdan and Jemal H. Abawajy

Abstract In this paper we propose Incremental Sequential PAttern Discovery using Equivalence classes (IncSPADE) algorithm to mine the dynamic database without the requirement of re-scanning the database again. In order to evaluate this algorithm, we conducted the experiments against three different artificial datasets. The result shows that IncSPADE outperformed the benchmarked algorithm called SPADE up to 20%.

Keywords Sequential pattern · Incremental · Updatable · Database

---

O. Adam (✉) · Z. Abdullah · A. Ngah
School of Informatics and Applied Mathematics, Universiti Malaysia Terengganu,
21030 Kuala Terengganu, Malaysia
e-mail: gsk2211@pps.umt.edu.my

Z. Abdullah
e-mail: zailania@umt.edu.my

A. Ngah
e-mail: amirnma@umt.edu.my

K. Mokhtar
School of Maritime Business and Management, Universiti Malaysia Terengganu,
21030 Kuala Terengganu, Malaysia
e-mail: kasypi@umt.edu.my

W.M.A.W. Ahmad
School of Dental Sciences, Universiti Sains Malaysia, 16150 Kubang Kerian,
Kelantan, Malaysia
e-mail: wmamir@usm.my

T. Herawan
Faculty of Computer Science and Information Technology, University of Malaya,
50603 Kuala Lumpur, Malaysia
e-mail: tutut@um.edu.my

P.J. Soh et al. (eds.), *Advances in Machine Learning and Signal Processing*,
Lecture Notes in Electrical Engineering 387, DOI 10.1007/978-3-319-32213-1_8

1 Introduction

The main goal of data mining is to discover the hidden patterns from a large amount of data [1]. There are many data mining problems have been introduced and one of them is sequential pattern mining [2]. Sequential pattern mining is an evolving data mining problem in various domain applications such as in marketing to find customer purchase patterns, web access patterns, DNA sequence analysis [3], etc. Nowadays, sequential pattern mining has received a great attention since it was first proposed by Agrawal and Srikant in 1995 [3]. The problem of sequential patterns mining is given a set of sequence where each sequence contains set of elements and each element consist of set of items, given a user specified minimum support threshold, and finally discover all frequent subsequence [2]. Up to date, there are many algorithms have been proposed for sequential pattern mining based on the property of Apriori algorithm. Apriori algorithm was introduced by [3] to find frequents sequence from given dataset through recursive and iterative procedures.

Apriori-based and Pattern growth [4] are the most common sequential pattern approaches. Pattern growth algorithms appeared in the early 2000s to solve the problem of generating candidates and test. It has few proposed algorithms compared to Apriori-based approach, the most common pattern growth algorithms are IncSpan [5], FreeSpan [6] and PrefixSpan [4]. Generally, Apriori-based approach algorithms can be categorized into horizontal and vertical data formats. Examples of horizontal data format algorithms are Apriori [3] and GSP [7] and vertical data format algorithms are SPADE [8] and SPAM [9]. The vertical format provides the advantage in generating the patterns and calculating their supports without performing costly I/O during database scans [10].

Although vertical algorithms have better performance in dense dataset as compared to horizontal algorithms, yet this performance is getting slow when it comes

N. Ahmad
Universiti Malaysia Pahang, Malaysia, 26300 Kuantan, Pahang, Malaysia
e-mail: noraziah@ump.edu.my

M.M. Deris
Faculty of Computer Science and Information Technology, Universiti Tun Hussein Onn Malaysia, 86400 Batu Pahat, Johor, Malaysia
e-mail: mmustafa@uthm.edu.my

A.R. Hamdan
Faculty of Information Science and Technology, Universiti Kebangsaan Malaysia, 43600 Bangi, Selangor, Malaysia
e-mail: arh@ftsm.ukm.my

J.H. Abawajy
School of Information Technology, Deakin University, Burwood Campus/221, Waurn Ponds, VIC 3216, Australia
e-mail: jemal.abawajy@deakin.edu.au

to a dynamic database [11]. A dynamic database is a database that is frequently updated and increased in term of number of records [12]. This update can occur in many different ways, such as APPEND to existing record (existing customer buys new items), INSERT new record (new customers buy items) and DELETE record (current records have been removed from the database) [5]. The idea of incremental sequential patterns is giving an updated database find the current frequent patterns [13, 14]. As a result to mine the latest patterns from updated database, most of the current algorithms have to rescan back from entire updated database [15]. Therefore, in this paper we propose an incremental sequential pattern algorithm IncSPADE is based on SPADE propriety with fixable data structure to mine updated database without having to start the mining process from the scratch once the database is updated. Experiments result shows that IncSPADE outperform SPADE by 20% in term of performance in most cases. IncSPADE algorithm constructs a tree map data structure (FINFT) stand for frequent infrequent tree map, that contains all frequent and infrequent sequences generated during the first time the algorithm start. Maintain such tree makes it easy to IncSPADE to mine updated database without rescanning the whole database.

The rest of the paper organized as follow: in Sect. 2 discuss related work in the field, Sect. 3 describe the methodology, Sect. 4 explains the obtained result and discussion and finally in Sect. 5 draw a conclusion and our further research direction

2 Related Works

Several algorithms have been proposed for mining incremental association rules, however only a little attention has been given to sequential pattern mining [16]. Most of the proposed solutions to mine incremental sequential pattern are Apriori-based and pattern growth approaches. SPADE was first proposed by [8]. It uses Eclat [17] approach which was proposed by the same author for mining frequents items. The key feature of SPADE is the layout of the database in a vertical id list format. SPADE breaks the search space into sub-lattices tree-equivalence classes—that can be processed independently in main memory [10]. It scans the database three time only or just once with some data preprocessing [8]. First SPADE scans the database to construct frequent 1-sequence. After frequent 1-sequence prefix tree equivalence classes have been set with root null, SPADE starts enumeration process to compute frequent k- sequence. The process of generating k-sequence done by using frequent k-1-sequence equivalence classes as prefix and generate k-sequence with simple join operation from k-1-sequence [18]. During sequence generation, SPADE prunes the items against minim support sub and if sequence TIDs less than minim support SPADE will not add to prefix tree and thus reducing the number of next generated sequences [19]. SPADE outperforms Apriori-all and GSP by an order of magnitude in most cases sequence [8].

Table 1 Sequence database

CID	TID	Items
1	10	A B
	20	B
	30	A B
2	20	A C
	30	A B C
	50	B
3	10	A
	30	B
	40	A
4	30	A B
	40	A
	50	B

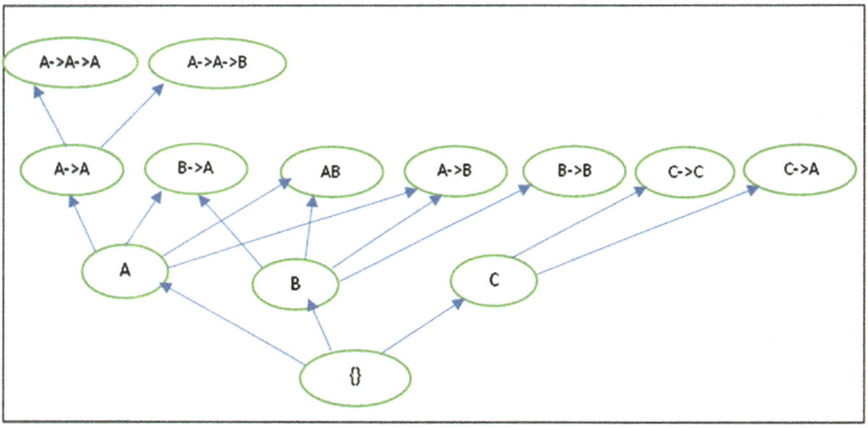

Fig. 1 Example of equivalence classes

Table 1 shows a sequence database and Fig. 1 shows the lattice tree constructed by SPADE for this database. Nodes marked in gray are infrequent sequences.

FASTUP algorithm [20], is one of the earliest work in incremental sequential pattern mining. FASTUP is an enhancement of GSP algorithm [7], FASTUP takes into account the previous result generated before start generating and examine candidates using generate and prune technique. The idea of having knowledge of previous result is to take advantages of the information related to sequence threshold to generated candidates, therefore, FASTUP avoids the issue of generating sequence depend on their support [16].

Comprehensive Incremental mining algorithm of Closed sequence (CISpan) proposed by [7]. It's an enhancement of the Clospan [21] algorithm [22]. The advantage of CISpan is the ability to remove sequence from the database in addition to database INSERTION. CISpan uses divide and conquer approach that makes INSERTION and DELETION operation are in depended from each other [23]. In INSERTION, the algorithm constructs an incremental lattice prefix tree to keep all frequents sequence that found in the updated database. However, removal of the sequence is done by updating the original database lattice tree. In the process of INSERTION CISpan only build a tree lattice for the newly inserted sequence, then CISpan merges the original prefix lattice with the incremental lattice result into updated frequent sequences [4].

3 Methodology

Up to our knowledge, there is not many works have been done in incremental sequential data mining. This section highlights some basic definitions and the proposed algorithm. IncSPADE is based on SPADE properties. IncSPADE is able to handle the updated database without reexamining the database from the scratch. There are two main phases of the IncSPADE, in the first phase IncSPADE behave just like SPADE. The second phase after the database has been updated; IncSPADE runs to generate the new frequents sequences.

3.1 Preliminary Definitions

Definition 1 (*sequence database*) A sequence is an ordered list of itemsets. Let *i* be $<i_1\ i_2 \ldots i_m>$ where i_j is an item, we donate a sequence *s* by $<s_1\ s_2 \ldots s_n>$ where s_j is an itemset.

Definition 4 (*support*) The *support* of a sequence *sa* in a sequence database *SDB* is defined as the number of sequences $s \in SDB$ such that $sa \sqsubseteq s$ and is denoted by *sup SDB* (*sa*).

Definition 3 (*prefix*) A sequence $sa = A_1, A_2, \ldots, A_n$ is a *prefix* of a sequence $sb = <B_1, B_2, \ldots, B_m>$, $\forall n < m$, iff $A_1 = B_1, A_2 = B_2, \ldots, A_{n-1} = B_{n-1}$ and the first $|A_n|$ items of B_n according to $\succ$. *lex* are equal to A_n.

Definition 5 (*sequential pattern mining*) Let *MinSupp* be a threshold set by the user and *SDB* be a sequence database. A sequence *s* is a *sequential pattern* and is considered *frequent* iff $supSDB(s) \geq MinSupp$.

Definition 2 (*Horizontal data format*) A sequence database in horizontal format is a database where each entry is a sequence. The horizontal data format was first introduced by Agrawal and Srikant upon proposing Apriori algorithm, its represents the items categorized into particular transactions. Table 2 shows a sample of horizontal sequence database.

Table 2 A sample of horizontal sequence database

SID	Sequence
1	<{A,B},{B},{A,B}>
2	<{A,C},{A,B,C},{B}>
3	<{A},{B},{A}>
4	<{A,B},{A},{B}>

Table 3 Vertical sequence database

A		B		C	
SID	Itemset	SID	Itemset	SID	Itemset
1	1,3	1	1,2,3	1	
2	1,2	2	2,3	2	1,2
3	1,3	3	2	3	
4	1,2	4	1,3	4	

Definition 3 (*Vertical data format*) A sequence database in vertical format is a database where each entry represents an item and indicates the list of sequences where the item appears and the position(s) where it appears. Table 3 shows the same as Table 1 database in the vertical format.

Definition 4 (*APPEND record*) An APPEND sequence database $DB\sim$ is a record that been appended to an existing record in original database, lets *DB* be original database and lets $S=<s_1, s_2, s_3, \ldots s_n>$ a sequence items in D, an updated database $D^{\sim}$ with $S\sim<i_1, i_2, i_3 \ldots i_n>$ final database $DB_F = DB \cap DB\sim$. Appended database has same numbers or fewer transactions as the original database.

Definition 5 (*INSERTION record*) An INSERTION sequence database $DB\sim$ is a records that been inserted to existing database, lets *DB* be original database and lets $S=<s_1, s_2, s_3, \ldots s_n>$ S= a sequence items in *DB*, an inserted database $DB\sim$ with $S\sim<i_1, i_2, i_3 \ldots i_n>$ final database $DB_F = DB \cup DB\sim$. INSERTION database has same or more transaction as the original database. Table 4 shows example of sequence database and its update in darker background that depict APPEND and INSERTION operations.

3.2 Incremental Sequential Mining

Let $|DB|$ be the number of data sequences in the original database and *MinSupp* is the minimum support. After the update of the original database $DB\sim$ a new records are added to original *DB*. The new records will be categorized by their *cid* in *DB*, each record will be merged with its corresponding *cid*. Thus final updated database $DB_F = DB \cup DB\sim$ where $DB\sim$ is the inserted dataset. In case of the *cid* record does not match with existing *cids* this records will be added as new *cid* and consider as INSERTION operation. Unlike APPEND database INSERTION requires overall

Table 4 Sequence database and its update

CID	TID	Items
	10	A B
1	20	B
	30	A B
	40	**C**
	20	A C
2	30	A B C
	50	B
	60	**A**
	10	A
3	30	B
	40	A
4	30	A B
	40	A
	50	B
	10	**AB**

changes in the frequent sequences tree structure, where some sequences might become infrequent due to change of the *MinSupp*.

IncSPADE is based on SPADE algorithm, its take an update database *DB*~ and run it against the previous obtained frequent and infrequent sequences tree, IncSPADE does not need to run the database from scratch again, it is only built prefix lattice tree structure for the newly appended sequences which did not appear previously on the original database. Using this idea, IncSPADE reduce time and memory space needed to mind the original and the updated database. Figures 2 and 3 presents IncSPADE model and IncSPADE algorithm, respectively.

1. **Scan database**

First step IncSPADE scans the original database to count the items and generate 1-freqnet sequence itemset.

2. **Create equivalence classes and enumerate**

After 1–frequent itemset generated, IncSPADE constructs a tree lattice search with root null and start enumeration process to generate k-frequents itemset from the prefix equivalence class k-1 frequent itemset.

3. **Frequent and none frequents sequence**

Once IncSPADE finished enumeration over the equivalence classes and prune sequence items against MinSupp, frequent and none frequent sequence.

4. **Scan updated DB**

This is a critical phase where IncSPADE scans the updated database to count the new sequence items. During this phase IncSPADE checks updated database size, if

```
DB = Original database
Start
Scan DB to find frequent and infrequent items
Generate frequents sequence
Construct frequent and infrequent tree map
UDB = updated database
Scan UDB
If checkItem(item)
 addToSearchTree(item);
Else
  CheckfrequentsItems(item)
    If isFrequent(item)
      updateItemTdis(item)
    else checkInfrequents(item)
       If ItemInfrequent(item)
         CheckItemTids(item)
          If (tisd > miusup)
             AddedItemToFrequent(item)
             RemoveItemFromInfrequent(item)
       else addToSearchTree(item)
       Endif
      Endif
    Endif
endif
forall New items do
  GenerateCandidate()
   If candidate tdis > minsup
    Add to frequent
   Else added to infrequent
   Endif
Endfor
Output = Generate frequents sequence
End
```

Fig. 2 IncSPADE algorithm

size is less or equal the original database, IncSPADE consider this update as APPEND else it will be considered as both APPEND and INSERT, MinSupp has to be recalculated then exams the existing frequent sequences against the new MinSupp.

5. **Enumerate and find 1-freqent**

After new items counted and process type identified, tree search will be constructed for the new item. IncSPADE enumerates to generating equivalence classes and prune the new items against frequents and none frequents itemset.

6. **Frequent and none frequents sequence**

At the end of enumerating process, IncSPADE come out will the new frequent and none frequents itemset which reflects the original and update databases with respect to MinSupp.

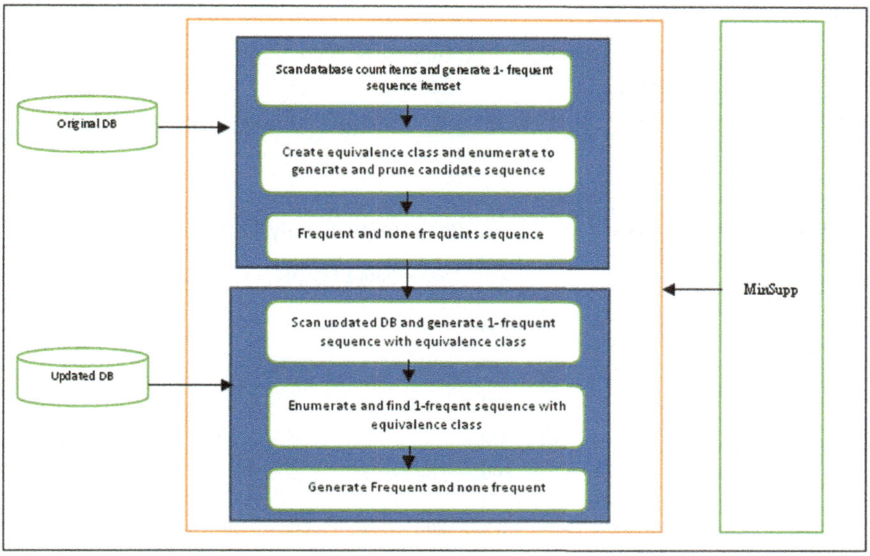

Fig. 3 IncSPADE model

4 Result and Discussion

The experiment was carried out on Intel® Core™ i3-3210 M CPU at 2.50 GHz speed with 1.00 GB RAM, running on Window 7 Home Premium 64bits. IncSPADE algorithm has been developed using Netbeans IDE 8 with Java JDK 1.8 and Java as a programming language.

Three artificial datasets was generated using our dataset generator. The datasets scenarios as follow: first dataset (DATA1) contains 20 items and 10 items per record, second dataset (DATA2) contains 30 items and 20 items per record. We split into at least two parts to stimulate database. The complete two datasets was run on SPADE algorithm with different MinSupp then continued to run the split datasets in IncSPADE. The datasets was split into two equal parts of 25,000 records for IncSPADE. The result shows that IncSPADE outperform SPADE in most cases by 20%. This is because IncSPADE used the previous knowledge to capture the frequent and none frequent sequences before determining the new frequent items. It helps IncSPADE to reduce the time complexity of generating and searching the new frequents items. Table 5 shows the datasets properties. Figures 4 and 5 depict the performance comparison between IncSPADE and SPADE for each dataset.

Table 5 Artificial datasets properties

Dataset	Number of records	Number of items	Items per record
Data 1	500,000	20	10
Data 2	500,000	30	20

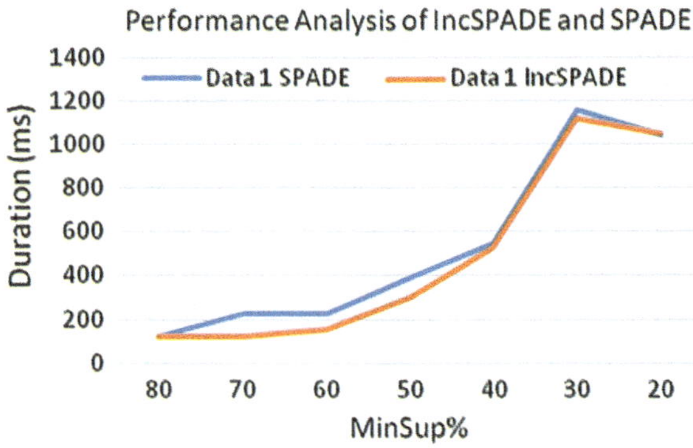

Fig. 4 Performance analysis of IncSPADE and SPAD

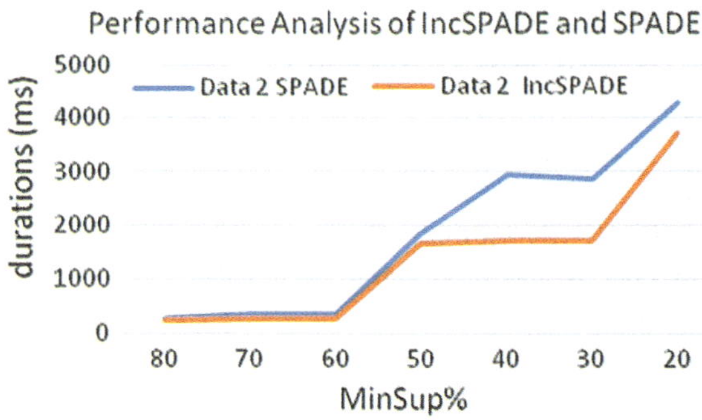

Fig. 5 Performance analysis of IncSPADE and SPADE

5 Conclusion

In this paper, we proposed IncSPADE algorithm to solve the issue of mining dynamic database without re-mine the database from scratch each time its updated. The result shows that IncSPADE was able to mine the updated database. In the near future, the algorithm will be enhanced and optimized to deal with the issue of constructing the frequent and infrequent tree.

References

1. Kumar V, Anupama C (2012) Mining association rules in student's assessment data. Int J Comput Sci Issues 9(5):211–216
2. Vishal SM (2014) A survey on sequential pattern mining algorithm. Int J Comput Sci Inf Technol 5(2):2486–2492
3. Agrawal R, Ramakrishnan S (1995) Mining sequential patterns. In: Proceedings of the 11th international conference on data engineering, pp 3–14
4. Pei J, Han J, Mortazavi-Asl B, Pinto H, Chen Q, Dayal U, Hsu MC (2001) Prefixspan: mining sequential patterns efficiently by prefix-projected pattern growth. In: Proceeding of 17th international conference on data engineering
5. Hong C, Yan X, Han J (2004) IncSpan: incremental mining of sequential patterns in large database. In: Proceedings of the 10th ACM SIGKDD international conference on knowledge discovery and data mining, pp 527–532
6. Han J, Dong G, Mortazavi-Asl B, Chen Q, Dayal U, Hsu M-C (2000) FreeSpan: frequent pattern-projected sequential pattern mining. In: Proceedings of the 6th ACM SIGKDD international conference on knowledge discovery and data mining, pp 355–359
7. Srikant R, Agrawal R (1996) Mining sequential patterns: generalizations and performance improvements. Adv Database Technol LNCS 1057:1–17
8. Zaki M (2001) SPADE: an efficient algorithm for mining frequent sequences. Mach Learn 42:31–60
9. Jay A, Gehrke J, Yiu T, Flannick J (2002) Sequential pattern mining using a bitmap representation. In: Proceedings of the 8th ACM SIGKDD international conference on knowledge discovery and data mining, pp 429–435
10. Fournier-Viger P, Gomariz A, Campos M, Thomas R (2014) Fast vertical mining of sequential patterns using co-occurrence information. LNAI 8443:40–52
11. Parthasarathy S, Zaki MJ, Ogihara M, Dwarkadas S (2002) Sequence mining in dynamic and interactive environments. Knowl Discov Bus Inf Syst 600:377–396
12. Lin MY, Lee SY (2004) Incremental update on sequential patterns in large databases by implicit merging and efficient counting. Inf Syst 29(5):385–404
13. Gupta M, Han J (2012) Approaches for pattern discovery using sequential data mining. Pattern discovery using sequence data mining: applications and studies, pp 137–154
14. Ezeife CI, Liu Y (2009) Fast incremental mining of web sequential patterns with PLWAP tree. Data Min Knowl Discov 19(3):376–416
15. Ezeife CI, Chen M (2004) Incremental mining of web sequential patterns using PLWAP tree on tolerance MinSupport. In: Proceeding of international database engineering and applications symposium, 2004, pp 465–469
16. Florent M, Poncelet P, Teisseire M (2003) Incremental mining of sequential patterns in large databases. Data Knowl Eng 46(1):97–121
17. Zaki M (2000) Scalable algorithms for association mining. Knowl Data Eng IEEE Trans 12 (3):372–390
18. Leleu M et al (2003) GO-SPADE: mining sequential patterns over datasets with consecutive repetitions. LNAI 2734:293–306
19. Mooney CH, John F (2013) Sequential pattern mining approaches and algorithms. ACM Comput Surv 45(2):2–39
20. Lin M-Y, Lee S-Y (2004) Incremental update on sequential patterns in large databases by implicit merging and efficient counting. Inf Syst 29(5):385–404
21. Wang J, Han J (2004) BIDE: efficient mining of frequent closed sequences. In: Proceedings of 20th international conference on data engineering, pp 79–90
22. He H, Wang D, Chen G, Zhang W (2014) An alert correlation analysis oriented incremental mining algorithm of closed sequential patterns with gap constraints. Int J Appl Math Inf Sci 8 (1L):41–46

23. Mallick B, Garg D, Grover PS (2013) Incremental mining of sequential patterns: progress and challenges. Intel Data Anal 17(3):507–530
24. Yuan D, Lee K, Cheng H, Krishna G, Li Z, Ma X, Zhou Y, Han J (2006) CISpan: (2008) comprehensive incremental mining algorithms of closed sequential patterns for multi-versional software mining. In: Proceeding of SDM'2008, pp 84–95

Choosing Geometric Dissimilarity Measure for Content Based Coral Reefs Image Retrieval

Wan Nural Jawahir Hj Wan Yussof, Muhammad Suzuri Hitam, Ezmahamrul Afreen Awalludin and Zainudin Bachok

Abstract The dissimilarity measure used by a Content-Based Image Retrieval (CBIR) system significantly affects its performance. Choosing the right dissimilarity measure is important especially when we have large low-level features to represent each image in the database. This paper presents the performance of various geometric distance measures for retrieval of images from a coral reefs database that consists of three groups of coral. Based on the results obtained by Precision-Recall graphs, there is no single distance measure that best for all queries. Therefore, Mean Average Precision is used to measure the overall performance, and the results showed that the top three best geometric distance measures for retrieving images from a coral database are the Squared Chord, City Block, and Canberra.

Keywords Geometric distance measures · Content-based image retrieval · Coral reefs

W.N.J. Hj Wan Yussof (✉) · M.S. Hitam
School of Informatics and Applied Mathematics,
Universiti Malaysia Terengganu, 21030
Kuala Terengganu, Terengganu Darul Iman, Malaysia
e-mail: wannurwy@umt.edu.my
URL: http://www.umt.edu.my

M.S. Hitam
e-mail: suzuri@umt.edu.my

E.A. Awalludin
Institute of Oceanography and Environment,
Universiti Malaysia Terengganu, 21030
Kuala Terengganu, Terengganu Darul Iman, Malaysia
e-mail: eafreen@gmail.com

Z. Bachok
School of Marine and Environmental Sciences,
Universiti Malaysia Terengganu, 21030 Kuala Terengganu
Terengganu Darul Iman, Malaysia
e-mail: zainudinb@umt.edu.my

P.J. Soh et al. (eds.), *Advances in Machine Learning and Signal Processing*,
Lecture Notes in Electrical Engineering 387, DOI 10.1007/978-3-319-32213-1_9

1 Introduction

Over the past few decades, the science of taxonomy has been suffering from dwindling numbers of experts [1]. 'When asking the question what coral is that?', not many people can identify the types of coral and most of them flicking through the pages of some books to know about that coral. As a result, most people get easily discouraged by the lack of any easy way to cut through the detail. To overcome this problem, development of an efficient content-based image retrieval (CBIR) system would provide people especially researchers and students with a powerful research tool and would allow them to pursue taxonomic studies in a better way. CBIR system aims at searching similar pictures from large image repositories, according to the users interest [2]. At the moment, it is worth noting that CBIR has been applied to diverse fields such as Medical [3, 4], Biometric [5], Forensic [6], etc. For coral reefs retrieval system, a user only provides a picture and then the system provides clues for assisting researchers to identify taxonomy level of the coral.

A typical CBIR system requires two important keys namely *features extraction* and *features similarity*. Typically, images in the repository are represented as a multi-dimensional feature vector extracted from a series of low-level features, such as color, texture or shape that is done through the first key. In the second key, a user may search a collection of images that are similar to a query image based on similarities of the corresponding image features. Various similarity/dissimilarity measures have been formulated as a function to estimate the similarity between pictures. To distinguish between similarity and dissimilarity measures is that the first measures of how similar two images are and the following measures of how different two images are.

While many researchers emphasize on features extraction as a first step for developing CBIR, in this study, we do the other way round for our CBIR system using a coral database. We adopt the classic features extraction techniques and focus on features similarity step. The study is inspired from [7–12] that have shown the performance of CBIR was also affected by the choice of similarity/dissimilarity measures. This paper presents a systematic comparison of seven geometric dissimilarity measures. It is a consequence of the study reported by [9] that geometric dissimilarity measures were better than information theoretic measures and statistic measure using Corel image collection. A study regarding similarity/dissimilarity measures using coral reefs image database has not yet reported. Therefore, we make an attempt to find out what is the most suitable dissimilarity measure for our coral reefs database that consists of three groups of coral—*Acropora* branching (ACB), *Acropora* submassive(ACS) and Coral foliose (CF).

The remainder of this paper is organized as follows. Section 2 describes seven dissimilarity measures used for the comparison. Section 3 presents a brief description of several features descriptors used for image representation. In Sect. 4, the discrimination powers of the geometric measure are determined and compared using images from a coral database. Finally, the conclusion is drawn in Sect. 5.

2 Dissimilarity Measures

Dissimilarity measure can be grouped into three categories; geometric measures, information theoretic measures, and statistic measures. These three categories are discussed further in [7–9, 13]. Geometric measures treat objects as vectors. Information theoretic measures are derivative from the Shannon's entropy theory [9] that treat objects as probabilistic distributions. Meanwhile, in statistic measures, comparing two objects are done in a distributed manner where the vector elements are assumed as samples [8]. In this study, we only look at the geometric measures. Euclidean distance, City Block, Chebyshev, Canberra, Squared Chord, Partial Histogram Intersection and Cosine Function Based are categorized as geometric measures that are considered in our study.

Let X denote the feature vector $(x_1, x_2, \ldots, x_n)$ representing the query image and Y the feature vector $(y_1, y_2, \ldots, y_n)$ representing an image in the database. The formulas are given as follows:

- Euclidean distance (L_2)

$$\mathrm{Euc}(X, Y) = \sqrt{\sum_{i=1}^{n} (x_i - y_i)^2} \tag{1}$$

- Cityblock distance (L_1)

$$\mathrm{City}(X, Y) = \sum_{i=1}^{n} |x_i - y_i| \tag{2}$$

- Chebyshev

$$\mathrm{Che}(X, Y) = \max_i |x_i - y_i| \tag{3}$$

- Canberra

$$\mathrm{Can}(X, Y) = \sum_{i=1}^{n} \frac{|x_i - y_i|}{x_i + y_i} \tag{4}$$

- Squared Chord

$$\mathrm{SC}(X, Y) = \sum_{i=1}^{n} (\sqrt{x_i} - \sqrt{y_i}) \tag{5}$$

- Partial Histogram Intersection

$$\mathrm{PHI}(X, Y) = 1 - \frac{\sum_{i=1}^{n}(\min(x_i, y_i))}{\min(|X|, |Y|)} \tag{6}$$

- Cosine Function Based

$$\mathrm{CFB}(X, Y) = 1 - \frac{X \cdot Y}{|X| \cdot |Y|} \tag{7}$$

3 Features Extraction

In image retrieval task, it is essential to capture important image properties concerning a set of meaningful features. This process is called features extraction. Shape, color, and texture are the most common features in underwater coral reefs images. However, only color and texture features are used in this study that incidentally are the indicators commonly used by marine scientists to identify components in a reef [12]. Three kinds of color features and two kinds of texture features are extracted from a coral image. All of the extracted features are concatenated to represent a set of feature vectors.

3.1 Colour Features

Color features are a reliable indicator as it has good separability for ACB and CF types of coral. Mostly ACB has yellowish color while most CF is greenish. For extraction of color features, once the color space is specified, a color feature can be extracted from images or regions. Some important color features have been proposed in the literature, including color histogram, color moments, and color auto-correlogram.

The HSV histogram is a method for describing the color content of an image using HSV color space. This method used quantization process to reduce the space required to store the histogram information and time to compare the histogram [14]. The output of the feature extraction is feature vector of length $8 \times 2 \times 2$ for each image.

Color moments are the distribution of color in an image that is interpreted as a probability distribution. Mean and standard deviation are two values that commonly used in color moments. The feature vector of color moments extracted from each image contains two moments for each RGB channel.

Color auto-correlogram represents how the spatial correlation between identical colors changes with its respective distance. Autocorrelogram is more stable to large appearance changes compared to other schemes that are based on purely local properties because correlogram takes into account the local spatial correlation and the global distribution of spatial correlation [15]. The feature vector of length 64 is extracted using color auto-correlogram for each image.

3.2 *Texture Features*

Using color features alone is not enough since ACS in our database comes with a variety of colors. As each type of coral has their repeating patterns, combining color with texture can help to differentiate between ACB, ACS, and CF used in this study.

Gabor wavelets are elements of a family of mutually similar Gabor functions [16]. The descriptor is optimal for measuring local spatial frequencies since it decomposes an image into sub-images using multiple resolutions and multiple orientations. In this study, we used six orientations and four resolutions to obtain a feature vector of length 48 with 6×4 mean square energy and 6×4 mean amplitude for each image.

Wavelet transforms decompose a signal into different scales with different levels of resolution using a family of basic functions obtained through translation and dilation of a mother wavelet [17]. In this study, we generated three level decomposition which results in $3 \times 3 \times 2 + 2 = 20$ components. In each of the components, wavelet features are constructed using mean and standard deviation that results in the feature vector of 2×20 for each image.

4 Experimental Results

To evaluate the performances of the geometric dissimilarity measures (GDM) described in the preceding section, we conduct experiments using a single image query to coral reef database. The database contains 90 images—30 images from ACB, 30 images from ACS and 30 images from CF. These coral reefs images were provided by the Institute of Oceanography and Environment (INOS), UMT. The images were taken in video format and for the purpose of this study, the video was transformed into still images and then were cropped and categorized manually. The video was taken from three meters depth from sea level using underwater camera GoPro. The examples of coral reefs images with different categories are shown in Fig. 1.

We use four randomly selected images from three types of coral reef for single-image queries. For each query, the retrieval results return 20 images according to similarity rankings and the query is repeated using other six GDM.

Fig. 1 Examples of coral reefs images. First row: *Acropora* branching (ACB), second row: *Acropora* submassive (ACS) and third row: Coral foliose (CF)

4.1 Precision-Recall Graph

Like other image retrieval tasks, the performance of our coral reefs retrieval is measured using the so-called precision and recall (PR) graph that is based on the number of retrieved images as cut-off values.

Precision and recall are defined as in Eqs. 8 and 9, respectively.

$$Precision = \frac{\text{No. of relevant images retrieved}}{\text{Total number of images retrieved}} \tag{8}$$

$$Recall = \frac{\text{No. of relevant images retrieved}}{\text{Total number of relevant images in the database}} \tag{9}$$

The PR graph is drawn by using data from a query that was randomly selected from each three groups of coral reefs collections. Three graphs represent ACB, ACS, and CF retrieval. The lines in the graphs represent the performance of seven GDM. To assure consistency, the randomly selected image is repeatedly used for different GDM.

4.1.1 PR for ACB

Figure 2 shows the retrieval results for ACB. Note that in the figure, the Partial Histogram Intersection line shows maximum precision until 0.73 recall. It is

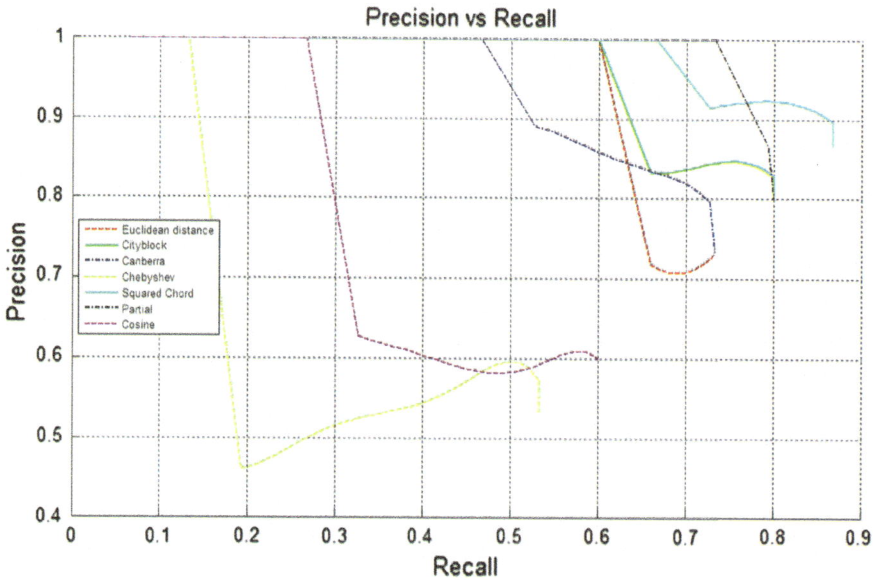

Fig. 2 Precision-recall graph for ACB

followed by the Squared Chord where its line keeps to the maximum precision until 0.67 recall. However, the maximum recall of Squared Chord is higher than the Partial Histogram Intersection. The Cityblock and Euclidean distance lines drop when they reach 0.6 recall. They are better than Canberra line that drops earlier at 0.47 recall. While other geometric measures are comparable to each other with precision at 1.0 till recall more than 0.45, this is not the case of Chebyshev and Cosine-based Function. Cosine-based Function line drops at recall 0.27 while Chebyshev line falls at 0.1 recall that is the worst amongst all. In this case, Chebyshev is least suitable compared to others while the Squared Chord is the most appropriate means for retrieving ACB.

4.1.2 PR for ACS

The PR lines for ACS in Fig. 3 are worse than ACB in Fig. 2 accept for the Chebyshev and Cosine-based Function. It maintains the maximum precision until 0.87 and at recall 0.93 its precision is still higher than 0.9. The Cosine-based Function shows better performance as compared to ACB graph where its line only drops after 0.53 recall. The rest lines drop before 0.5 recall with the worst goes to Canberra where the precision drastically falls to 0.53. The line belongs to Euclidean distance shows fluctuated and lasts at 0.8 recall while Partial Histogram Intersection, Squared Chord, Cityblock and Cosine-based Function last at 0.73

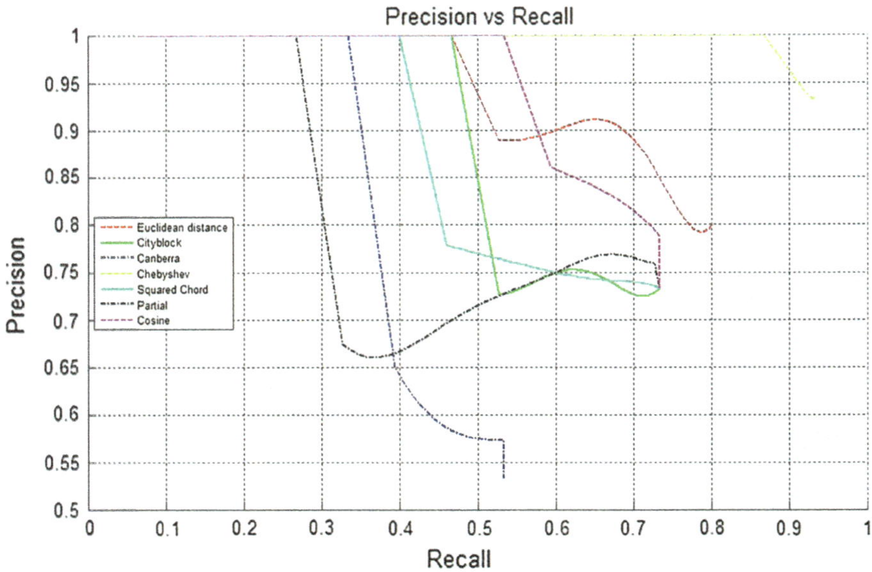

Fig. 3 Precision-recall graph for ACS

recall. As a conclusion, Canberra is not suitable for retrieving ACS. On the other hand, the most appropriate GDM for this case is Chebyshev.

4.1.3 PR for CF

Figure 4 presents the PR graph for CF. As shown in the graph, the best result is obtained by the Euclidean distance and Squared Chord. They maintain 1.0 precision until 0.8 recall, and the lowest precision they obtain are 0.93. The Cityblock also maintains 1.0 precision until 0.8 recall, but then the precision drastically drops to 0.83. The Canberra cuts earlier than the above measures and lasts at 0.93 recall. The Partial Histogram Intersection and Cosine-based Function have the final precision similar to the Cityblock but both of them drop earlier. The worst amongst all is the Chebyshev where its lowest precision is 0.57 at recall 0.6. This means the Chebyshev is not suitable for retrieving CF. One can choose either the Euclidean distance or the Squared Chord in this case.

4.2 Mean Average Precision

Even though PR graph is very informative, it is often a desire to boil the information from a PR graph down to a few numbers, or even a single number for better

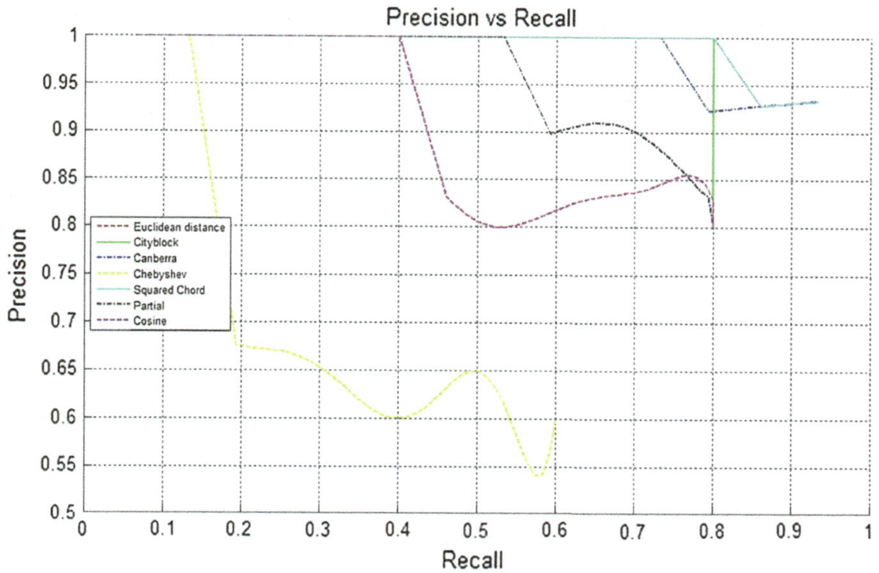

Fig. 4 Precision-recall graph for CF

discrimination and stability. The most standard way is to calculate the Mean Average Precision (MAP). MAP provides a single-figure measure of quality across recall levels. For a single information need, the first step is to compute an Average Precision (AP) that is the average of the precision value obtained for the set of N returned images.

$$\text{AP} = \frac{\sum_{n=1}^{N} Precision}{N} \quad (10)$$

Then, MAP for M different queries under the same category can be computed as follows:

$$\text{MAP} = \frac{\sum_{m=1}^{M} AP}{M} \quad (11)$$

In this experiment, we calculate AP and MAP for each dissimilarity measure, by performing a single-image query on 12 images; four images from ACB; four images from ACS and four images from CF. The average precision for ACB, ACS, and CF for all GDM are shown in Table 1.

Once we have obtained AP for all the queries for each category, the MAP values are just obtained by averaging the results from Table 2. For ACB, the best geometric measure is the Squared Chord with 0.925 MAP followed by the City Block and the Canberra with MAP values 0.907 and 0.882, respectively. The Chebyshev performs better for ACS with the highest MAP 0.856. It follows by the City Block with MAP

Table 1 Average precision for ACB, ACS, and CF

GDM	ACB				ACS				CF			
	Q1	Q2	Q3	Q4	Q1	Q2	Q3	Q4	Q1	Q2	Q3	Q4
Euc	0.907	0.917	0.981	0.678	0.952	0.789	0.254	0.913	0.743	0.986	0.670	0.814
City	0.937	0.974	1.000	0.719	0.947	0.871	0.376	0.933	0.791	0.972	0.782	0.883
Can	0.912	0.972	0.966	0.678	0.888	0.758	0.385	0.926	0.794	0.980	0.844	0.932
Che	0.593	0.382	0.579	0.504	0.996	0.754	0.701	0.974	0.597	0.685	0.473	0.726
SC	0.970	0.996	0.996	0.738	0.914	0.805	0.342	0.909	0.878	0.986	0.900	0.856
PHI	0.967	0.965	0.986	0.800	0.855	0.691	0.271	0.879	0.345	0.941	0.652	0.502
CFB	0.723	0.981	0.974	0.870	0.974	0.791	0.221	0.908	1.000	0.898	0.565	0.960

Table 2 Mean average precision

GDM	ACB	ACS	CF	Overall MAP
Euc	0.871	0.727	0.803	0.800
City	0.907	0.782	0.857	0.849
Can	0.882	0.739	0.888	0.836
Che	0.515	0.856	0.619	0.663
SC	0.925	0.743	0.905	0.857
PHI	0.674	0.670	0.760	0.681
CFB	0.872	0.724	0.856	0.817

value 0.782. The third rank goes to the Squared Chord with 0.743 MAP. For CF, the highest MAP is obtained by the Squared Chord with 0.905 MAP. The Canberra and the City Block are in the second, and the third rank with the MAP values 0.888 and 0.857, respectively.

With the MAP values, one can determine the overall performance for each of the geometric measures by calculating the MAP mean values. In overall, the top three best GDM for retrieving images from a coral database are the Squared Chord, City Block, and Canberra.

5 Conclusion

In conclusion, a different group of coral entails a different kind of dissimilarity measure but the Squared Chord, the City Block and the Canberra give the best overall performance. Even though the Chebyshev is the worst geometric measure but it has shown the best result in retrieving ACS. Having a good judgment is hard with this study. Therefore, future study will be devoted to both features and similarity/dissimilarity measures so that the best combination techniques will be selected for an effectiveness of coral reefs images retrieval system.

Acknowledgments We would like to thank Institute of Oceonography and Environment (INOS) for kindly sharing the coral reefs images.

References

1. Rodman James E, Cody Jeannine H (2003) The taxonomic impediment overcome: NSF's partnerships for enhancing expertise in taxonomy (PEET) as a model. Syst Biol 52(3):428–435
2. Datta R, Joshi D, Li J, Wang JZ (2008) Image retrieval: ideas, influences, and trends of the new age. ACM Comput Surv (CSUR) 40(2):5
3. Faria AV, Oishi K, Yoshida S, Hillis A, Miller MI, Mori S (2015) Content-based image retrieval for brain MRI: an image-searching engine and population-based analysis to utilize past clinical data for future diagnosis. NeuroImage Clin 7:367–376

4. Jiji GW, DuraiRaj PJ (2015) Content-based image retrieval techniques for the analysis of dermatological lesions using particle swarm optimization technique. Appl Soft Comput 30:650–662
5. Iqbal K, Odetayo MO, James A (2012) Content-based image retrieval approach for biometric security using colour, texture and shape features controlled by fuzzy heuristics. J Comput Syst Sci 78(4):1258–1277
6. Zhou J, Abdel-Mottaleb M (2005) A content-based system for human identification based on bitewing dental X-ray images. Pattern Recogn 38(11):2132–2142
7. Zhang D, Lu G (2003) Evaluation of similarity measurement for image retrieval. In: Proceedings of the 2003 international conference on neural networks and signal processing, vol 2. IEEE, pp 928–931
8. Hu R, Ruger S, Song D, Liu H, Huang Z (2008) Image retrieval: ideas, influences, and trends of the new age. ACM Comp Surv 40(2):5:1–5:60
9. Haiming L, Dawei S, Rüger S, Hu R, Uren V (2008) Comparing dissimilarity measures for content-based image retrieval. ACM Comp Surv 40(2):5:1–5:60
10. Cho HC, Hadjiiski L, Sahiner B, Chan HP, Helvie M, Paramagul C, Nees AV (2011) Similarity evaluation in a content-based image retrieval (CBIR) CADx system for characterization of breast masses on ultrasound images. Med Phys 38(4):1820–1831
11. Collins J, Okada K (2012) A comparative study of similarity measures for content-based medical image retrieval. In: CLEF (Online Working Notes/Labs/Workshop)
12. Marcos MSA, Soriano M, Saloma C (2007) Low-level color and texture feature extraction of coral reef components. Sci Diliman 15(1)
13. Rubner Y, Puzicha J, Tomasi C, Buhmann JM (2001) Empirical evaluation of dissimilarity measures for color and texture. Comput Vis Image Underst 84(1):25–43
14. Huang J, Zabih R (1998) Combining color and spatial information for content-based image retrieval. In: Proceedings of ECDL
15. Stricker MA, Orengo M (1995) Similarity of color images. In: IS&T/SPIE's symposium on electronic imaging: science and technology. International Society for Optics and Photonics, pp 381–392
16. Lee TS (1996) Image representation using 2D Gabor wavelets. IEEE Trans Pattern Anal Mach Intell 18(10):959–971
17. Tsai SJS (2002) Power transformer partial discharge (PD) acoustic signal detection using fiber sensors and wavelet analysis, modeling, and simulation. Doctoral dissertation, Virginia Polytechnic Institute

Block Compressive Sensing (BCS) Based Multi-phase Reconstruction (MPR) Framework for Video

Mansoor Ebrahim and Wai Chong Chia

Abstract In this paper, a Multi-phase Reconstruction (MPR) framework that uses certain key frames to produce some Side Information (SI) to improve the reconstruction quality of the non-key frames is proposed. After a sequence of frames has been encoded using Block Compressive Sensing (BCS) and transmitted to the host workstation, some SI is produced by first aligning the key frames to the non-key frames. The aligned frames are then fused together using Wavelet to exploit the spatial and temporal correlations between them, and to generate a set of predicted non-key frames. Next, the difference between the initially reconstructed and the predicted non-key frames at the measurement level is calculated. The difference is then decoded to recover a set of residual frames. The reconstruction of the final non-key frames is completed by adding the residual frames to the predicted non-key frames. The experimental results show that the proposed framework is able to outperform other frameworks by 1.5–3.0 dB at lower sub-rates.

1 Introduction

Sensor nodes in a Visual Sensor Network (VSN) could now couple with low cost cameras to create various new applications in the field of tracking, surveillance, and disaster monitoring. However, the use of cameras also brings with it a set of new challenges, because capturing and encoding videos increases the amount of data that needs to be processed and transmitted significantly. Indirectly, this also increases the computational burden and memory requirement of the sensor nodes. Because the sensor nodes are mostly powered by batteries, energy consumption is also one of the critical issues that needs to be taken into consideration. In order to

M. Ebrahim (✉) · W.C. Chia
Faculty of Science and Technology, Sunway University, 46150 Bandar Sunway, Petaling Jaya, Selangor, Malaysia
e-mail: 12032389@imail.sunway.edu.my

W.C. Chia
e-mail: waichongc@sunway.edu.my

P.J. Soh et al. (eds.), *Advances in Machine Learning and Signal Processing*, Lecture Notes in Electrical Engineering 387, DOI 10.1007/978-3-319-32213-1_10

overcome the aforementioned issues, compressing the videos before transmission takes place is a suitable solution. In standard video compression framework, Motion Estimation and Compensation (ME/MC) has been widely adopted to achieve this. However, the use of ME/MC requires significant amount of memory for processing that also leads to increase in energy consumption [1].

Recently, Compressive Sensing (CS) [2] has emerged to be one of the better solutions for low-power application such as the VSN. The basic idea of CS is to sample and represent a signal with a few non-zero coefficients [2]. One of the leading edge of this is the single-pixel camera [3] that directly reduces the sampling and number of data that will be streaming out. This reduces the amount of data that has to be processed and transmitted by the sensor node, but increases the complexity of reconstructing the original signal. In other words, CS helps to create a simple-encoder complex-decoder paradigm that shifted the computational burden to the decoder.

In this paper, a Multi-Phase Reconstruction (MPR) framework is proposed. Assuming that after a sequence of frames has been encoded using Block Compressive Sensing (BCS) and transmitted to a host workstation, the MPR framework is applied to reconstruct the frames. It uses certain key frames to produce some side information that can be used to improve the reconstruction quality of the non-key frames. The aim is to predict the non-key frames from the key frames by exploiting the correlations in them. First, image registration is applied to align the key frames to the non-key frames. The aligned frames are then fused together using Wavelet. Such approach is able to generate an approximation of the non-key frames in shorter time, when compared to MC/ME. The paper is organized as follows. The background of CS and related works are described in Sect. 2. This is followed by a detailed explanation of the proposed MPR framework in Sect. 3. All the simulation results are presented and discussed in Sect. 4. Finally, the paper is concluded in Sect. 5.

2 Background

2.1 Compressive Sensing

CS allows high prospect of signal recreation by using minimum number of unsystematic estimations, provided that the signal/image is sparse. Consider that we want to recover a real-valued signal $\mathbf{x}$ with length $\mathbf{N}$ from $\mathbf{M}$ measurements ($M \ll N$), the signal must has some sparse representations in the transformation domain $\boldsymbol{\Psi}$ with random measurement matrix $\boldsymbol{\Phi}$. Then, the set of measurements $\mathbf{y}$ is given as:

$$y = \Phi x \tag{1}$$

It is also assumed that $\boldsymbol{\Phi}$ is orthonormal i.e. $\boldsymbol{\Phi}\,\boldsymbol{\Phi}^{T} = \mathbf{I}$. Where, $\mathbf{I}$ is the identity matrix. Nevertheless, to recover $\mathbf{x}$ from such small measurements is not directly possible, i.e. inverse projection of $\hat{x} = \mathbf{\Phi}^{-1}y$ is ill-posed [4]. But since $\mathbf{x}$ has some sparse representations in $\mathbf{\Psi}$, $\mathbf{x}$ can be reconstructed from the sparse representations $\hat{x} = \mathbf{\Psi}\mathbf{x}$ by solving the ℓ_0 optimization problem that can be expressed as:

$$\hat{x} = \mathrm{argmin}_{\hat{x}} \|\hat{x}\|_{l2}, \quad s.t.\ y = \mathbf{\Phi}\mathbf{\Psi}^{-1}\hat{x} \tag{2}$$

However, solving the ℓ_0 constrained optimization problem is difficult. The Iterative Hard Thresholding (ITH) algorithm [5] was proposed to resolve the aforementioned problem by substituting it with an unconstrained optimization problem using Lagrangian multiplier ℓ_1 and ℓ_2:

$$\hat{x} = \mathrm{argmin}_{\hat{x}} \|\Psi x\|_{l1} + \lambda\|y - \Phi_x x\|_{l2} \tag{3}$$

In order to speed-up the process of solving Eq. (3) for images, the Block Compressive Sensing (BCS) [6] is proposed to handle this using a block-based basis. In order words, an image is first divided into small $\mathbf{B} \times \mathbf{B}$ independent blocks, and each block is then individually sampled based on the same measurement matrix $\mathbf{\Phi}$ with a constrained (block-diagonal) structure. In addition to this, [6] also proposed to solve Eq. (3) using a Projective-based Landweber (PL) algorithm. The BCS was later enhanced by [7], that cast the reconstruction based on contourlets and wavelet transforms. Moreover, a smoothing filter is incorporated into the PL iteration to find a better CS reconstruction that achieves both sparsity and smoothness. This scheme is known as BCS-SPL in the remaining part of the paper. Beside the PL algorithm, [8] attempted to solve Eq. (3) by using the Total Variation (TV) minimization by Augmented Lagrangian and Alternating Direction algorithm (TV-AL3) [8]. The TV-AL3 is a modified version of [9]. It resolves the computational complexity issue by using splitting and alternating approaches. The scheme that makes use of the block-based processing similar to BCS, but resolves Eq. (3) using the TV-AL3 algorithm is referred to as BCS-TV-AL3 throughout the paper.

2.2 Related Work

In [10], a CS-based video reconstruction approach that recovers each frame within a video sequence independently using 2D Discrete Wavelet Transform (2D-DWT) is proposed. However, the temporal correlations between consecutive frames were not taken into consideration. An alternative approach is to exploits the temporal correlations by makes use of 3D-DWT and reconstructs a group of frames all at once [10]. But the increase in dimensionality also lead to the increase in memory requirement and computational burden. [11], applied coded aperture mask designs to each frame and attempted to solve multiple frame altogether to exploit the correlations between consecutive frames.

In addition to above, some of the CS-based video reconstruction approaches also make use of Motion Estimation and Compensation (ME/MC), which is widely used in traditional video compression scheme. In [12], a multi-scale video reconstruction approach is presented. It uses an iteration mechanism between Motion Estimation (ME) and sparsity based reconstruction of the frames themselves. The framework is based on the LIMAT method (use Motion Compensation (MC) to improve sparsity in the 3D-DWT) for standard video compression.

In [13] a residual based reconstruction approach, which is known as Modified-CS-Residual, is presented. It deals with the reconstruction problem of sparse signals by using some side information. The side information is generated by using Least Mean Squares or Kalman filtered based prediction methods.

Another reconstruction algorithm based on the above mentioned principle is presented in [14]. The scheme is known as k-t FOCUSS. The framework employs on ME/MC based predictions and residual encoding to optimal the samples allocation between the prediction and residual encoding steps. k-t FOCUSS assumes that there exists multiple key frames, and then CS reconstruction is performed by taking residuals between each non-key frame and a bidirectional (ME/MC) prediction for each of the key frames.

In [15], a ME/MC based framework is presented. The framework incorporated MC/ME into the reconstruction process of BCS-SPL for video and referred as MC-BCS-SPL. Initially block-based random CS measurements are applied frame by frame for the video sequence. Then, the decoder incorporates the reconstruction from a ME/MC-based residual; the proposed MC-BCS-SPL framework alternatively reconstructs frames of the video sequence and their corresponding motion fields, using one to improve the quality of the other in an iterative fashion.

3 Proposed Multi-phase Reconstruction (MPR) Framework

The inter-frame correlations and redundancies are usually exploited and removed using conventional ME/MC in traditional video compression scheme. But in this case, we aim to achieve this by using the proposed Multi-Phase Reconstruction (MPR) framework. In contrast to the traditional video compression scheme, the proposed MPR framework shifts all the complex processing to the host workstation (decoder).

The overall architecture of the proposed framework is shown in Fig. 1, where a sensor node $\mathbf{S}$ is recording a video. Specifically, we consider a set of $\mathbf{J}$ consecutive frames from a video sequence that we termed as a Group of Pictures (GoP). Moreover, we also assume that the current GoP is tailed by another GoP. The GoP consists of a key frame $\mathbf{F_K}$ (the first), and $(\mathbf{J-1})$ non-key frames $\mathbf{F_{NK}}$. The $\mathbf{F_K}$ and $\mathbf{F_{NK}}$ are encoded at sub-rate of $\mathbf{M_K}$ and $\mathbf{M_{NK}}$ respectively, with $\mathbf{M_K} > \mathbf{M_{NK}}$. Each video frame $\mathbf{F_{x,}}$ is first divided into a small block of size $\mathbf{B} \times \mathbf{B}$, where $\mathbf{x}$ represents the frame number within the GoP i.e. $\mathbf{0} \leq \mathbf{x} \leq \mathbf{J-1}$, and $\mathbf{F_0}$ is equivalent to $\mathbf{F_K}$.

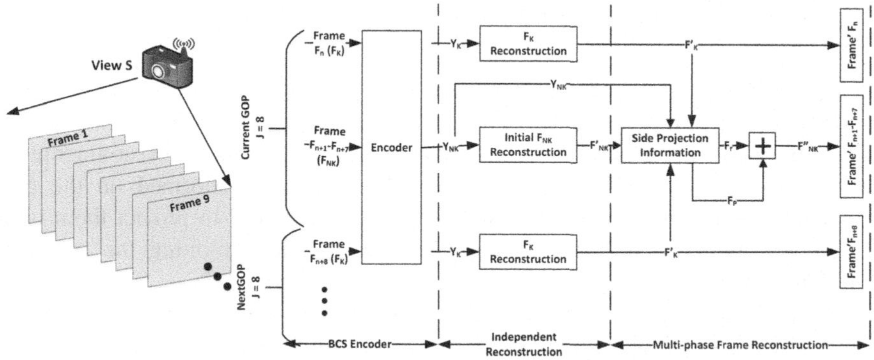

Fig. 1 System architecture for video reconstruction using the proposed MPR framework

Each block within a frame will then be sampled with respect to the sampling matrix $\mathbf{\Phi_x}$ to produce a set of measurements ($\mathbf{Y_x}$) as defined in Eq. (4)

$$Y_x = \Phi_x F_x.. \tag{4}$$

The frames are encoded and transmitted independently. The measurements ($\mathbf{Y_x}$) of each frame received by the host workstation are first decoded independently in a frame by frame manner by solving Eq. (3) using BCS-TV-AL3 till a complete GoP is obtained. Once the GoP is obtained, the proposed MPR framework is then applied to the GoP to exploit the temporal correlations between the frames.

As illustrated in Fig. 1, the first frame (key frame) of the current and next GoP serve as the reference frames for the MPR framework to generate side information for improving the quality of $(\mathbf{J-1})$ non-key frames $\mathbf{F_{NK}}$ of the current GoP. Generally, the proposed MPR framework can be divided into three major phases as shown in Fig. 2. The detail explanation of each phase is provided in the following subsections.

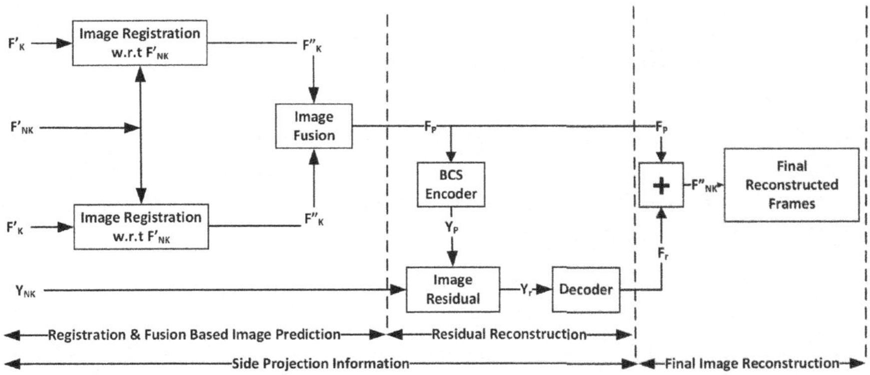

Fig. 2 Proposed multi-phase reconstruction (MPR) framework

3.1 Phase 1—Image Prediction

As shown in Fig. 2, this phase consists of two key steps: (i) registration [16] and (ii) fusion [17] in order to generate a prediction of (**J−1**) non-key frames ($\mathbf{F'_{NK}}$) within the GoP. First, intensity based image registration (requires less amount of pre-processing and able to achieve better alignment) is performed on the two independently reconstructed key frames $\mathbf{F'_K}$ within the GoP to project them onto the same plane of $\mathbf{F'_{NK}}$ i.e. aligning $\mathbf{F'_K}$ to $\mathbf{F'_{NK}}$ and exploiting the temporal correlation between them.

The registration process begins with the pre-processing of **F′K** and **F′NK** by using phase correlation to find the gross alignment between the two frames to estimate an initial transformation matrix (I-tform). Next, affine transformation is used to align **F′K** w.r.t. **F′NK** to give transformed **F′K** that is called **F″KT**. Then, the Image Similarity Metric (ISM) using mutual information is performed to evaluate the accuracy of the registration. The ISM returns a metric value by comparing the **F″KT** to the **F′NK**. Finally, the optimizer (gradient descent) states the method to exploit the ISM in order to produce a final registered image **F″K**. Details of intensity based image registration can be found in [16].

Second, both the registered $\mathbf{F''_K}$ are decomposed into their respective approximation (**A**) and detail (**D**) coefficients maps with 5 levels of decomposition using symlet 4-tap filter. The **A** and **D** coefficients in the two decomposition maps are then be fused together using point-to-point operations.

For each **A** coefficient from the same coordinate of the two decomposition maps, the magnitudes are compared and the coefficient with higher value is picked as the output in the fused map. Whereas for the **D** coefficients, the average magnitude of the two **D** coefficients from the same coordinate of the two decomposition maps is computed. The average value then serves as the output in the fused map. After fusing all the **A** and **D** coefficients from the two decomposition maps, inverse transformation is applied on the fused map to reconstruct the predicted frames $\mathbf{F_P}$. The registration and fusion approach estimates the object motions and creates predicted frames $\mathbf{F_P}$.

3.2 Phase 2—Residual Reconstruction

After the predicted frames $\mathbf{F_P}$ are generated, the projection of $\mathbf{F_P}$ onto the measurement basis $\mathbf{Y_P} = \mathbf{\Phi_x I_P}$ is performed. Then, the difference between the given measurements $\mathbf{Y_x}$ and $\mathbf{Y_P}$ is determined as expressed in Eq. (5) and the output is known as the residual measurement $\mathbf{Y_r}$.

$$Y_r = Y_X - Y_P \tag{5}$$

To obtain the residual frames $\mathbf{F_r}$, the residual measurements are then decoded by solving Eq. (3) using BCS-TV-AL3 reconstruction.

3.3 Phase 3—Final Frame Reconstruction

In order to produce the final reconstructed frames $\mathbf{F''_{NK}}$ with in the GoP, the $\mathbf{F_r}$ and $\mathbf{F_P}$ are added together. It is a normal point-to-point addition that is expressed in Eq. (6). By doing so a uniformity in terms of image measurements ($\mathbf{Y}$) is achieved i.e. the measurements computed for $\mathbf{F''_{NK}}$ is to some extent equal to the measurements $\mathbf{Y_{NK}}$.

$$F''_{NK} = F_r + F_P \tag{6}$$

After the key frames $\mathbf{F'_K}$ ($\mathbf{F_0}$ and $\mathbf{F_J}$ from $\mathbf{Y_0}$ and $\mathbf{Y_J}$) are reconstructed using BCS-TV-AL3, they are used as the reference frames for the reconstruction of the non-key frames $\mathbf{F'_{NK}}$ between them. The MPR framework produces the non-key frame $\mathbf{F''_1}$ from $\mathbf{Y_1}$, $\mathbf{F_0}$ and $\mathbf{F_J}$ in the same way as $\mathbf{F''_2}$ is produced from $\mathbf{Y_2}$, $\mathbf{F_0}$ and $\mathbf{F_J}$. The process continues for all the remaining non-key frames. We expect the reconstruction quality to drop when reconstructing non-key frames that are far from the key frames. Hence, the reconstruction quality may deteriorates more as the GoP size (**J**) increases.

4 Experimental Results

The proposed MPR framework coupled with BCS-TV-AL3 is referred to as MPR-TV. It is applied to a set of standard test video sequences (i.e. grayscale CIF [18] frames of size 352 × 288). However, only the results obtained from the **Mother-Daughter** (300 frames), and **Mobile** (300 frames) are presented in the main text, because we were able to observe the same trends in other datasets that we have tested. The evaluation is carried out by recording the Peak Signal to Noise Ratio (PSNR) at different sub-rates. Due to the random $\mathbf{\Phi}$, the image quality may vary. Hence, all PSNR values represent an average of 5 independent trials.

The GoP size (**J**) is fixed at **8** frames and the block size of **16** × **16** rather than 32 × 32 and 64 × 64 is adopted, because smaller block size leads to less memory usage [14]. All the non-key frames within a GoP are encoded at lower sub-rates (0.05, 0.1, 0.15, 0.2, 0.25, 0.3) with the key frames encoded at a fixed sub-rate of 0.5.

4.1 Proposed MPR Framework with BCS-TV-AL3

In this subsection the performance of the proposed MPR-TV framework is compared to the independent BCS-TV-AL3. In Table 1 the PSNR results are averaged over only the non-key frames attained for each video sequence. It can be seen that the proposed framework shows an average gain of 2 dB - 5 dB over independent BCS-TV-AL3,

Table 1 PSNR (dB) achieved by using the conventional BCS-TV-AL3 and the proposed framework to encode various video sequences

Mother-Daughter						
Sub-rate	**0.05**	**0.1**	**0.15**	**0.2**	**0.25**	**0.3**
BCS-TV-AL3 [19]	27.44	29.80	31.51	33.81	35.12	36.54
MPR-TV	32.00	34.26	35.81	37.65	38.63	39.61
Gain	**4.56**	**4.46**	**4.3**	**3.84**	**3.51**	**3.07**
Mobile						
Sub-rate	**0.05**	**0.1**	**0.15**	**0.2**	**0.25**	**0.3**
BCS-TV-AL3 [19]	17.23	18.40	19.48	20.59	21.54	22.40
MPR-TV	20.59	21.68	22.55	23.44	24.20	24.96
Gain	**3.36**	**3.28**	**3.07**	**2.85**	**2.66**	**2.56**

and the gain is more prominent in low-motion video Mother-Daughter. The PSNR of the first 100 non-key frames encoded at sub-rates of 0.1 is also shown in Fig. 3.

It is also noticed that the performance gain decreases when the sub-rate increases. As mentioned earlier, $\mathbf{F_K}$ are the reference frames that are transmitted at a higher sub-rate than that of $\mathbf{F_{NK}}$. Hence, $\mathbf{F_K}$ produces a larger set of measurements, which superimposes the correlated smaller set of measurements encompasses by $\mathbf{F_{NK}}$. Thus, reducing the prediction errors of $\mathbf{F_{NK}}$ that occurs due to smaller set measurements and produce an improved version of $\mathbf{F_{NK}}$.

4.2 *Visual Result Comparison*

Since MPR performs better at lower sub-rate, it is important to ensure that the sub-rate used is sufficient to produce visually recognizable frame. Shown in Fig. 4

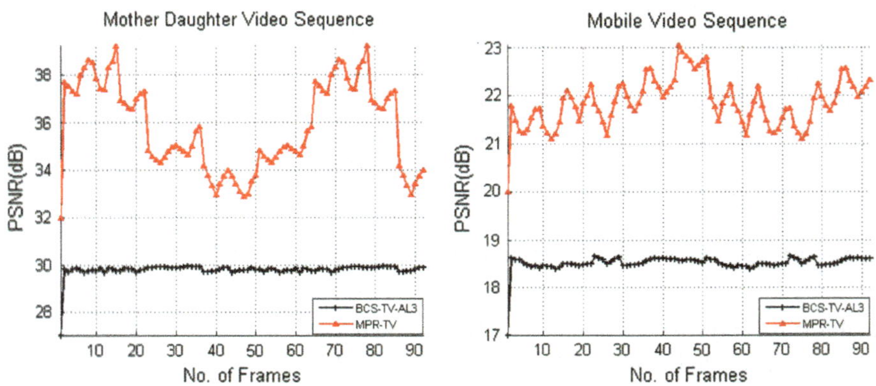

Fig. 3 PSNR (dB) of proposed MPR-TV and BCS-TV-AL3 @ 0.1 for first 100 non-key frames of different video sequence

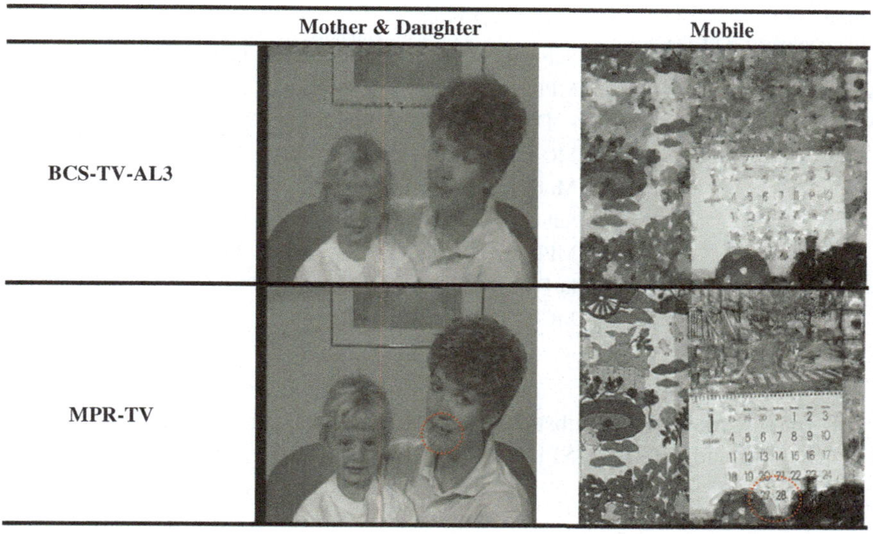

Fig. 4 Visual results of the reconstruction for center frame of 1st GoP at sub-rates = 0.1 for different video sequence using independent BCS-TV-AL3 and proposed MPR-TV framework

are centre frame (**x = 4**) of the first GoP reconstructed by using MPR-TV and BCS-TV-AL3 at sub-rate of 0.1. It should be reminded that the reference frames $\mathbf{F_K}$ used in the reconstruction of $\mathbf{F_{NK}}$ are reconstructed at a sub-rate of 0.5.

By comparing the visual results presented in Fig. 4, it can be noticed that the frame reconstructed by using the proposed MPR framework looks much sharper. It also performs better for low-motion video (Mother-Daughter) due to more accurate frame prediction and residual reconstruction. For video contains fast moving objects (Mobile), the MPR is exposed to some noise as highlighted by the red dotted circle.

4.3 Comparison of Proposed MPR Framework with Other Reconstruction Schemes

The proposed MPR framework is compared with k-t FOCUSS [14], and MC-BCS-SPL [15] as their implementations were readily available at the time of writing. All the simulation results that we obtained for the first 100 frames are summarized in Table 2, presented in terms of gain i.e. proposed final reconstruction over independent reconstruction.

The results for k-t FOCUSS and MC-BCS-SPL were obtained after modifying their available code [20, 21] respectively, with regard to the experimental setup described in Sect. 4. From the simulation results, it can be seen that the proposed

Table 2 Performance gain (dB) comparison of the proposed MPR framework with k-t FOCUSS [14] and MC-BCS-SPL [15] for various video sequences

Mother and Daughter						
Sub-rate	**0.05**	**0.1**	**0.15**	**0.2**	**0.25**	**0.3**
MPR-TV	**5.96**	**5.83**	**5.69**	**5.57**	**5.41**	**5.27**
kt-Focuss [20]	1.08	1.87	2.65	3.34	3.99	4.75
MC-BCS-SPL [21]	2.17	3.04	3.77	4.30	4.85	5.14
Mobile						
Sub-rate	**0.05**	**0.1**	**0.15**	**0.2**	**0.25**	**0.3**
MPR-TV	**3.73**	**3.59**	**3.47**	**3.32**	3.19	3.06
kt-Focuss [20]	0.66	1.02	1.89	2.52	3.12	3.86
MC-BCS-SPL [21]	0.96	1.82	2.15	2.95	**3.81**	**4.55**

MPR framework provides substantial gain at lower sub-rates when compared with k-t FOCUSS and MC-BCS-SPL.

4.4 Execution Time

All the schemes are implemented using Matlab (R2014a) running on a computer with an Intel(R) Xeon(R) E5-1620 3.6 GHz CPU and 8 GB RAM. We measured the average execution time required to reconstruct a single frame at various sub-rates. The results show that the average execution time of the proposed MPR-TV framework, kt-Focuss [14], and MC-BCS-SPL [15], ranges from **6.39–12.06 s, 24.78–30.01 s, and 63.54–69.51 s** respectively. At lower sub-rate, all the three schemes take longer time to find a better reconstruction due to small number of received measurements. Overall, the proposed MPR-TV framework is approximately **2–3 times** shorter than others. This is due to the less complex BCS-TV-AL3 and the simplified process of predicting the non-key frames using the proposed MPR framework. However, it is important to note that all the implementations above have not been optimized for execution time.

5 Conclusion and Future Work

In this paper, a new framework for video reconstruction encoded independently with BCS is developed. The framework is capable of exploiting the correlations present within the video frames. The simulation results prove that the proposed framework outperform the conventional BCS-TV-AL3 framework by a margin of 3 to $\sim$5 dB at different sub-rates for different set of video sequences.

The Proposed scheme will be extended in future for multi-view video scenario in order to exploit not only the temporal, but also the inter-view correlations to improve reconstruction quality of video in multi-view environment.

References

1. Ebrahim M, Chong CW (2014) A comprehensive review of distributed coding algorithms for visual sensor network (VSN), In press, Int J Commun Networks Inf Secur (IJCNIS), 6(2):104–117
2. Donoho DL (2006) Compressed sensing. IEEE Trans Inf Theory 52(4):1289–1306. doi:10.1109/TIT.2006.871582
3. Duarte MF, Davenport MA, Takhar D, Laska JN, Sun T, Kelly KF, Baraniuk RG (2008) Single pixel imaging via compressive sampling. IEEE Signal Process Mag 25(2):83–91. doi:10.1109/MSP.2007.914730
4. Candes E, Romberg J (2007) Sparsity and incoherence in compressive sampling. Inverse Prob 23(3):969–985. doi:10.1088/0266-5611/23/3/008
5. Blumensath T, Davies ME (2009) Iterative hard thresholding for compressed sensing. Appl Comput Harmonic Anal. Elsevier, 27(3):10. doi:10.1016/j.acha.2009.04.002
6. Gan L (2007) Block compressed sensing of natural images. In: Proceedings of the international conference on digital signal processing. Cardiff, UK, pp 403–406, July 2007. doi:10.1109/ICDSP.2007.4288604
7. Mun S, Fowler JE (2009) Block compressed sensing of images using directional transforms. In: Proceedings of the international conference on image processing. Cairo, Egypt, pp 3021–3024, November 2009. doi:10.1109/ICIP.2009.5414429
8. Li C (2013) Compressive sensing for 3d data processing tasks: applications, models and algorithms. PhD thesis, Rice University, Houston, Texas, United States. doi:http://hdl.handle.net/1911/70314
9. Chambolle A, Lions PL (1997) Image recovery via total variation minimization and related problems, Numerische Mathematik Electronic Edition. Springer, 76(2):21. doi:10.1007/s002110050258
10. Wakin MB, Laska JN, Duarte MF, Baron D, Sarvotham S, Takhar D, Kelly KF, Baraniuk RG (2006) Compressive imaging for video representation and coding. In: Picture coding symposium—PCS 2006. Beijing, China, April 2006
11. Marcia R, Willet R (2008) Compressive coded aperture video reconstruction. In: Proceeding of European Signal Processing Conference (EUSIPCO)
12. Park JY, Wakin MB (2009) A multiscale framework for compressive sensing of video. In: Proceedings of the picture coding symposium, pp 1–4, May 2009
13. Lu W, Vaswani N (2009) Modified compressive sensing for real-time dynamic MR imaging. In: Proceedings of the international conference on image processing (ICIP'09). IEEE, Cairo, Egypt, pp 3045–3048. doi:10.1109/ICIP.2009.5414208
14. Jung H, Sung K, Nayak KS, Kim EY, Ye JC (2009) k-t FOCUSS: a general compressed sensing framework for high resolution dynamic MRI. Magn Reson Med 61(1):14. doi:10.1002/mrm.21757
15. Mun S, Fowler JE (2011) Residual reconstruction for block-based compressed sensing of video. In: Storer JA, Marcellin MW (eds) Proceedings of the IEEE data compression conference. Snowbird, UT, pp 183–192, March 2011
16. Mathworks (2014) Automatic Registration. Retrieved 20 April 2014 from http://www.mathworks.com/help/images/-automatic-registration.html
17. de Zeeuw P (1998) Wavelet and image fusion. CWI research STW (March 1998), Retrieved 20 April 2014 from Mathworks wfusimg: www.mathworks.com/help/wavelet/ref/wfusimg.html
18. YUV Video Sequences, http://trace.eas.asu.edu/yuv/. Retrieved 15 Jan 2015
19. TVAL3, Courtesy Rice University, from http://www.caam.rice.edu/~optimization/L1/TVAL3/
20. k-t-FOCUSS. Version 1, http://bisp.kaist.ac.kr/research_02.html. Retrieved 15 April 2015
21. Mc-bcs-spl Version 1.0-1, www.ece.msstate.edu/~fowler/BCSSPL//. Retrieved 15 April 2015

PAT and Px Code Sidelobe Reduction Using Wavelet Neural Network

Fayad Mohammed Ghawbar, Mustafa Sami, Nor Shahida Mohd Shah and Yasin Yousif

Abstract Pulse compression is a significant aspect for improving the radar detection and range resolution. To improve the range detection, the pulse width is increased to overcome the transmitter maximum peak power limitations. However, pulse compression is accompanied with time sidelobes that can mask the small targets. The Wavelet Neural Network (WNN) is a new technique used for pulse compression sidelobe reduction. In this paper, Morlet function is applied as an activation function for WNN and the backpropagation (BP) is implemented for training the networks. The WNN is applied based on PAT and Px polyphase codes. The performance of WNN is evaluated in terms of Signal to Noise Ratio (SNR) and the computational complexity. The simulation results indicate that the WNN has higher Peak Sidelobe Level (PSL) than the Auto Correlation Function (ACF) with more than 100 dB and higher PSL than the Neural Network (NN) with more than 100 dB.

1 Introduction

In radar systems, Pulse compression is a very important research topic in which it allows to utilize a long pulse for achieving a high energy and to get a short-range resolution at the same time. As a result, maximum range detection and high range resolution can be obtained simultaneously. This compression can be employed using either phase or frequency modulation to get a widened bandwidth signal rather than using an amplitude modulation, which is rarely used [1]. To obtain a

F.M. Ghawbar (✉) · M. Sami · N.S.M. Shah · Y. Yousif
Communication Department, Faculty of Electrical and Electronic Engineering,
Universiti Tun Hussein Onn Malaysia, 86400 Parit Raja, Batu Pahat, Johor, Malaysia
e-mail: fay_mh600@yahoo.com

M. Sami
e-mail: mustafa_sami87@yahoo.com

P.J. Soh et al. (eds.), *Advances in Machine Learning and Signal Processing*,
Lecture Notes in Electrical Engineering 387, DOI 10.1007/978-3-319-32213-1_11

short pulse, the received signal is passed through a match filter to compress the long pulse to a period $1/B$ where B is the modulated bandwidth [2]. However, pulse compression has some limitations during the compression process in which it generates large sidelobes. These generated sidelobes are objectionable since small target can be masked by a large target and that may create false targets [2, 3].

Polyphase codes are derived consistently detecting a frequency modulation pulse suppression waveform signal. Detection is applied either by a local oscillator and is called single sideband detection at the band edge of the waveform or by Inphase I and Qdate at the Nyquist rate. These codes are harmonically based on an incremental phase [4]. Polyphase codes have some advantages over analog pulse compression waveforms. The ability to obtain low sidelobe without weighting is the main advantage of polyphase codes. In addition, they are significantly doppler shift tolerant and they can be applied easily without having reflections. More advantages can be found in this research [5]. Frank codes are considered a classical family of polyphase codes that are closely related to the stepped frequency of a linear chirp and barker codes [1, 6]. It was previously introduced by Robert L. Frank (1963) [7]. After the frank code has been proposed other researchers have updated and developed this code into P-codes [8, 9]. Generally, frank codes can be described by declaring that the longer frank codes somehow correspond the chirp code and the shorter frank codes somehow correspond the barker codes [1].

Wavelet Networks are a new model of networks that are a combination of the sigmoid neural network and wavelet analysis [10]. Wavelet Neural Networks (WNN) were previously proposed by Zhang and Benveniste in 1992 as an alternative to both feedforward neural networks and wavelet decomposition using backpropogation (BP) algorithm for training the networks [11]. This new type of hybrid network would help to decrease the weaknesses associated with neural networks and wavelet analysis while still preserve the good characteristics of each method.

There are many types of research and applications have implemented Wavelet networks in different fields achieving great results such as, Engineering, digital signal processing, time-series prediction, control, signal classification, static, dynamic, nonlinear modeling, etc. [10]. Jin Zhao et al. presented a new WNN to ratify the application of this network in multi-variable functional approximation using a multi-input and multi- output feedforward wavelet neural network. In addition, wavelet basis is used as an activation function in the hidden layer instead of the sigmoid function and that is the only difference with the standard neural network. As a result, the difference in activation function results in a better performance with WNN compared to the standard NN [12]. After all, from reviewing and studying many types of research on pulse compression, there are still many weaknesses in sidelobe reduction techniques for pulse compression. Therefore, the WNN is a newly adopted method in polyphase pulse compression used for sidelobe reduction. In this paper, a new approach of sidelobe pulse reduction implementing the Px and PAT code, is proposed and the wavelet Neural Network is applied. The simulation of this approach is performed to obtain low sidelobes for Px and PAT codes under different code elements.

This paper is organized as follows. Section 2 discusses on the Px and PAT codes. In Sect. 3 the WNN and its learning algorithm is presented. Section 4 summarizes the simulation and results for the NN and WNN including their performance evaluation under different factors. Finally, the conclusion of this paper is provided in Sect. 5.

2 Px and PAT Code

Px and PAT codes are classes of polyphase codes of perfect square length derived for improving previous polyphase codes performance. Px code is a modified class of frank code having a dc frequency in the middle of the pulse rather than the beginning in frank code [13]. It is considered the best choice with a significant performance compared to previously mentioned pulse compression polyphase codes (i.e., frank, P1, P2, P3, P4 codes) [14]. The Px code was presented by Rapajic and Kennedy in 1998 [15] to have similar peak sidelobes of Frank code but with a utilized lower integrated sidelobe level (ISL) [13]. The mathematical expression of Px code is represented by the following equation.

$$\emptyset_{l,m} = \begin{cases} \frac{2\pi}{L}\left[\frac{L+1}{2} - l\right]\left[\frac{L+1}{2} - m\right] & L\,even \\ \frac{2\pi}{L}\left[\frac{L+1}{2} - l\right]\left[\frac{L}{2} - m\right] & L\,odd \end{cases} \quad 1 \leq l, m \leq L \tag{1}$$

where: L is the code length and l,m are the matrix indexes.

Figure 1 shows the ACF amplitude of the 25 and 100 element Px code that has a peak sidelobe of 18.19 and 29.8 dB respectively.

The PAT code is a new model of polyphase sequences of square length $N = L^2$. PAT code was presented by Daniele Petrolati et al. in 2012 [16]. The generation of PAT codes derives based on a significant method called Spectral Domain Synthesis

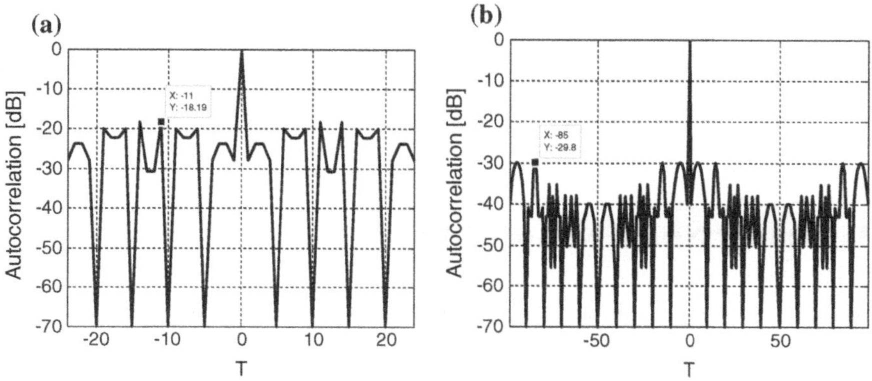

Fig. 1 ACF output of the Px code **a** 25 element **b** 100 element

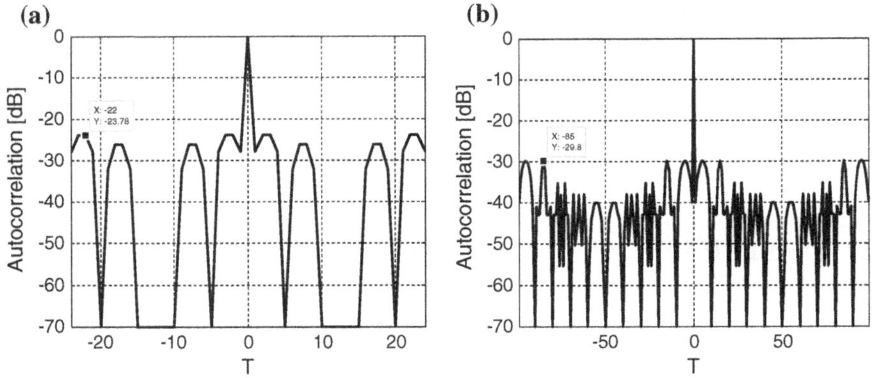

Fig. 2 ACF output of the PAT code **a** 25 element **b** 100 element

(SDS) which results in a simple code generation formula. In addition, PAT codes reduced the limitations related to Px code and previously mentioned polyphase codes and it has a better merit factor compared to odd lengths of Frank code and a similar merit factor compared to even lengths of Px code [16]. The mathematical formula representation of PAT code is shown as below:

$$\emptyset_{l,m} = \emptyset_{l-1,m} - \frac{2\pi}{L}\left[\left(\frac{L+1}{2} - 1\right)m + 0.5\right] 1 \leq l, m \leq L \tag{2}$$

With the initial condition

$$\emptyset_{0,L} = 0 \tag{3}$$

Figure 2 shows the ACF amplitude of the 25 and 100 element PAT code that has peak sidelobes of 23.78 and 29.8 dB respectively. By comparing the results from Figs. 1 and 2 of Px and PAT code, it is shown that for odd lengths, the PSL of PAT code is larger than the Px code. However, for even lengths, the PAT code preserves the same PSL of Px code.

The results in Figs. 1 and 2 show the performance comparison of the ACF of Px and PAT codes, which confirm that the 25 element PAT code almost avoids the deep nulls and peaks resulting in a sharper PAT code ACF.

3 Wavelet Neural Network (WNN)

Wavelet analysis (WA) is a mathematical tool and analysis method used in a variety of research and applications. It is currently implemented in time series analysis, intensity, and time position. [10]. The WNN structure is described simply as shown in Fig. 3

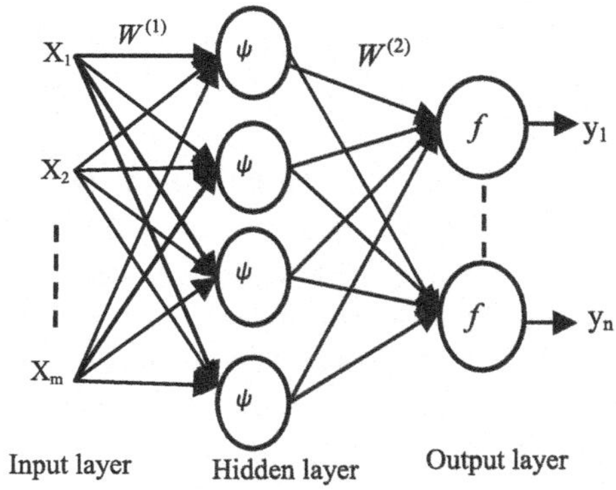

Fig. 3 The structure of wavelet neural network

The structure of the WNN can be presented as follows:

Step 1: Initialize the parameters of WNN

The matrix between input layer and hidden layer: $w^{(1)} = \left(w_{jk}^{(1)}\right)_{p \times m}$

The matrix between hidden layer and output layer: $w^{(2)} = \left(w_{ij}^{(2)}\right)_{n \times p}$

Dilation vector of the hidden layer neuron: $a_j = (a_1, a_1, \ldots a_p)$

Translation vector of the hidden layer neuron: $b_j = (b_1, b_1, \ldots b_p)$

where *m, p* and *n* are the WNN nodes in the input layer, hidden layer and output layer respectively. The initialization of these parameters can be applied randomly.

Step 2: Compute the forward pass to generate the network's output by using the following formula:

$$\psi_{a_j,b_j}\left(\sum_{k=1}^{m} w_{jk}^{(1)} x_k\right) = \psi_{a_j,b_j}(net_j^{(1)}) = \frac{1}{\sqrt{a_j}}\psi\left(\frac{net_j^{(1)} - b_j}{a_j}\right) \tag{4}$$

ψ is a special wavelet function named a mother wavelet and $net_j^{(1)}$ is the output of the hidden layer. The output of the *i*th node of output layer is:

$$y_i = f\left(\sum\nolimits_{j=1}^{p} w_{ij}^{(2)} \psi_{a_j,b_j}(net_j^{(1)})\right) = f(net_j^{(2)}) \tag{5}$$

where X is the input vector of WNN and can be defined as $X = (x_1, x_2, \ldots, x_m)$. Once X is obtained, the output of the *j*th node in the hidden layer can be computed.

Step 3: calculate the total error of each output by taking the difference of the output vector and the target vector:

$$E = \frac{1}{2}\sum_{q=1}^{Q}\sum_{i=1}^{n}\left(d_{qi} - y_{qi}\right)^2 \tag{6}$$

where E is the total error, Q is the number of training samples for each sample q, d_q is the target vector and it is defined as $D_q = (d_{q1}, d_{q2}, \ldots, d_{qn})$, and y_{qi} is the output vector and it is represented by $Y_q = (y_{q1}, y_{q2}, \ldots, y_{qn})$. Achieving the minimum total error E of each output is the main purpose of the WNN.

Step 4: compute the partial derivatives of each parameter $\frac{\partial E}{\partial w_{jk}^{(1)}}, \frac{\partial E}{\partial w_{ij}^{(2)}}, \frac{\partial E}{\partial a_j}$ and $\frac{\partial E}{\partial b_j}$

$$\frac{\partial E}{\partial w_{ij}^{(2)}} = -\sum_{q=1}^{Q}\left(d_{qi} - y_{qi}\right).y_{qi}.\left(1 - y_{qi}\right).\psi_{a_j,b_j}\left(net_{qj}^{(1)}\right) \tag{7}$$

$$\frac{\partial E}{\partial w_{jk}^{(1)}} = -a_j^{-1}\sum_{q=1}^{Q}\left[\psi'_{a_j,b_j}\left(net_j^{(1)}\right).x_{qk}.\sum_{i=1}^{n}\left(d_{qi} - y_{qi}\right).y_{qi}.\left(1 - y_{qi}\right)w_{ij}^{(2)}\right] \tag{8}$$

$$\frac{\partial E}{\partial a_j} = -a_j^{-1}\sum_{q=1}^{Q}\left\{\left[\frac{a_j^{-1}}{2}\psi_{a_j,b_j}\left(net_{qj}^{(1)}\right) + \frac{net_{qj}^{(1)} - b_j}{a_j^2}\psi'_{a_j,b_j}\left(net_{qj}^{(1)}\right)\right].\sum_{i=1}^{n}\left(d_{qi} - y_{qi}\right).y_{qi}.\left(1 - y_{qi}\right)w_{ij}^{(2)}\right\} \tag{9}$$

$$\frac{\partial E}{\partial b_j} = -a_j^{-1}\sum_{q=1}^{Q}\left[\psi'_{a_j,b_j}\left(net_j^{(1)}\right).\sum_{i=1}^{n}\left(d_{qi} - y_{qi}\right).y_{qi}.\left(1 - y_{qi}\right)w_{ij}^{(2)}\right] \tag{10}$$

Step 5: update the Parameters of WNN by setting the learning rate and momentum as $\eta = 0.2$ and $\alpha = 0.99$ as shown in the Eqs. (11–14).

$$w_{ij}^{(2)}(t+1) = w_{ij}^{(2)}(t) - \eta\frac{\partial E}{\partial w_{ij}^{(2)}} + \alpha\left[w_{ij}^{(2)}(t) - w_{ij}^{(2)}(t-1)\right] \tag{11}$$

$$w_{jk}^{(1)}(t+1) = w_{jk}^{(1)}(t) - \eta\frac{\partial E}{\partial w_{jk}^{(1)}} + \alpha\left[w_{jk}^{(1)}(t) - w_{jk}^{(1)}(t-1)\right] \tag{12}$$

$$a_j(t+1) = a_j(t) - \eta\frac{\partial E}{\partial a_j} + \alpha\left[a_j(t) - a_j(t-1)\right] \tag{13}$$

$$b_j(t+1) = b_j(t) - \eta\frac{\partial E}{\partial b_j} + \alpha\left[b_j(t) - b_j(t-1)\right] \tag{14}$$

4 Simulation and Results

This section illustrates the performance of the Wavelet Neural Network for radar signal pulse compression. Two different codes of Px and PAT with different lengths of 25 and 100 elements are implemented. It is known that the Polyphase has complex numbers consist of the real and imaginary part. There are two methods to solve such complex numbers have been implemented previously by several studies such as [17]. The first method is by dealing with the real and imaginary parts separately using two training networks, one for the real and one for the imaginary. The second method is by using the free parameters as complex numbers and implementing a modified algorithm called the complex BPA for the training purpose [18]. However, there are some weaknesses in these methods exist in its computational complexity and hardware requirements. Furthermore, using two neural networks is difficult to obtain similar powerful training results for the real and imaginary parts separately.

In this paper, a new approach is implemented which it allows to use one network that ensures employing both real and imaginary parts. Therefore, one network is used for each length, 25 and 100 in which each of these networks contributes to both Px and PAT codes. After obtaining the autocorrelation of Px and PAT codes, input data is converted from the rectangular into the polar form which consists of the amplitude and the phase. Then, the amplitude is trained based on the desired signal for the WNN. The desired signal is a vector contains zeros for all values except for the middle-value point. For example, if the code length is 25, the desired signal $d =$ [zero(1,24),25, zero(1,24)]. Once the training is completed, the newly trained values are converted back to the complex form. This would result in zero imaginary for all vector values except for the desired middle point.

This WNN consists of one layer for input (m), hidden (p) and the output layer (n). The input and the output layer consist of one node while the hidden layer consists of three neurons based on this following equation [19].

$$N_h \geq (2^* N_i) + 1 \tag{15}$$

where N_h represents the number of neurons in the hidden layer and N_i represents the number of neurons in the input layer. The Morlet function is used as an activation function in the hidden layer while the sigmoid function is represented in the output layer. The following equation shows the Morlet function.

$$\psi(t) = \cos(1.75t)exp\left(-\frac{t^2}{2}\right) \tag{16}$$

Each Network for the 25 and 100 elements are trained for 200 epoch based on their training learning algorithm. After the training is completed. The WNN can be directly applied for radar pulse compression. The performance of the WNN can be evaluated based on various factors as the following

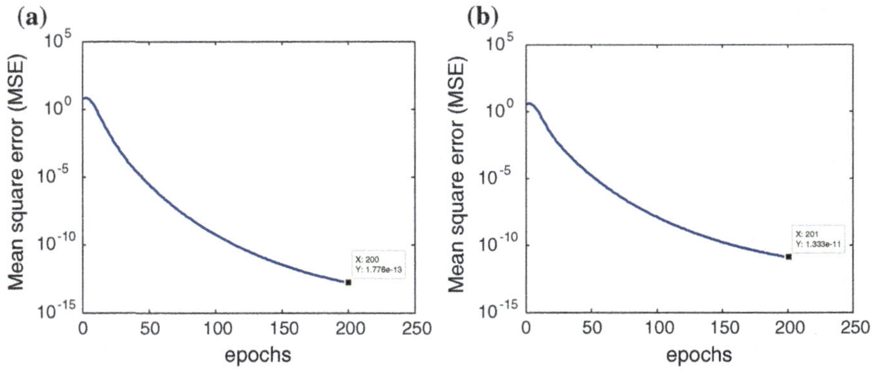

Fig. 4 MSE for **a** 100 element **b** 25 elements

Table 1 Convergence speed

Method	MSE	
	25 elements	100 elements
NN	2.24e-9	3.16e-9
WNN	1.33e-11	1.77e-13

4.1 Convergence Performance

Figure 4 shows the MSE for the 25 and 100 elements. From this figure, it is observed that the WNN for the 100 elements has a lower MSE than the 25 elements. It is also found out that MSE decreases with the code length due to the increase in neurons numbers of the input layer and that results in more interconnections between the adjacent layers. However, the increasing number of interconnections results in more hardware requirements. Table 1 shows the MSE values for WNN and NN for both code lengths networks. It is clearly shown that WNN has a lower MSE values compared to the NN for both lengths. Therefore, the WNN has a higher convergence speed than the NN.

4.2 PSL Performance

PSL is defined as the measure of sidelobe with respect to the mainlobe [20]. Table 2 displays the PSL results for Px and PAT code with free noise case. It is clearly observed that the PAT code has a lower sidelobe than the Px code in the ACF at the 25 elements. However, at the 100 elements both Px and PAT code exhibits the same sidelobe with 29.8 dB. That illustrates the PAT code has a similar PSL of Px at even lengths while it has a better PSL performance at odd lengths. The PAT code has a higher sidelobe reduction than the Px code for the 25 elements while it obtains

Table 2 PSL for PAT and Px codes for 25 and 100 elements

Method	PSL (dB)			
	Px code		PAT code	
	25 elements	100 elements	25 elements	100 elements
ACF	18.19	29.8	23.78	29.8
NN	109.32	124.82	112.50	124.90
WNN	176.81	255.13	192.85	255.59

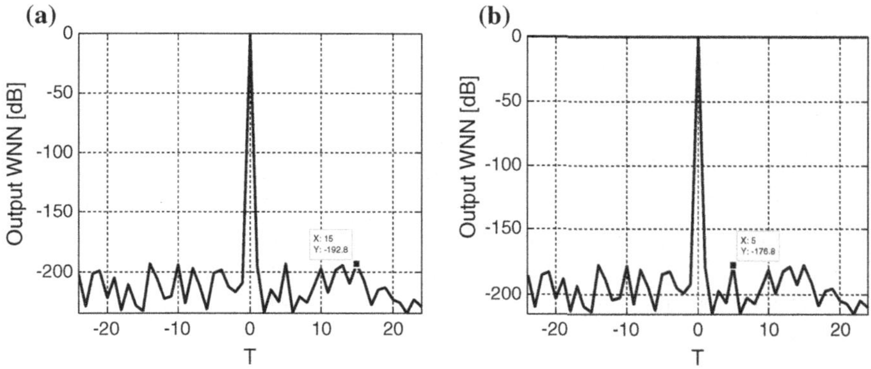

Fig. 5 The WNN output for the 25 elements **a** PAT code **b** Px code

a similar PSL at the 100 elements. Figure 5 displays the WNN PSL results for the 25 elements of PAT and Px codes. Furthermore, Table 2 compares the PSL output results of NN and WNN and demonstrates that WNN has the highest PSL magnitude over other approaches.

4.3 Noise Performance

Noise is an important metric to test the performance of any network. It might mask the target signal if it is very high resulting in a very low detection. Therefore, the ability of noise rejection is verified using the white additive Gaussian noise for the simulation. Different Signal to Noise Ratio (SNR) conditions are applied ranging from 0 to 20 dB for the ACF and WNN. Table 3 and 4 display the PSL output for the Px and PAT codes of the 25 and 100 elements. It is evidently shown that increasing the SNR results in increasing the PSL magnitudes as well. Results from Table 3 for the 25 elements shows that PAT code has the highest PSL magnitudes compared to the Px code since it has the least ACF sidelobe. Furthermore, Table 4 shows the PSL output results of the 100 elements for PAT and Px codes. It is observed that the PSL outputs for both Px and PAT codes are almost similar.

Table 3 PSL for Px and PAT code with 25 elements at different SNR

Method		PSL (dB)				
		SNR = 0 dB	SNR = 5 dB	SNR = 10 dB	SNR = 15 dB	SNR = 20 dB
PX	ACF	10.395	11.567	13.830	14.138	16.164
	WNN	95.981	113.525	125.447	130.999	151.166
	NN	68.180	74.680	87.182	95.001	98.480
PAT	ACF	12.396	15.433	17.088	19.623	20.692
	WNN	110.120	122.556	134.820	149.329	162.377
	NN	70.750	84.990	89.300	96.670	100.340

Table 4 PSL for Px and PAT code with 100 elements at different SNR

Method		PSL (dB)				
		SNR = 0 dB	SNR = 5 dB	SNR = 10 dB	SNR = 15 dB	SNR = 20 dB
PX	ACF	18.502	19.815	22.330	24.729	27.758
	WNN	172.669	179.473	186.755	195.742	200.767
	NN	86.320	88.760	95.350	111.850	118.860
PAT	ACF	17.950	19.815	22.230	24.280	27.530
	WNN	169.149	178.505	186.009	194.770	200.762
	NN	87.620	89.480	95.600	112.330	119.090

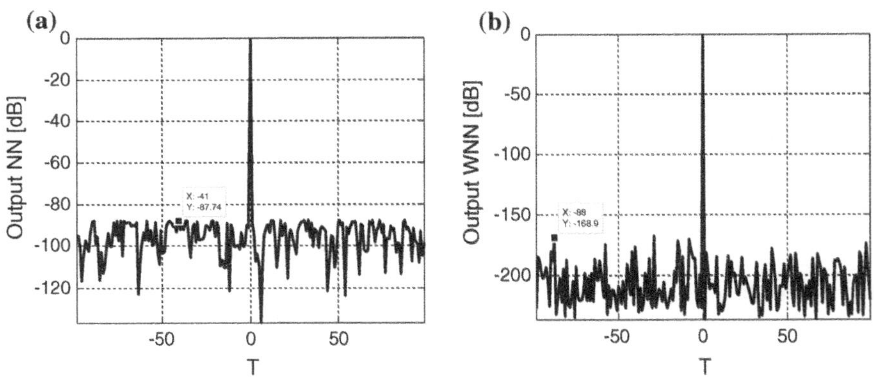

Fig. 6 The outputs of the 100 elements PAT code at 0 dB **a** NN, **b** WNN

Figure 6 shows the outputs of the NN and WNN for the PAT code 100 elements at 0 dB. In addition, Tables 3 and 4 compare the performance of the WNN and the NN. The results show that the WNN has a better sidelobe reduction with noisy condition as it has a better noise rejection capability.

Table 5 CPU time (s) utilization for training

Method	ACF	NN	WNN
CPU time	No training	4.60	1.190

4.4 Computational Complexity

To evaluate the performance of a network, computational complexity is an important factor to be considered. Therefore, the CPU time utilization during the training for each approach in the same processor is shown in Table 5. The CPU time utilization results show that WNN has an efficient speed compared to the NN. In addition, it is observed that WNN has a lower number of operations than the NN approach. Thus, the WNN method is less complexity compared to the NN.

5 Conclusion

In this paper, the WNN is adopted for radar pulse compression. Morelt function is used as an activation function and BP is used for training the networks. The WNN is applied based on PAT and Px codes. The simulation results indicate that the performance of WNN based pulse compression is much better than NN. The results reveal that the WNN has higher convergence speed than the NN with a lower MSE of 1.77e-13 for the 100 element. In addition, the WNN presents better PSL values with noise condition and exhibits less computational complexity than NN. Furthermore, this paper shows that the PAT code is more efficient than Px polyphase code and it has a better sidelobe reduction. The efficiency of PAT code has improved the performance of both WNN and NN. Though the WNN is applied for the PAT and Px codes, it is still can be implemented for any other polyphase codes.

References

1. Bogler PL (1990) Radar principles with applications to tracking systems. Wiley, New York
2. Skolnik MI (1980) Introduction to radar systems, 2nd edn. McGraw Hill, New York
3. Hussain A, Khan ZH, Khalid A, Iqbal M (2014) A comparison of pulse compression techniques for ranging applications. In: Computational intelligence for decision support in cyber-physical systems. Springer Singapore, pp 169–191
4. Darwich T (2007) High Resolution Detection systems using low sidelobe pulse compression techniques. Ph.D. Thesis, University of Louisiana
5. Kretschmer FF Jr, Lewis BL (1982) Polyphase pulse compression waveforms (No. NRL-8540). Naval Research Lab, Washington, DC
6. Wirth WD (1992) Compression of polyphase codes with Doppler shift. In: Radar 92. International conference. IET, pp 469–472
7. Frank RL (1963) Polyphase codes with good nonperiodic correlation properties. IEEE Trans Inf Theory 9(1):43–45

8. Lewis BL, Kretschmer FF Jr (1981) A new class of polyphase pulse compression codes and techniques. IEEE Trans Aerosp Electron Syst AES-17(3):364–372
9. Lewis BL, Kretschmer FF (1982) Linear frequency modulation derived polyphase pulse compression codes. IEEE Trans Aerosp Electron Syst AES-18(5):637–641
10. Alexandridis AK, Zapranis AD (2014) Wavelet neural networks: with applications in financial engineering, chaos, and classification. Wiley
11. Acosta FMA, Vesin J (1992) Nonlinear prediction of time series using radial wavelet networks. In: Proceedings of the IEEE-SP international symposium time-frequency and time-scale analysis. Victoria, BC, pp 189–192
12. Zhao J, Chen W, Luo J (2005) Feedforward wavelet neural network and multi-variable functional approximation. In: Computational and information science. Springer, pp 32–37
13. Levanon N, Mozeson E (2004) Radar signals. John Wiley and Sons, NJ
14. Jamil M, Zepernick H, Pettersson MI (2008) Performance assessment of polyphase pulse compression codes. In: IEEE 10th International symposium on spread spectrum techniques and applications 2008, ISSSTA '08, pp 166–172
15. Rapajic PB, Kennedy RA (1998) Merit factor based comparison of new polyphase sequences. IEEE Commun Lett 2(10):269–270
16. Petrolati D, Angeletti P, Toso G (2012) New piecewise linear polyphase sequences based on a spectral domain synthesis. IEEE Trans Inf Theory 58(7):4890–4898
17. Padaki AV, George K (2010) Improving performance in neural network based pulse compression for binary and polyphase codes. In: 2010 12th International conference on computer modelling and simulation (UKSim). Cambridge, pp 278–283
18. Leung H, Haykin S (1991) The complex back propagation algorithm. IEEE Trans Signal Process 39(9):2101–2104
19. Kishan M, Chilukuri K, Sanjay R (1997) Elements of artificial neural net-works. MIT press, Bradford
20. Nathanson FE, Reilly JP, Cohen MN (1991) Radar design principles-signal processing and the environment. NASA STI/Recon Technical, Report

Detecting Neighbor Discovery Protocol-Based Flooding Attack Using Machine Learning Techniques

Firas Najjar, Mohammad M. Kadhum and Homam El-Taj

Abstract Neighbor Discovery Protocol (NDP) is stateless and lacks of authentication which exposes it to flooding attacks. Securing NDP is critical due to the large deployment of open network. Commonly existing solutions for securing NDP violate its design principle in terms of overhead and complexity. Other solutions suffer from high false positive alerts which affects solution trustiness. This paper aims to investigate the use of machine learning mechanism for detecting NDP flooding attacks. It was found that the advantage of using machine learning is that the detection can be done without relying on attack signatures they can learn broader definitions of attack attributes.

1 Introduction

The huge growth of Internet users led to the exhaustion of the existing Internet Protocol v4 (IPv4) addresses [1]. To overcome this issue, Internet Assigned Number Authority (IANA) [2] has started to allocate IP addresses using Internet Protocol version 6 [3] which provides a massive number of IP addresses. Although IPv6 was built with security in mind and as a successor of IPv4, it inherits security weakness from IPv4 protocol. Neighbor Discovery Protocol (NDP) [4] is the main supported protocol for IPv6 used to enable IPv6 node to discover each other's

F. Najjar (✉) · M.M. Kadhum
National Advanced IPv6 Centre (NAv6), Universiti Sains Malaysia (USM), 11800 Pulau Pinang, Malaysia
e-mail: firas@nav6.usm.my

M.M. Kadhum
Telecommunications Research Lab, School of Computing, Queen's University, Kingston, ON K7L 3N6, Canada
e-mail: kadhum@nav6.usm.my; kadhum@cs.queensu.ca

H. El-Taj
Community Collage, University of Tabuk, Tabuk, Saudi Arabia
e-mail: h.eltaj@ut.edu.sa

P.J. Soh et al. (eds.), *Advances in Machine Learning and Signal Processing*, Lecture Notes in Electrical Engineering 387, DOI 10.1007/978-3-319-32213-1_12

presence. NDP has no authentication or registration mechanism; therefore, it is exposed to attacks.

IPv6 networks provide the ability for any connected node (router or host) to configure its IP address and start communicating with other nodes without any registration or authentication and that the other node must response to it without any validation. Therefor the attacker can easily flood the network with any NDP fake messages and most hosts must blindly accept and process all these messages. Consequently, the system resources of the hosts will be exhausted, which may eventually lead to freeze them all, enforcing rebooting to clear the memory of thousands of fake addresses.

Different approaches proposed to prevent and monitor NDP misuse, spoofing, and denial of server attacks (DoS). However, most of them violate the design principle of NDP in term of overhead, complexity, and high rate of false positive alarms. Thus, proposed solutions on NDP must preserve the original design without any modification while improving its security.

One of the latest technologies for monitoring and preventing computer cyber threats is Intrusion Detection and Prevention System (IDPS) which become an essential component of computer security. The role of IDPS is to detect any sign of a possible incident that violates the system polices, warn the system administrator, and try to stop or slow down the detected violation. Moreover, IDPS contains ruled-based algorithms with learning algorithms which recognize detected complex patterns to help making intelligent decisions or predictions when it faces new or previously unseen network behavioral [5].

Machine learning algorithms in IDPS are used to recognize valid, novel, potentially useful and meaningful detected network behavioral using non-trivial mechanisms. Therefore, machine learning algorithms become an effective approach for IDPS for detecting novel and known attacks.

The rest of this paper is organized as follows: Sect. 2 presents background of NDP and IDPS. Section 3 describes related works. Section 4 identifies the common techniques of machine learning. The evaluation of machine learning techniques is described in Sect. 5. Finally, Sect. 6 concludes the research work presented in this paper.

2 Background

This section provides background of NDP with IDS technologies.

2.1 NDP

NDP provides a stateless mechanism that gives the ability for connected nodes to configure their IP addresses, configure their gateway, and start communication with neighbor nodes without any authentication or registration inside the local site [6]. Consequently, the attacker can claim himself as any node inside the network and launches various attacks.

Even NDP includes IPsec [7] in its original specification, to secure NDP messages, there are no instructions introducing the use of IPsec, and how the automatic exchanging of keys is done. Therefore, manual configuration of security associations must be done, which makes it impractical for most purposes [8].

IPv6 NDP allows nodes to identify its neighbors on the same LAN, and advertise its existence to other neighbors. To complete its functions, NDP uses the below ICMPv6 [9] messages:

- **Router Advertisement** (RA) messages are originated by routers and sent periodically, or sent in response to Router Solicitation. Routers use RAs to advertise their presence and send specific parameters such as MTU, Router Prefix, lifetime for each prefix, and hop limits.
- **Router Solicitation** (RS) messages are originated by hosts at the system startup to request a router to send RA immediately.
- **Neighbor Solicitation** (NS) messages are originated by hosts, which attempt to discover the link-layer addresses of other nodes on the same local link, originated in Duplicate Address Detection DAD Algorithm, or originated to verify the reachability of a neighbor.
- **Neighbor Advertisement** (NA) messages are sent to advertise the changes of the host MAC address and IP address, or solicited response to NS messages.
- **Redirect messages** are used to redirect traffic from one router to another.

2.2 *IDPS*

In the past, Intrusions were detected manually. It was done by reading and analyzing all systems logs trying to detect anomalies or attacks. Such process takes a lot of time, effort and needs specialized trained employees to make the detection. Therefore, detection process should be done automatically [10].

Not every detected anomaly can be treated as threats; even though it may exhibit characteristics that are similar to threats ones [11]. It also has the potential to translate into significant critical and actionable information [12].

Anomaly based systems attempt to map events to the point where they learn what is normal and then detect an unusual occurrence that indicates an intrusion. National Institute of Standards and Technology (NIST) [13] defines (IDSs) as the process of monitoring the events occurring in a computer system or network and analyzing them for signs of possible incidents, which are violations or imminent threats of violation of computer security policies, acceptable use policies, or standard security practices.

Moreover, NIST defines Intrusion prevention system (IPS) as the process of performing intrusion detection and attempting to stop detected possible incidents.

This means that IPS is an active IDS that performs all IDS activity and try to stop intrusions. In this paper, IDPS acronym is used instead of IDS and IPS acronyms. IDPS methodologies can be categorized into:

- **Misuse-based**, (also denoted as signature-based) search for predefined patterns, or signatures within the captured data. This methodology is very effective in detecting predefined threats and has a small number of false positive alert. However, it has no ability to detect new threats or undefined one.
- **Anomaly-based**, where normal behavior of the target system is defined. It generates anomaly alert whenever any deviation is detected. Usually anomaly-based IDSP suffers from high number of false positive alert.
- **Strict Anomaly Detection**, "not use" is alternative name for it, introduced by Sasha and Beetle firstly introduce Strict Anomaly Detection in 2000 [14] to detect any violation of system rules. The best use of strict anomaly detection in environments where legitimate activities is well defined [15].

3 Related Work

The lack of authentication limitation makes NDP prone to attacks [16–19] despite the integration of IPsec within IPv6 to secure it. However, IPsec needs manual configuration which makes it limited to small network with known hosts [20].

Solutions proposed to overcome NDP limitation can be divided according to the purpose; preventing NDP attacks and monitoring NDP attacks which are IDPS, for preventing NDP attacks. Common exist preventing solutions have made changes on the original design of the protocol which increases the complexity of the protocol. On the other hand, IDPS solutions intend to monitor the NDP without any modification to the original design of the protocol. These approaches detect any violation of predefined normal behavior of the protocol, alert system administrators and try to stop intruders.

SEND [21] and Cryptographically Generated Address CGAs [22] are examples of preventing NDP attacks. These solutions are the best choice for securing IPv6 networks where IPsec is considered to be an impractical. However, SEND and CGA have not been widely implemented or deployed due to their high complexity and issues that reportedly holding back some vendors; these issues including intellectual property claims and concerning licensing terms [20]. Another example, Hassan et al. [23] proposed the use of a digital signature to secure IPv6 neighbor discovery protocol which has less complexly than CGA. However, the proposed solution cannot detect DAD and Neighbor Unreachability Detection attacks. Shah et al. [24] proposed highly randomized technique for address generation that safeguards node's privacy and asserts address uniqueness on the link.

The main limitation of the solutions proposed to prevent NDP attacks is the complexity that increases the overhead of the protocol as there extra added functions. In contrast, IDPS solutions did not change or increase the complexity of

NDP. The main role of these solutions is to alert the system administrator of any violation of NDP normal behavioral.

There are two types of monitoring solutions: passive and active mechanism which generates extra packets for additional analysis. NDPmon [25] is an example of passive mechanism that tracks changes in MAC-IP pairings; any changes trigger alerts to the system administrator. The main drawback is that training phase must be free of any compromised node; otherwise, the whole detection process fails. Another example presented by Najjar et al. [15] where a finite state machine is built to model the normal behavioral of NDP. Strict Anomaly Detection was used to detect any violation.

On the other hand, active mechanisms [26, 27] use probe packets for additional observations. Bansal [28] uses Multicast Listener Discovery (MLD) probing to reduce the traffic, while Kumar [29] proposed a host-based IDPS which verified any changes made on its neighbor cache by sending NS probes. Main limitation of the active technique is that the generated traffic can be used by the attackers to perform DoS attacks by flooding the nodes with fake MAC-IP address pairs.

4 Dataset

Datasets are an essential part of evaluating and testing the machine learning solutions. In order to utilize a machine learning technique to detect NDP anomalies, NDP dataset must successfully capture the normal behavioral. Failing in capturing normal behavior of the protocol affects the accuracy of the machine learning technique.

The most common benchmark datasets lack of IPv6 data flow, which makes them useless to detect NDP flooding attack. Dataset presented in [30] is selected in testing and evaluations as it successfully captures the normal behavioral of NDP, and capture RA and NS flooding attacks which make it good selection choice. Table 1 summarizes the summation of captured packets inside the dataset.

Table 1 Captured packets summarisation and duration in NDP Dataset

Class	Duration	Packet	RS	RA	NS	NA	RD	IP	MAC
Normal	24 h	2991	13	460	1159	1070	289	13	7
Flood_RA	25 s	79771	0	45678	34093	0	0	45691	45685
Flood_NS	23 s	101759	0	0	89279	12373	107	89292	89286

4.1 NDP Features Definition

Strict anomaly detection methodology is used to observe any violation in normal NDP messages flows. As the characteristics of anomalies flows differ from normal flows, the events variables that are used to distinguish between normal packets flow and abnormal ones are defined as NDP features set. Table 2 illustrates the number of NDP messages that characterizes the legitimate flow. These values are extracted from the protocol constants to define the legitimate NDP messages flows, any violation of these values indicate a prohibited event that NDP features set must be characterized. Figure 1 illustrate the process of generating the new features. The output of this process generates the following eight new features from the dataset:

- **Duration** is the time in seconds for counting NDP messages, number of MAC addresses and number of IP addresses used to generate these messages. In this paper, 3 s was chosen as the duration time for counting the packets. However, 3 s is long time for flooding attacks; therefore, if the number of counted type packets is over the threshold, then the new duration become 0.1 s as shown in Fig. 1.
- **Number of MAC** address is the incremental count of MAC address in the network. This feature is very useful in detecting spoofing MAC addresses because some attacks violate the system polices not NDP, such as the number of MAC addresses connected to specific ports.
- **Number of IPs** is the incremental count of IP addresses. It is unlike number of MAC addresses feature because IPv6 permits to obtain more than one IP address for each MAC address. Hence, the attacker can generate fake NDP messages using different IP address using a legitimate MAC address. Such situation can occur when there is a physical security on switch ports.
- **Number of RS** messages is the counted number of RS messages within duration time. Table 2 shows the permitted number of RS messages.
- **Number of RA** messages is the counted number of RA messages within duration time. Table 2 shows the maximum permitted number RA messages.

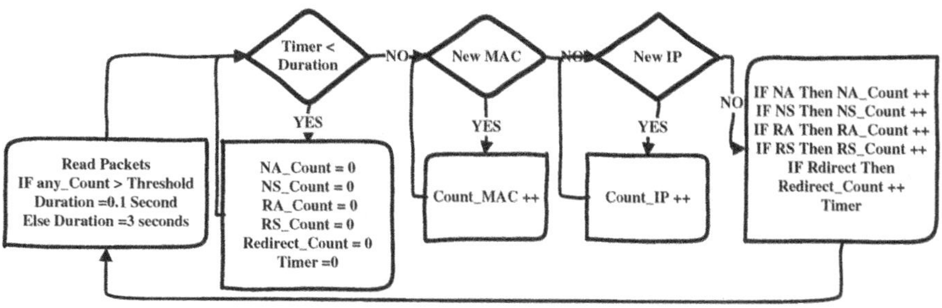

Fig. 1 Process of generating new features from Dataset

Table 2 Number of NDP Messages based on Protocol Constant

Message	Protocol constant	Description
RS	3 Transmissions/8 s	Maximum number of RS packets for each IP
RA	3 Transmissions/8 s	Maximum number of RA packets for each subnet
NS	3 Transmissions/3 s	Maximum number of NS packets for each IP
NA	3 Transmissions/Threshold	Maximum number of NA packets for each IP

- **Number of NS** messages is the counted number of NS messages within duration time. Table 2 shows the maximum permitted number NS messages.
- **Number of NA** messages is the counted number of NA messages within duration time. Table 2 shows the maximum permitted number NA messages.
- **Number of redirect** messages is the counted number of redirect messages within duration.

5 Machine Learning

In this section, most of popular machine learning techniques are highlighted and discussed. In some attack datasets, a single attribute can do the whole job while other attributes are redundant and irrelevant [31]. In another attack datasets, attributes must contribute equally and independently to detect the attack. A third scenario, an attack can be simply detected using logical structure, with selecting a few numbers of attributes using a decision tree method [32]. Therefore, the aim of this paper is to compare different machine learning techniques to detect NDP flooding attacks.

ZeroR Method. ZeroR is the simplest classification method which usually used as a benchmark for other classification methods. For classification, ZeroR depends on the class type attribute and ignores all other attributes. In classification, ZeroR constructs frequent table of the classes and select the most frequent class for classification [32].

One Rule Method. One Rule method (OneR) or 1-level decision tree is a simple and accurate method which classifies using a single attribute to create a classification rule. Frequency table is constructed for each attribute against the class. Then, the rule with the smallest total error is chosen [33].

Naive Bayesian. Naive Bayesian is another simple technique which uses all attributes and deals with them equally. Moreover, it assumes all attributes are statistically autonomous. Although this assumption is not realistic for most real-world datasets, it works well in practice [34].

Decision Trees. Decision tree uses a tree structure to build a classification model. The final result of the model is a tree with branches (decision nodes) and leaf nodes. Iterative Dichotomiser 3 (ID3) was the basis of the decision tree proposed by J. R. Quinlan [35] which uses top-down and greedy search through the space of possible branches with no backtracking. Entropy and Information Gain used in ID3 to construct a decision tree. This paper used C4.5 [35] algorithm in the evaluation experiments, which it is an extension to ID3.

Nearest Neighbor. The nearest neighbor algorithm (KNN) [36] is a pattern recognition statistical method which classifies according to the nearest *k* object in the dataset. It uses similarity or distance metrics to choose the nearest objects. KNN is an example of lazy learning techniques where the method does nothing until the prediction is made. This algorithm is one of the highly accurate machine learning algorithm that involves no learning cost and builds a new model for each test. The testing may become costly if the number of instances in the input dataset increases.

Support vector machine. Support vector machine (SVM) is one of the most robust and accurate methods in all well-known data mining algorithms. It performs classification for finding the hyperplane which maximizes the margins between classes [37].

6 Experiments Evaluation

WEKA tool [38] was used in testing and evaluating the different machine learning techniques on detecting NDP RA and NS flooding attacks. Testing and evaluation have been performed using a Dell XPS Intel Core i7 processor, and 8 GB RAM, with Windows 10 pro operating system.

Testing and evaluation process went through different phases. First phase is applying preprocessing procedures to remove irrelevant and redundant instances. WEKA-RemoveDuplicate function was utilized in this phase. The total number of instances was reduced from 1639 instances to 510 instances, 1129 instances were duplicated. Thus, 1129 repeated behavior was removed (sending RA every 3 s).

Second phase is applying machine learning techniques (described in Sect. 4). Default setting was chosen for most techniques; except SVM, where Linear Kernel was chosen.

The experimental results which are illustrated in Table 3 show that most techniques successfully detect flooding attacks which means that the generated features successfully characterize different behaviors of NDP. Benchmark technique detected 43 % of the attacks. This result is affected by removing redundant instances. As without removing duplicate instance, the detection rates become 73 %, and it will increase if the scenarios change, since Normal is the main characteristics of networks.

OneR managed to detect 90 % of instances correctly using one feature selection. This result indicates that there is a high difference between the captured behaviors, since the dataset only contain two flooding attacks along with the normal behavior. Most of 10 % misclassified instances as a consequence of the automatic generation of new IP addresses which triggered by the faked prefix insides the fakes RA. The rest techniques classify 100 % instances. The results showed that the generated features explicitly distinguish between NDP behaviors; hence, successfully classify all instances using different approaches.

Table 3 Comparison between machine learning techniques in detecting NDP flooding attacks

<table>
<tr><th>Classifier</th><th>Time in seconds</th><th>Percentage of detection accuracy (%)</th><th>Confusion Matrix</th></tr>
<tr><td>ZeroR</td><td>0</td><td>43.52</td><td>a b c <– classified as
0 79 0 | a = Normal
0 222 0 | b = Flood_RA
0 209 0 | c = Flood_NS</td></tr>
<tr><td>OneR</td><td>0.01</td><td>90.00</td><td>a b c <– classified as
28 0 51 | a = Normal
0 222 0 | b = Flood_RA
0 0 209 | c = Flood_NS</td></tr>
<tr><td>Naive Bayesian</td><td>0.02</td><td>100</td><td>a b c <– classified as
79 0 0 | a = Normal
0 222 0 | b = Flood_RA
0 0 209 | c = Flood_NS</td></tr>
<tr><td>C4.5</td><td>0.08</td><td>100</td><td>a b c <– classified as
79 0 0 | a = Normal
0 222 0 | b = Flood_RA
0 0 209 | c = Flood_NS</td></tr>
<tr><td>KNN</td><td>0</td><td>100</td><td>a b c <– classified as
79 0 0 | a = Normal
0 222 0 | b = Flood_RA
0 0 209 | c = Flood_NS</td></tr>
<tr><td>SVM</td><td>1.08</td><td>100</td><td>a b c <– classified as
79 0 0 | a = Normal
0 222 0 | b = Flood_RA
0 0 209 | c = Flood_NS</td></tr>
</table>

7 Conclusion

Securing NDP is a major task since the large implementation of Internet in public environments where no trust between users is established. Attackers can easily flood the network with a huge number of NDP messages. Accordingly, the connect device is frozen and disconnected. Most of proposed solutions for securing NDP increase the protocol overhead and complexity; therefore, they are not widely implemented or deployed. To avoid increasing the complexity of NDP, IDSP is the best choice which has the ability to detect, notify, and stop detected protocol anomalies. In order to detect anomalies, IDPS must utilize effective tools, such as machine learning techniques. This paper utilizes machine learning techniques to detect NDP flooding attacks. The experimental results show that most of machine learning techniques can accurately detect NDP flooding attack. Thus, the complexity of flooding attack recognition is very low. This implies that flooding behavioral is much different from normal behavioral; that is why the detection rates are very high for most techniques. In addition, NDP implementation created according to well-known defined specifications which illustrated in the Request for Comments (RFC 4861, 4862), normal behavioral of the protocol can be accurately modelled, and any violation can be treated as anomaly. Therefore, Strict Anomaly detection methodology is the optimum method for detecting NDP anomalies due to its simplicity in modelling the protocol normal behavioral. The best technique can be utilized with strict anomalies detection is C.45, since it generates classifiers expressed as a decision trees based on certified rules.

Acknowledgments This work was supported by National Advanced IPv6 Centre (NAv6), Universiti Sains Malaysia.

References

1. Postel J (1981) Internet protocol: DARPA Internet program protocol specification. RFC 791, September
2. Huston G (2015) GeoHuston. http://www.potaroo.net/tools/ipv4/index.html
3. Deering S, Hinden R (1998) Internet protocol, Version 6 (IPv6) specification, RFC 2460, December
4. Narten T, Nordmark E, Simpson W, Sliman H (2007) Neighbor Discovery for IP version 6 (IPv6). RFC 4861 September
5. Bhattacharyya DK, Kalita JK (2013) Network anomaly detection: a machine learning perspective. CRC Press
6. Narten T, Thomson S, Jinmei T (2007) IPv6 stateless address autoconfiguration
7. Kent S, Seo K (2005) Security architecture for the internet protocol, RFC4301, December
8. Frankel S, Graveman R, Pearce J, Rooks M (2010) Guidelines for the secure deployment of IPv6, vol 800. NIST Special Publication
9. Conta A, Gupta M (2006) Internet control message protocol (icmpv6) for the internet protocol version 6 (ipv6) specification
10. Allen J, Christie A, Fithen W (2000) State of the practice of intrusion detection technologies
11. Sommer R, Paxson V (2010) Outside the closed world: on using machine learning for network intrusion detection. In: 2010 IEEE symposium on security and privacy (SP). IEEE
12. Chandola V, Banerjee A, Kumar V (2009) Anomaly detection: a survey. ACM Comput Surv (CSUR) 41(3):15
13. Scarfone K, Mell P (2007) Guide to intrusion detection and prevention systems (IDPS). National Institute of Standards and Technology (NIST), Feb 2007
14. Sasha, Beetle (2000) A strict anomaly detection model for IDS. Phrack Magazine Volume 0xa Issue 0x38, May1

15. Najjar F, Kadhum M, El-Taj H (2015) Neighbor discovery protocol anomaly detection using finite state machine and strict anomaly detection, Proceedings of the 4th International Conference on Internet Applications, Protocols and Services (NETAPPS2015), 978-967-0910-06-2
16. Nikander P, Kempf J, Nordmark E (2004) IPv6 Neighbor Discovery (ND) trust models and threats, RFC 3756, May
17. Gashinsky I, Jaeggli J, Kumari W (2012) Operational neighbor discovery problems. No. RFC 6583
18. Najjar F, El-Taj H (2015) Ipv6 change threats behavior. Int J Adv Comput Sci Appl 6(1)
19. Jankiewicz E, Loughney J, Narten T (2011) Ipv6 node requirements. No. RFC 6434
20. Bouras C, Karaliotas A, Ganos P (2003) The deployment of IPv6 in an IPv4 world and transition strategies. Internet Res 13(2):86–93
21. Kempf E, Sommerfeld J, Zill B, Arkko B, Nikander P (2005) Secure neighbor discovery (SEND). No. RFC 3971
22. Bagnulo M, Arkko J (2006) Cryptographically Generated Addresses (CGA) Extension Field Format. RFC 4581, October
23. Hassan R, Ahmed AS, Osman NE (2014) Enhancing security for IPV6 neighbor discovery protocol using cryptography. Am J Appl Sci 11(9):1472
24. Shah JL, Parvez J (2015) Optimizing security and address configuration in IPv6 SLAAC. Proc Comput Sci 54:177–185
25. Beck F, Cholez T, Festor O, Chrisment I (2007) Monitoring the neighbor discovery protocol. In: The second international workshop on IPv6 Today-Technology and Deployment-IPv6TD 2007
26. Barbhuiya FA, Bansal G, Kumar N, Biswas S, Nandi S (2013) De- tection of neighbor discovery protocol based attacks in IPv6 network. Netw Sci 2:91–113
27. Barbhuiya F, Biswas S, Hubballi N, Nandi S (2011) A host based DES approach for detecting ARP spoofing. In: Symposium on computational intelligence in cyber security. IEEE, pp. 114–121
28. Bansal G, Kumar N, Nandi S, Biswas S (2012) Detection of NDP based attacks using MLD. In: Proceedings of the fifth international conference on security of information and networks. ACM
29. Kumar N, Bansal G, Biswas S, Nandi S (2013) Host based IDS for NDP related attacks: NS and NA Spoofing. In: India Conference (INDICON). IEEE, pp 1–6
30. Najjar F, Kadhum M (2015) Reliable Behavioral Dataset for IPv6 Neighbor Discovery Protocol Investigation. In: International conference on IT convergence and security
31. Tang J, Alelyani S, Liu H (2014) Feature selection for classification: a review. Data Classif Algorithms Appl 37
32. Witten IH, Frank E (2005) Data mining: practical machine learning tools and techniques. Morgan Kaufmann
33. Holte RC (1993) Very simple classification rules perform well on most commonly used datasets. Mach Learn 11(1):63–90
34. Domingos P, Pazzani M (1997) On the optimality of the simple Bayesian classifier under zero-one loss. Mach Learn 29:103–130
35. Quinlan JR (2014) C4. 5: programs for machine learning. Elsevier
36. Fix E, Hodges JL (1951) Discriminatory analysis, nonparametric discrimination. USAF School of Aviation Medicine, Randolph Field, Tex., Project 21-49-004, Rept. 4, Contract AF41(128)-31, February
37. Chang C-C, Lin (2011) LIBSVM: a library for support vector machines. ACM Trans Intell Syst Technol (TIST) 2(3):27
38. Hall M et al (2009) The WEKA data mining software: an update. ACM SIGKDD Explor Newsl 11(1):10–18

MFCC Global Features Selection in Improving Speech Emotion Recognition Rate

Noor Aina Zaidan and Md Sah Salam

Abstract Feature selection is one of the important aspects that contribute most to the emotion recognition system performance apart from the database and the classification technique used. Based on the previous finding, Mel Frequency Cepstral Coefficients (MFCC) are said to be good for emotion recognition purpose. This paper discusses the use of MFCC features to recognize human emotion on Berlin database in the German language. Global features are extracted from MFCC and tested with three classification methods; Naive Bayes, Artificial Neural Network (ANN) and Support Vector Machine (SVM). We investigate the capabilities of MFCC global features using 13, 26 and 39-dimensional cepstral features in recognizing emotions from speech. The result from the experiment will be further discussed in this paper.

Keywords Speech emotion recognition · MFCC · Global features · Coefficient number

1 Introduction

Human conversation is more effective in the presence of emotion. Current technology shows that the existence of two-way communication between human and machine is possible. Human emotions can be understood by the intelligent machine, but information will not be successfully delivered correctly if the emotion is misinterpreted. To develop computer applications that can provide a natural response to the user, detailed studies have been carried out in recent years.

N.A. Zaidan (✉) · M.S. Salam
Faculty of Computing, Universiti Teknologi Malaysia,
81310 Skudai, Johor, Malaysia
e-mail: ainazaidan@gmail.com

M.S. Salam
e-mail: sah@utm.my

P.J. Soh et al. (eds.), *Advances in Machine Learning and Signal Processing*,
Lecture Notes in Electrical Engineering 387, DOI 10.1007/978-3-319-32213-1_13

Automatic Speech Emotion Recognition (SER) requires applications that can create a natural human-computer interaction where there is such a concept for the computer to understand the emotions of human from speech. Human speech is said to be one of the most effective methods of communication that came with the intentions and emotions [1]. If the computer is given the ability to identify human emotions in the same way as a human does, communication is bound to be more effective.

Human's speeches have a variety of emotional states, for example, happy, anger, sad, fear, surprise, disgust, and so on. Many factors that should be considered for a machine to understand human emotion contained in a speech. There are lots of feature selection and classification method used in recognizing emotion from human speech.

Emotional speech features that contribute to an emotion are various. It is vital to design a system with suitable features that capable to describe different emotions [2]. From a survey done by Ayadi [2], prosodic features such as energy and pitch carry the most emotional information of an utterance. Moreover, spectral feature like Mel- Frequency Cepstral coefficients (MFCC) is also said to be among the most popular feature extraction method and widely used by researchers based on previous finding [3, 4].

Prosodic and spectral feature are widely used for emotion recognition task. Some of the researchers use both local and global feature extracted from the speech signal for their automatic emotion classification. Schuller et al. [5] used global statistic framework extracted from pitch and energy related features for their experiment. Origlia et al. [6] presented new features set consist of both local and global prosodic features. The prosodic global feature like min, max, mean and standard deviation for fundamental frequency and intensity were extracted and tested for their research. Huang et al. [7] proposed a new framework by combining global wavelet packet decomposition and local traditional prosodic utterance-level features. Other than that, Ravi et al. [8] investigate the multidimensional MFCC features that offer better accuracy for stuttered speech recognition task. The past research motivated us to investigate the capabilities of MFCC global features using 13, 26 and 39-dimensional cepstral features in recognizing emotions from speech.

Beside feature selection method, classification technique is also an important process that contributes to the emotion recognition system performance. The frequently used methods in speech emotion recognition field are SVM, ANN, Hidden Markov Model (HMM), Gaussian Mixture Model (GMM) and K-Nearest Neighbors (K-NN). Each classification method has its own advantages and limitations. There is no final decision about which method is the best to use to classify emotion [2].

This paper is organized into 6 sections. Section 2 describes the MFCC feature extraction and its process. Section 3 contains the emotional speech database used in the experiment. Section 4 discusses the classification method used to run the experiment. The initial result and discussion are explained in Sect. 5 and finally, the conclusions of the obtained result and the future work are presented in Sect. 6.

2 MFCC Feature Extraction

Emotion in speech is represented by a large number of parameters and the changes of the parameter may cause changes in emotion recognition. It is crucial to do a proper feature selection as it is believed to influence the classification performance. MFCC is a technique where spectral information of the speech signal can be represented.

In several pattern recognition processes that involve human voice, cepstral coefficients are believed to be robust to used as a set of features [8]. Each number of coefficients holds a specific value for each sound frame. Normally, in speech analysis, the MFCC coefficients number that frequently used is in between the sufficient range of 8–14 [9]. In this paper, MFCCs with a number of coefficients 13, 26 and 39 were extracted from the speech signal. The coefficient number is used to capture information of the time-varying spectral envelope [10]. The MFCC are analyzed and processed as follows [11]:

1. Divide signal into frames.
2. Compute the FFT to obtain the amplitude spectrum.
3. Apply the Mel filter.
4. Take the logarithm to map the spectrum onto Mel scale.
5. Take the Discrete Cosine Transform of Mel log-amplitudes.

Figure 1 illustrates the processed above:

MFCC spectral feature has a huge total number of feature vectors for emotional speech. If the system is relying on the hisgh recognition accuracy with less time consuming, MFCC features will have the disadvantage compared to other prosodic feature: pitch, fundamental frequency, formant, in term of its process speed. Compared to prosodic features, a spectral feature like MFCC carry a huge number of digital information and require lots of process time. To overcome this limitation, it is recommended to use the global feature as a second filter to the MFCC features extracted to save the duration of processing time. The global features that suitable to use in such case are the value of Min, Max, Mean, Median, Mode, Standard Deviation and Variation. These values will be extracted from the MFCC feature from each frame rows and stored as final features for MFCC.

Current automatic speech recognition (ASR) systems normally use appended MFCC features, delta, and double-delta cepstral features [12]. MFCC coefficient

Fig. 1 Steps to compute MFCC

number is usually selected from MFCC (13 dimensional), MFCC + delta (26 dimensional), and MFCC + delta + delta (39 dimensional). MFCC features are extracted using 13 coefficient number and its first and second derivative (Δ's and ΔΔ's), resulting in 26 and 39 coefficient number representing each frame. Multidimensional feature vector were constructed for each speech frame by most feature extraction method [8]. In this paper, three experiments were run using global MFCC with 13, 26 and 39 coefficients. The global features as mentioned before were extracted from each MFCC features vector by row. The entire feature extraction process is performed by MATLAB software.

Figure 2 illustrate on how to find global feature in MFCC features that have been extracted earlier using coefficient number 13. Each frame (1–13) contains a window of features vector according to the signal length. The total of the feature vector is considered as (row x column): (13 frames x signal length). The first step is to find the global feature in the first frame until the last frame. Then, the global feature will be reshaped according to the WEKA (The Waikato Environment for Knowledge Analysis: an open source software that provide an algorithm for machine learning and pre-processing tools [13]) ARFF file format and feed into the classification technique used. Refer to Fig. 3 below for output global feature after reshape.

Figure 3 illustrate the output feature vector to be feed into WEKA where the input file consists of global feature extracted from MFCC using 13, 26 and 39 coefficient numbers. A total of 322 emotional speech recording file ranging from 46 files for each of the seven emotions were used for the experiment. The final feature vector for experiment 1 will be counted as (Min1, Max1, Mean1, Median1, Mode1, Std1, and Var1) until (Min13, Max13, Mean13, Median13, Mode13, Std13, and Var13), make it a total of $(7 \times 13 \times 322)$ feature vectors. Feature extraction process was repeated for 26 and 39 dimensional (The final feature vectors for experiment 2 will be counted as (Min1, Max1, Mean1, Median1, Mode1, Std1, and Var1) until (Min26, Max26, Mean26, Median26, Mode26, Std26, and Var26), make it a total of $(7 \times 26 \times 322)$ feature vectors and experiment 3 will be counted as (Min1, Max1, Mean1, Median1, Mode1, Std1, and Var1) until (Min39, Max39, Mean39, Median39, Mode39, Std39, and Var39), make it a total of $(7 \times 39 \times 322)$ feature vectors.

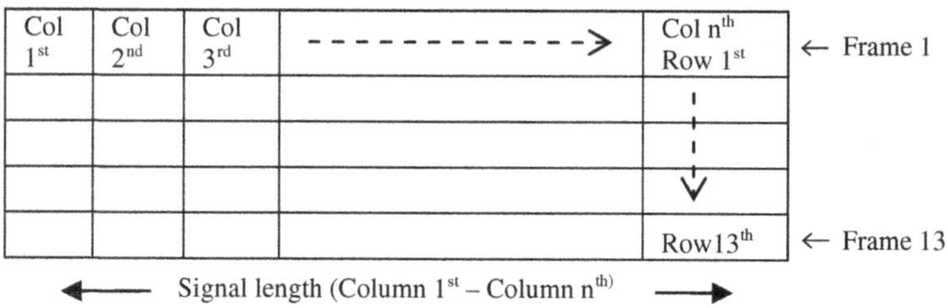

Fig. 2 Finding global feature in MFCC window frame (Frame 1–Frame 13)

Min 1	Max 1	Mean 1	Median 1	Mode 1	Std 1	Var 1	-----▶	Min 13	Max 13	Mean 13	Median 13	Mode 13	Std 13	Var 13
2														
↓														
322														

Fig. 3 Output global feature after reshape (coefficient 13)

3 Emotional Speech Database

The database used in this experiment is Berlin database of emotional speech (EMO-DB). EMO-DB is open source and contains reliable emotional speech data to get start with [14]. It is well-known and used by many researchers in speaker independent case [15] and also for the benchmarking purpose. The acted data is recorded in the German language with duration in between 1 until 8 s each sentence and covers seven emotions: happiness, sadness, fear, anger, disgust, boredom, and neutral. 46 recording files (.wav) from each emotion were selected to run the experiment make it a total of 322 recording files. Feature extraction technique needs the labeling name for each emotion class as a reference to store the data for each class. Labeling process is important to track the emotional class for the final stage of the process: classification and recognition.

4 Classification and Recognition

SVM method is said to have a good capability in classifying emotions and accessible to build the hierarchical classification method [16]. In circumstances where the training data should be limited, SVM is recommended to use due to its good classification performance [17].

Another classifier, ANN Neural network with input data from several neighbouring frames could be used for identifying emotional speech [18]. It learns and understands the data and changes the value of the weights, based on back-propagation from the target output layer. The learning process involves the network architecture and the weights are updated so that the data can be classified in their own class.

After going through the features extraction process, the global features were used as input data for the classification process. A general process flow involved is shown in Fig. 4 below.

In this experiment, three classifiers were selected for performance comparisons which are: Naïve Bayes, ANN, and SVM. These classifiers are provided in WEKA classification tools. The performance of the classifiers is validated using 10 fold validation method. The default parameters for each classifier in WEKA were used to carry out the experiment: ANN parameter with learning rate 0.3, momentum rate

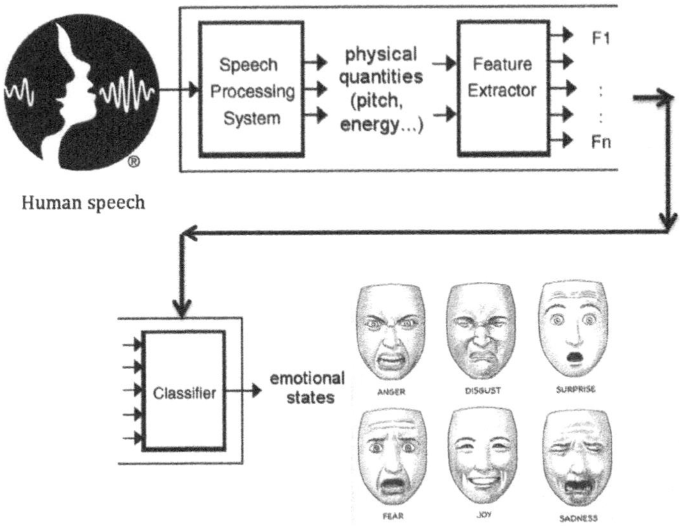

Fig. 4 Emotion recognition process flow

0.2, training time (number of epochs) 500, validation threshold 20 and SVM with linear kernel: K(x, y) = <x, y> function. In this work, the efficiency in using the global features selected from MFCC to classify seven emotions was observed. Three experiments were run to test the effect of coefficient number used using the global features:

A. *Experiment 1—Global feature from MFCC with 13 coefficient number.*
B. *Experiment 2—Global feature from MFCC with 26 coefficient number.*
C. *Experiment 3—Global feature from MFCC with 39 coefficient number.*

5 Result and Discussion

Experiment 1, 2 and 3 were carried out using global features extracted from MFCC series of feature vector using a number of coefficients 13, 26 (MFCC + delta) and 39 (MFCC + delta + delta) respectively. Three classifiers were used: Naive Bayes, ANN, and SVM for classification of seven emotions and overall accuracy were computed. Tables 1, 2 and 3 below show the results of emotion recognition rate.

The result from Table 1 shows that the selected classifier can distinguish seven emotions from one another with the average recognition rate of above 55 %. Naive Bayes gives the lowest result with 57.1 % followed by SVM with 67.7 % and ANN gives the highest result with 70.2 %.

From Tables 2 and 3 above, we can see that result from an experiment that use 26 and 39 coefficient numbers provides the average recognition rate of above 45 %.

Table 1 Emotion Recognition Rate (%) using 13 coefficient number

Emotion/classifier	Happiness	Neutral	Anger	Sadness	Fear	Boredom	Disgust	Average classified	Process time (s)
Naive Bayes	45.7	63.0	67.4	78.3	30.4	52.2	63.0	57.14	0.04
ANN	69.6	67.4	80.4	84.8	65.2	54.3	69.6	70.19	44.76
SVM	71.7	58.7	76.1	80.4	71.7	52.2	63.0	67.69	1.56

Table 2 Emotion Recognition Rate (%) using 26 coefficient number

Emotion/classifier	Happiness	Neutral	Anger	Sadness	Fear	Boredom	Disgust	Average classified	Process time (s)
Naive Bayes	47.8	41.3	60.9	97.8	19.6	21.7	58.7	49.69	0.02
ANN	63.0	65.2	82.6	91.3	58.7	56.5	65.2	68.93	157.31
SVM	71.7	69.9	80.4	87.0	60.9	54.3	69.9	70.59	1.04

Table 3 Emotion recognition rate (%) using 39 coefficient number

Emotion/classifier	Happiness	Neutral	Anger	Sadness	Fear	Boredom	Disgust	Average classified	Process time (s)
Naive Bayes	43.5	45.7	54.3	93.5	19.6	26.1	58.7	48.77	0.03
ANN	65.2	63.0	82.6	78.3	60.9	58.7	71.7	68.63	385.66
SVM	73.9	73.9	78.3	82.6	58.7	50.0	65.2	68.94	1.73

Fig. 5 13 coefficient number

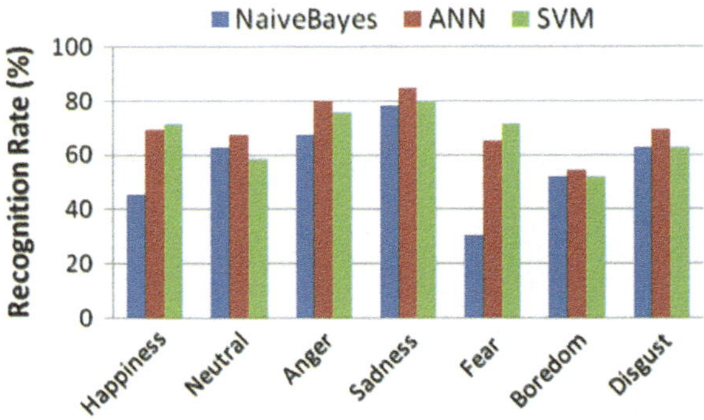

For experiment 2, Naive Bayes gives the lowest average result with 49.7 % followed by ANN with 68.9 % and SVM gives the highest result with 70.6 %. For experiment 3, Naive Bayes gives the lowest average result with 48.8 % followed by ANN with 68.6 % and SVM gives the highest result with 68.9 %. It shows that the use of coefficient 13, 26 and 39 give small difference results in average recognition rate and a huge difference in processing time. Figures 5, 6 and 7 below visualize the result from the experiments.

Figure 5 shows a moderately consistent accuracy compared to Figs. 6 and 7. As we can see in the figures above, the use of coefficient 13 provides slightly small differences in recognition per emotion and provide consistent result compared to the coefficient 26 and 39. This is probably because MFCC extracted using coefficient 13 carry the untainted emotional features without any modification. The first and second derivation of the feature may alter the characteristic of the original feature that cause changes in emotion recognition accuracy. Global features like maximum, minimum, mean, median, mode, standard deviation and variation that extracted from MFCC features using 13, 26 and 39 coefficient number were proven to give a fine result.

According to Figs. 6 and 7, MFCC delta and delta-delta features do carry more emotion information and able to recognize emotion adequately for emotion recognition task. The use of 26 and 39 coefficients number with bigger dimension may give relevant features that may lead to higher accuracy for the SVM classifier. But the redundancy of feature may also give lower classification accuracy for Naive Bayes and ANN classifier.

Based on the result of the three experiments as illustrated in Fig. 8 above, we examine that in comparison to different MFCC coefficient number, 13, 26 and 39; the use of MFCC global feature with coefficient number 13 provides better recognition accuracy by using Naive Bayes and ANN classifier. On the other hand, coefficient number 26 provides better recognition accuracy with SVM classifier while coefficient number 39 provides an insignificantly different result on both classifiers ANN and SVM. Naive Bayes give the lowest result, SVM method give satisfying result due to its high recognition accuracy and less processing time while ANN gives a good result but acquire long processing time. Hence, MFCC global feature with coefficient number 26 and SVM method is suggested for future work due to its better accuracy and less processing time.

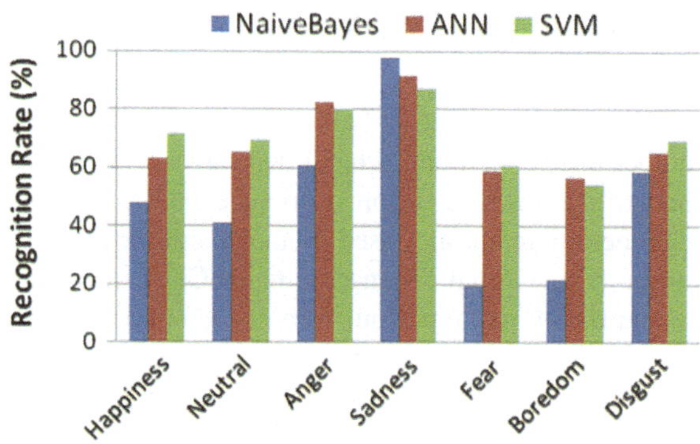

Fig. 6 26 coefficient number

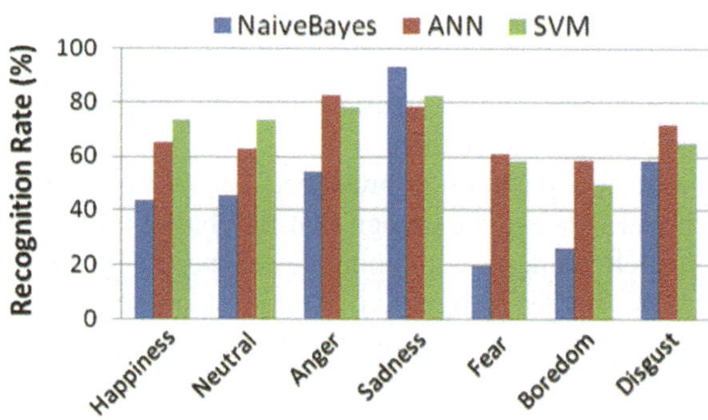

Fig. 7 39 coefficient number

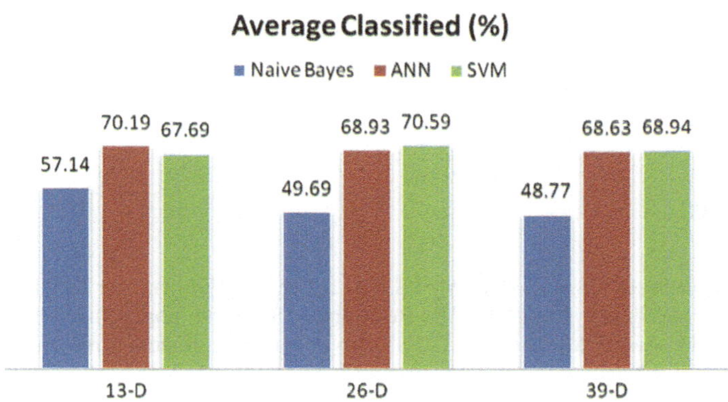

Fig. 8 Average emotion classification result

6 Conclusion and Future Work

This paper studied the capabilities of global MFCC spectral features extracted using 13 coefficient number and its first and second derivative (Δ's and ΔΔ's): 26 and 39 coefficient number, to recognize emotion from speech. Based on the experimental result, the different coefficient number used do make significant different in emotion recognition accuracy rate. Global feature from MFCC carries enough emotional information to recognize human emotion from speech and capable of recognizing emotion above 50 % accuracy. Average recognition result from experiment 1 using 13 coefficient numbers is slightly higher than experiment 2 and 3 for Naive Bayes and ANN classifier. However, global features from MFCC extracted using 26 coefficient number gives the highest recognition rate at 70.59 % accuracy using SVM classifier. Even though MFCC global features with coefficient number 13, 26, and 39 are carrying enough emotional information to represent emotion in human speech but, selection of global feature from MFCC with coefficient number 26 is reliable to use in emotion recognition task due to its better accuracy and less processing time. To increase the recognition accuracy, appending another potential MFCC global feature is recommended.

It is believed that a combination of prosodic, spectral and wavelet features from emotional speech can increase the recognition rate based on the previous finding. Our ongoing research work is combining few types of prosodic, spectral and wavelet feature to get higher recognition rate and study about the contribution of certain features towards certain emotion. By the use of global feature along with other features will increase the efficiency of the emotion recognition accuracy. Bear in mind, it is important to reduce the process time to get better emotion recognition rate in real time application, so avoiding the use of very large emotional features is of utmost important.

Acknowledgments Special thanks to Ministry of Education (MOE) and Research Management Centre (RMC), Universiti Teknologi Malaysia providing financial support of this research in FRGS Vot number R.J130000.7828.4F253.

References

1. Panda B, Padhi D, Dash K, Mohanty PS (2012) Use of SVM classifier & MFCC in speech emotion recognition system. Int J Adv Res Comput Sci Softw Eng 2(3):225–230
2. El Ayadi M, Kamel M, Karray F (2011) Survey on speech emotion recognition: features, classification schemes, and databases. Pattern Recogn 44(3):572–587
3. Kishore KVK, Satish PK (2013) Emotion recognition in speech using MFCC and wavelet features. In: 2013 IEEE 3rd international advance computing conference (IACC), pp. 842–847
4. Zhou Y, Sun Y, Zhang J, Yan Y (2009) Speech emotion recognition using both spectral and prosodic features. In: International conference on information engineering and computer science, pp 1–4, Dec

5. Schuller B, Rigoll G, Lang M (2003) Hidden Markov model-based speech emotion recognition. In: ICME '03 Proceedings of the 2003 International Conference on Multimedia and Expo—Vol 2, pp 401–404 (2003)
6. Origlia A, Galatà V, Ludusan B (2010) Automatic classification of emotions via global and local prosodic features on a multilingual emotional database. In: Proceedings of the 5th international conference on speech prosody
7. Huang Y, Zhang G, Li X (2011) Improved emotion recognition with novel global utterance-level features. Appl Math Inf Sci 5(2):147–153
8. Ravi Kumar KM, Ganesan S (2011) Comparison of multidimensional MFCC feature vectors for objective assessment of stuttered disfluencies. Int J Adv Netw Appl 860:854–860
9. Loughran R, Walker J, O'Neill M, O'Farrell M (2008) The use of mel-frequency cepstral coefficients in musical instrument identification. In: Music conference, Belfast
10. Pradier MF (2011) Emotion recognition from speech signals and perception of music. Thesis, University of Stuttgart
11. Zhongzhe X (2008) Recognition of emotions in audio signals. Thesis
12. Kumar K, Kim C, Stern RM (2011) Delta-spectral cepstral coefficients for robust speech recognition. In: IEEE international conference on acoustics, speech and signal processing (ICASSP), pp 4784–4787
13. Hall M, Frank E, Holmes G, Pfahringer B, Reutemann P, Witten IH (2009) The WEKA data mining software: an update. SIGKDD Explor 11(1):10–18
14. Burkhardt F, Paeschke A, Rolfes M (2005) A database of German emotional speech. In: Interspeech, pp 3–6
15. Neiberg D, Laukka P, Ananthakrishnan G (2010) Classification of affective speech using normalized time-frequency cepstra. In: Fifth international conference on speech prosody
16. Qi-Rong M, Zhan Y-Z (2010) A novel hierarchical speech emotion recognition method based on improved DDAGSVM. Comput Sci Inf Syst 7(1):211–222
17. Shen P, Changjun Z, Chen X (2011) Automatic speech emotion recognition using support vector machine. In: International conference on electronic and mechanical engineering and information technology, pp 621–625, Aug
18. Safdarkhani MK, Mojaver SP, Atieghechi S, Riahi MS (2012) Emotion Recognition of Speech Using ANN and GMM. Aust J Basic Appl Sci 6(9):45–57

A Parametric Study of Compact UWB Antenna with Multiple Notched-Band Functions

Md. Moinul Islam, Mohammad Tariqul Islam and Mohammad Rashed Iqbal Faruque

Abstract In the paper, a parametric study of compact UWB antenna is conducted with multiple notched-band functions. Three notch bands at 3.6, 7.5, and 8.3 GHz frequencies have been attained with three slots on the radiating patch whereas two notch bands at 5.2 and 5.8 GHz frequencies have been acquired with two slots on the partial ground plane. This proposed antenna covers impedance bandwidth (VSWR < 2) from 3.06 to 10.67 GHz with notch band 3.4–3.8 GHz (WiMAX), 5.1–5.35 GHz (WLAN), 5.6–6 GHz (WLAN), 7.15–7.65 GHz (X-band satellite communication) and 8.05–8.65 GHz (ITU 8-GHz band) whereas the overall antenna size is 25 mm × 31 mm. The parametric studies are executed depending on the surface current flow and design parameters of slots.

1 Introduction

The Federal Communications Commission (FCC) approved the 3.1–10.6 GHz frequency band for commercial applications [1, 2]. However, attentions and challenges with UWB is increased rapidly. There are some coexisting narrow bands such as WiMAX, WLAN, X-band and ITU-band. These narrow bands can interfere

Md.M. Islam (✉) · M.R.I. Faruque
Space Science Centre, Universiti Kebangsaan Malaysia (UKM),
43600 Bangi, Selangor D.E, Malaysia
e-mail: mmoiislam@yahoo.com

M.R.I. Faruque
e-mail: rashed@ukm.edu.my

Md.M. Islam · M.T. Islam
Department of Electrical, Electronic and Systems Engineering,
Universiti Kebangsaan Malaysia (UKM), 43600 Bangi, Selangor D.E, Malaysia
e-mail: tariqul@ukm.edu.my

Md.M. Islam
Department of Software Engineering, Daffodil International University,
102 Shukrabad, Dhanmondi, Dhaka, Bangladesh

P.J. Soh et al. (eds.), *Advances in Machine Learning and Signal Processing*,
Lecture Notes in Electrical Engineering 387, DOI 10.1007/978-3-319-32213-1_14

with the UWB systems. For removing these difficulties, multiple band-notched are extremely applied in UWB communication systems. In order to attain the feature, different design configurations are reported in the literature with modified radiating element or/and the ground on the planar UWB antennas [3–10]. Quintuple [3, 4], quadruple [5, 6], triple [7], dual [8, 9] and single [10] notched frequencies have been attained with UWB antennas applying different design configurations. A compact disc antenna has been reviewed with quintuple notched band frequency for UWB applications using slots and stubs [3]. But, our presented antenna is more compact than that. A monopole UWB antenna is reviewed with five notched band functions [4], but in this design, the parametric studies on the surface current flow and design parameters of slots are not available. An UWB antenna is explained using a modified H-shaped resonator with quadruple notched bands [5], but this antenna dimension is larger than our proposed design. A compact monopole antenna is narrated with quadruple notched band using lateral L-shaped slot [6]. In the way, triple notched bands are presented in the UWB antennas by using two elliptic single complementary split-ring slots and locating two rectangular split-ring resonators. However, dual notched bands are proposed with applying etching a partial annular slot [8], one semi-circular with one rectangular and partial concentric annular slot with another rectangular slot [9]. A single notched frequency is also observed by loading a pair of SRRs [10], but this antenna dimension is very large.

The paper explains a parametric study of compact UWB antenna conducted with multiple band-notched functions. The parametric studies are executed depending on the surface current flow and design parameters of slots. This parametric analysis ensures that the presented antenna bears significant values to maintain multiple band-notch functions.

2 Antenna Design

By adopting various design procedures and techniques of ultra-wideband antenna in [3–10], a monopole line-fed UWB antenna has been designed, which is illustrated in Fig. 1. The presented antenna is made using FR4 material of 1.6 mm thick, conducting 0.02 loss tangent, and 4.6 dielectric constant. By using slots, multiple notched bands are attained at 3.6, 5.2, 5.8, 7.5, and 8.3 GHz respectively. The HFSS is applied for all the simulations. The following are the optimum values: W_S = 3.9 mm, L = 31 mm,, W = 25 mm, L_S = 4.2 mm, L_{GND} = 16 mm, g = 0.29 mm, R = 7 mm, W_F = 2.25 mm, SB_2 = 5.3 mm, SB_1 = 3.5 mm, SB_3 = 9.8 mm, SC_1 = 1.75 mm, SB_T = 0.4 mm, SC_2 = 4.9 mm, SC_3 = 6.5 mm, SD_1 = 2.1 mm, SC_T = 0.4 mm, SD_2 = 3.6 5 mm, SD_T = 0.4 mm, SE_1 = 1.5 mm, SD_3 = 6.5 mm, SE_2 = 2.4 mm, SE_T = 0.4 mm, SE_3 = 6.5 mm, SF_1 = 2.35 mm, SF_3 = 6.5 mm, SF_2 = 1.4 mm, SF_T = 0.4 mm.

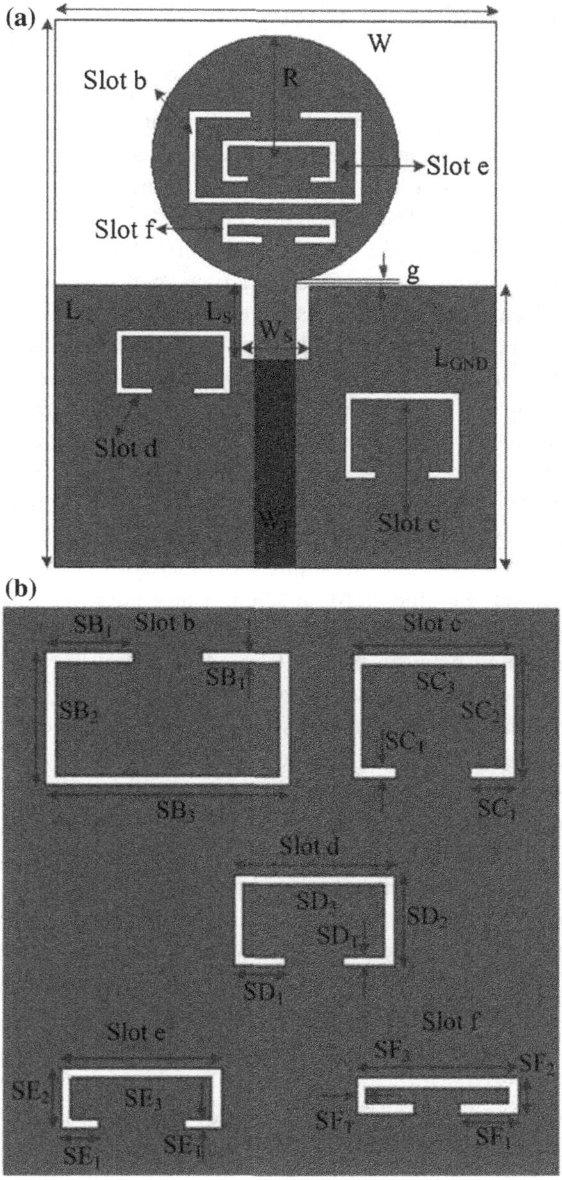

Fig. 1 The presented **a** antenna layout **b** the slots design specifications

3 Results and Discussion

Initially, we are observing the surface current of both the patch and the ground plane of the proposed antenna. Our target is to design and fabricate five band-notched antennas for UWB applications. However, we have looked the place where current flows low (low citation). Then we have cut C-shaped slots and achieved the desired notched band individually. The slots are responsible for creating notch bands, where the density of the current flow is enhanced around the neighboring region. The variations between initial surface current and surface current after cutting slots have been shown in Fig. 2.

In order to understand the mutual coupling between the band-notched structures, the parametric study has been carried out by changing one parameter at a time and fixing the others. For simplicity, only the parameters SB_3, SE_3, and SF_T are changed where the optimized parameters are $SB_3 = 9.8$ mm, $SC_3 = SD_3 = SE_3 = SF_3 = 6.5$ mm, and $SF_T = 0.4$ mm. As shown in Figs. 3 and 4, the first, the second, the fourth and the fifth notched bands shift to lower frequency as the SB_3 and SE_3 increase or decrease but the third has no effect where the length of the corresponding slots increases or decreases. As shown in Fig. 5, the second and the third notched bands have no effects due to the fact that the band-notched structures of adjacent frequencies are etched in different place and the coupling between them is very weak. From the results, we can also conclude that the notch frequency is controllable by changing the length of the slot.

The parametric analysis by varying the distance between different slots has been performed. It can be realized from the Fig. 6 that two notch bands (2nd and 4th) are shifted when these two slots are closer to feed line and ground between these reported slots whereas all notch bands are shifted while these two slots are in symmetric fashion. Therefore, we have achieved the desired results while asymmetric fashion in the referred position. It can be realized from the Fig. 7 that four notch bands are shifted using distance 0.6 mm between these reported slots whereas three notch bands are shifted using distance 0.4 mm between them. Therefore, 0.8 mm distance is the optimized value. It can be observed from Fig. 8 that 2.05 mm is the optimum distance between slot e and slot f.

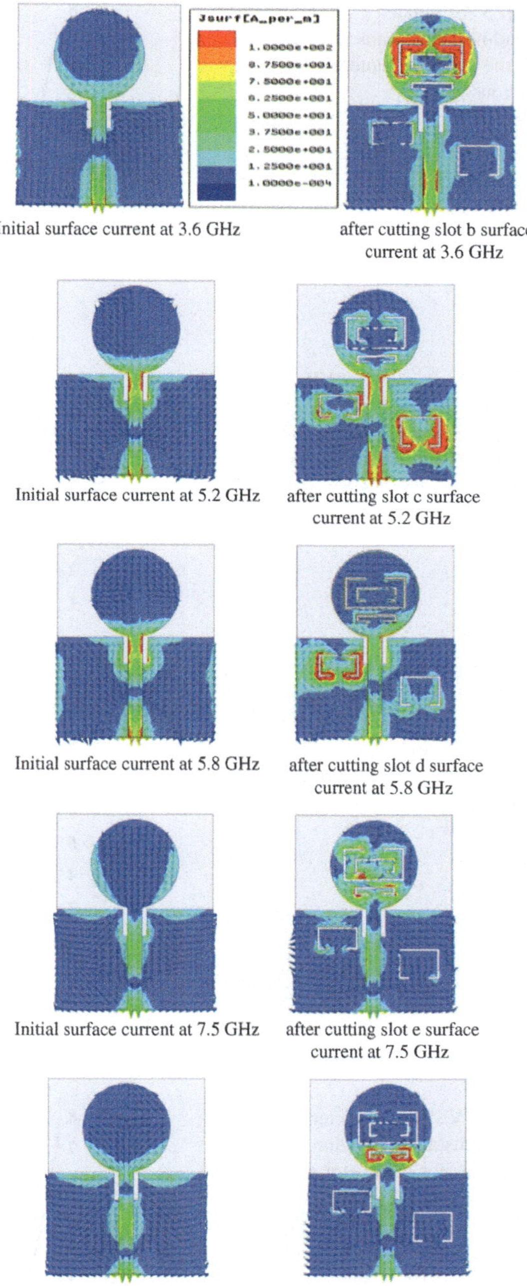

Initial surface current at 3.6 GHz

after cutting slot b surface current at 3.6 GHz

Initial surface current at 5.2 GHz

after cutting slot c surface current at 5.2 GHz

Initial surface current at 5.8 GHz

after cutting slot d surface current at 5.8 GHz

Initial surface current at 7.5 GHz

after cutting slot e surface current at 7.5 GHz

Initial surface current at 8.3 GHz

after cutting slot f surface current at 8.3 GHz

Fig. 2 Effect of surface current distribution before and after cutting slots

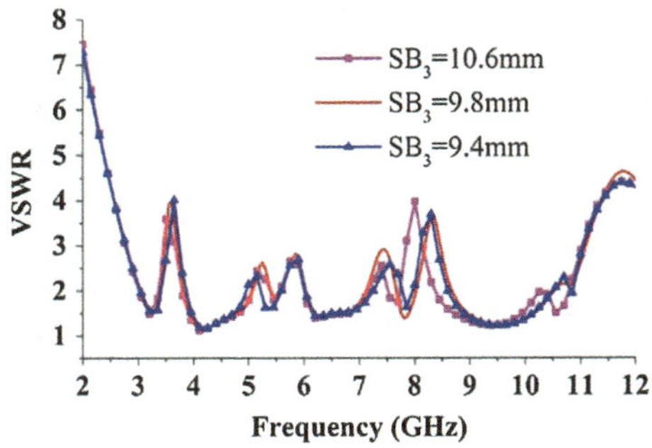

Fig. 3 Simulated band-notched characteristics of the proposed antenna for various SB3

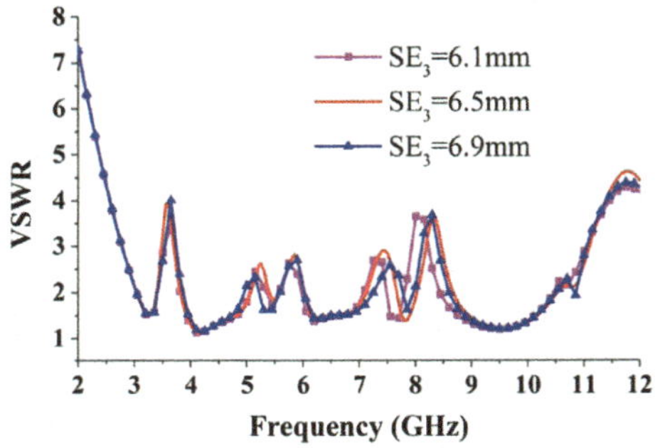

Fig. 4 Simulated VSWRs for different values of R_1. Simulated band-notched characteristics of the proposed antenna for various SE_3

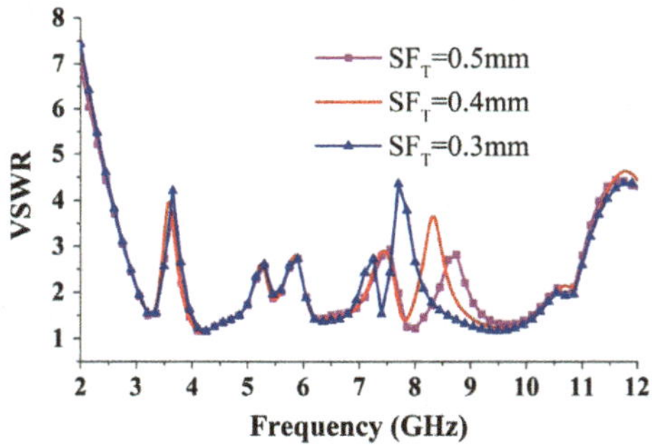

Fig. 5 Simulated band-notched characteristics of the proposed antenna for various SF_T

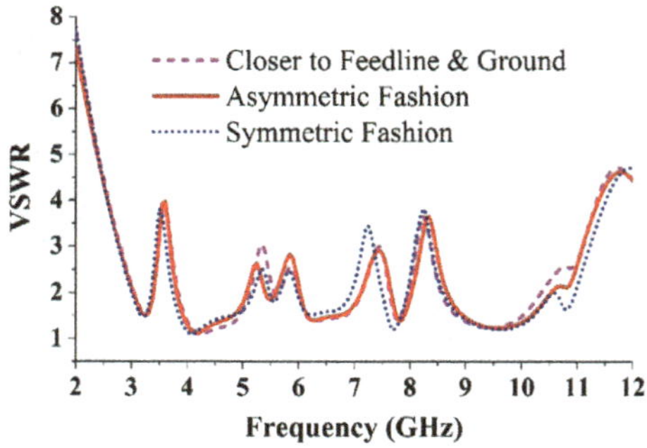

Fig. 6 VSWR characteristics on the distance variation between slot c and slot d

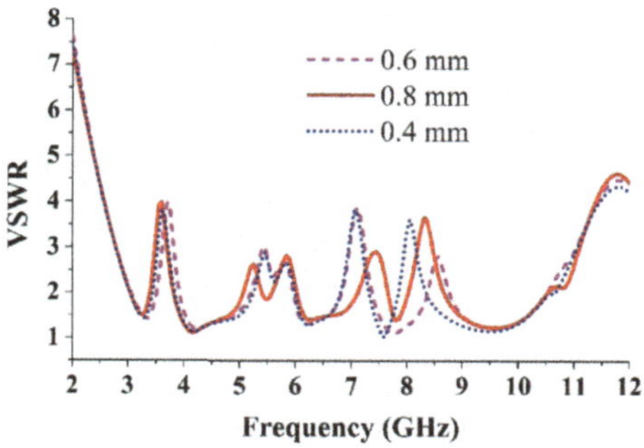

Fig. 7 VSWR characteristics on the distance variation between slot b and slot f

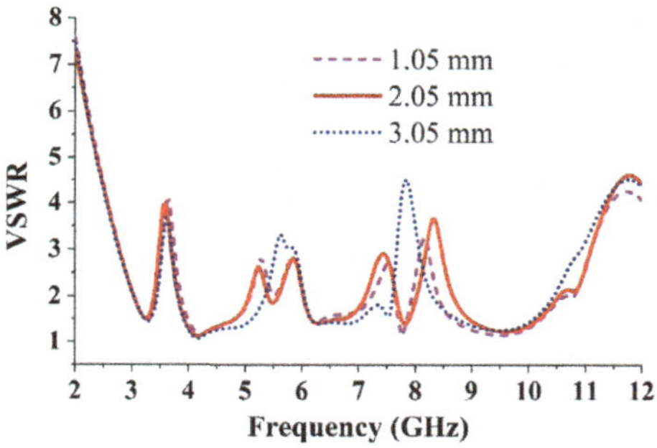

Fig. 8 VSWR characteristics on the distance variation between slot e and slot f

4 Conclusions

A parametric study is performed with multiple band-notched functions for UWB applications. Three slots are used on the patch and two slots are on the ground plane to attain multiple notched bands. This proposed antenna delivers the impedance bandwidth (VSWR < 2) from 3.06 to 10.67 GHz with multiple notched band whereas the antenna size remains unchanged. The parametric studies are executed depending on the surface current flow and design parameters of slots. This parametric analysis ensures that the presented antenna bears significant values in case of multiple band-notch functions.

References

1. Azim R, Islam MT, Misran N (2010) Printed planar antenna for wideband applications. J Infrared Millimeter Terahertz Waves 31(8):969–978
2. Islam M, Moniruzzaman M, Misran N, Shakib M (2009) Curve fitting based particle swarm optimization for UWB patch antenna. J Electromagn Waves Appl 23(17–18):2421–2432
3. Jie X, Shen D, Zhang X, Wu K (2012) A compact disc ultrawideband (UWB) antenna with quintuple band rejections. IEEE Antennas Wirel Propag Lett 11:1517–1520

4. Islam MM, Islam MT, Samsuzzaman M, Faruque MRI (2015) Five band-notched ultrawide band (UWB) antenna loaded with C-shaped slots. Microwave Opt Technol Lett 57(6):1470–1475
5. Sung Y (2013) Quad band-notched ultrawideband antenna with a modified H-shaped resonator. Microwaves Antennas Propag IET 7(12):999–1004
6. Wu ZH, Wei F, Shi X-W, Li W-T (2013) A compact quad band-notched UWB monopole antenna loaded one lateral L-shaped slot. Prog Electromagnet Res 139:303–315
7. Sarkar D, Srivastava KV, Saurav K (2014) A compact microstrip-fed triple band-notched UWB monopole antenna. IEEE Antennas Wirel Propag Lett 13:396–399
8. Azim R, Islam MT (2013) Compact planar UWB antenna with band notch characteristics for WLAN and DSRC. Prog Electromagnet Res 133:391–406
9. Azim R, Islam M, Mandeep J, Mobashsher A (2012) A planar circular ring ultra-wideband antenna with dual band-notched characteristics. J Electromagn Waves Appl 26(14–15):2022–2032
10. Siddiqui JY, Saha C, Antar YM (2014) Compact SRR loaded UWB circular monopole antenna with frequency notch characteristics. IEEE Antennas Propag Trans 62(8):4015–4020

Parallel Implementation of Morphological Operations on Binary Images Using CUDA

Jun Ming Koay, Yoong Choon Chang, Shahirina Mohd Tahir and Sankaraiah Sreeramula

Abstract Morphology is a common technique used in image processing because it is a powerful tool with relatively low complexity. Albeit simple, morphological operations are typically time consuming due to the fact that the same operations are repeated on every pixel of an image. Since the processing of the pixels of an image is an embarrassingly-parallel process, the morphological operations can be carried out in parallel on Nvidia graphic cards using Compute Unified Device Architecture (CUDA). However, most of the existing CUDA work focuses on the morphological operations on grayscale images. For binary image, it can be represented in the form of a bitmap so that a 32-bit processor will be able to process 32 binary pixels concurrently. With the combination of the bitmap representation and van Herk/Gil-Werman (vHGW) algorithm, the performance of the proposed implementation in term of computation time improves significantly compared to the existing implementations.

Keywords Morphology · Binary images · Bitmap · Parallel · CUDA · GPU · vHGW

J.M. Koay (✉)
Faculty of Engineering, Multimedia University, 63100 Cyberjaya, Selangor, Malaysia
e-mail: kjm_coffee@hotmail.com

Y.C. Chang
Faculty of Engineering and Science, Universiti Tunku Abdul Rahman, 43000 Kajang, Selangor, Malaysia
e-mail: ycchang@utar.edu.my

S.M. Tahir · S. Sreeramula
MIMOS Bhd, Technology Park Malaysia, 57000 Kuala Lumpur, Malaysia
e-mail: shahirina.mtahir@mimos.my

S. Sreeramula
e-mail: sankaraiah.ramula@mimos.my

P.J. Soh et al. (eds.), *Advances in Machine Learning and Signal Processing*, Lecture Notes in Electrical Engineering 387, DOI 10.1007/978-3-319-32213-1_15

1 Introduction

Morphology is a common technique used in image processing, since it is a simple and powerful tool. In general, morphology can be applied on both grayscale and binary images. The two main operations in morphology are erosion and dilation. In the context of grayscale images, erosion is the process of scanning through the neighbouring pixels under consideration and assigning the smallest value to the central pixel. On the other hand, dilation is the complete opposite process of erosion, where instead of finding the smallest pixel value, the largest pixel value is assigned to the central pixel. By combining the erosion and dilation operations, more complicated morphological operations such as opening and closing can be performed.

In order to process an input image using morphological operator, an extra input which is known as structuring element is also required. The structuring element selects the neighbouring pixels that need to be considered during the morphological operations. In fact, the structuring element can have any type of shapes and sizes. If an erosion operation is performed using a 3×3 structuring element, the pixel value of all the nine pixels inside the window are compared with each other and the smallest value is assigned to the central pixel.

Although morphological operations are straightforward, its process is time consuming because the same operations need to be repeated on every pixel of an image. Hence, there is a need to speed up the operations for real-time applications. Since the operations on the pixels are independent of each other, the pixels can be processed in parallel in order to speed up the morphological operations.

There are a few researchers that have implemented the morphological operations on parallel programming platforms like Nvidia's Compute Unified Device Architecture (CUDA). However, most of the implemented work is focused only on grayscale images [2], [5]. Although the grayscale implementation of morphological operations can also be used to process binary images, it is not an optimised way to handle binary images because an 8-bit grayscale pixel has 256 possible values while a binary pixel can only have two possible values. Hence, special optimisation techniques that utilise the bi-level nature of the binary pixels are proposed in this paper in order to further speed up the computation time of the morphological operations.

The rest of the paper is organised as follows. Section 2 discusses related work on the implementation of morphological operations using CUDA. In Sect. 3, the proposed CUDA implementation is explained in detail. Section 4 compares the performance of the proposed CUDA implementation with the existing implementations. Finally, Sect. 5 concludes the proposed work.

2 Related Work

2.1 Decomposition of Structuring Element

For a morphological operation using a rectangular structuring element with a size of $x \times y$, the basic operations need to be repeated for xy times to obtain the output of one single pixel. Hence, for an image with n number of pixels, the total number of operations that need to be executed is nxy. Nevertheless, the computational complexity can easily be reduced by using a technique known as structuring element decomposition [4]. By using this technique, a rectangular structuring element can be decomposed into one-dimensional horizontal and vertical structuring elements. This means that instead of using a rectangular structuring element to perform the morphological operations, the same output can be obtained by using a horizontal structuring element followed by a vertical structuring element. This optimises the computational complexity from nxy to $n(x + y)$. By using this technique, the speedup obtained for a 5×5 structuring element is approximately 2.5 times.

2.2 vHGW Algorithm

One of the drawbacks of the conventional way of implementing morphology is the computation time increases as the size of the structuring element becomes larger. In order to solve this problem, van Herk [7] and Gil and Werman [3] proposed a method known as van Herk/Gil-Werman (vHGW). By using the vHGW algorithm, the computation time remains relatively constant regardless of the structuring element size. For an erosion operation on a grayscale image using a horizontal structuring element of size p, the steps in the vHGW algorithm are explained as follow.

1. The image rows are split into multiple horizontal segments of size p. In addition, aprons with a size of $(p-1)/2$ are appended to both the left and right ends of the segments to form segments with a total size of $2p-1$, which is shown in Fig. 1.
2. Starting from the midpoint of the segments, a cumulative comparison is carried out for the elements on the left half of the segment and the minimum values are

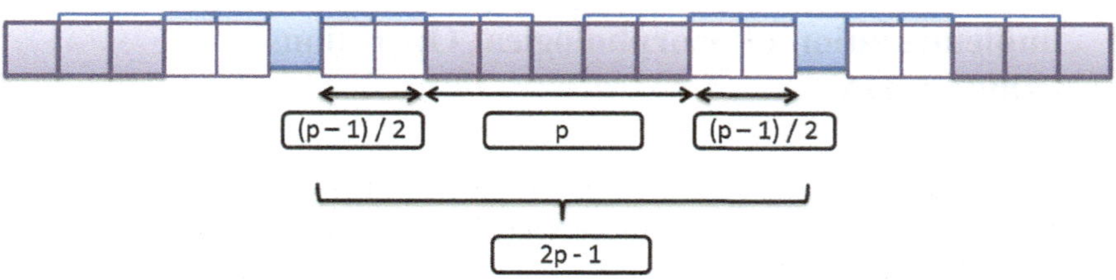

Fig. 1 First step of vHGW algorithm: dividing the image rows into multiple segments

stored inside a new array known as suffix array. The same operation is repeated for the elements on the right half of the segment and the results are stored inside another array known as prefix array.

3. The value of the elements in the suffix array are compared with the value of the elements in the prefix array to obtain the final minimum values.

2.3 CUDA Implementation of the vHGW Algorithm

In general, for most of the morphological operations, the pixels of an image can be processed independently. Hence, it offers an opportunity to speed up the processing by parallelising the implementation on a many-core platform. In this case, graphics processing units (GPUs) are well suited to perform the morphological operations in a parallel manner due to its highly parallel structure. There are various computing platforms that are available for the implementation of general purpose computing on a GPU, such as openCL, openGL and CUDA. For the proposed research work, CUDA is chosen as the computing platform because it is a matured platform in the field of computer vision, with support from computer vision libraries like OpenCV.

Various work has been done on the implementation of the morphological operations using CUDA. Domanski, Vallotton and Wang [2] implemented the vHGW algorithm using CUDA and observed that the CUDA version of the algorithm achieves a speedup of 13 to 33 times compared to the serial implementation. Domanski et al. also showed that the computation time of the vHGW CUDA implementation remains relatively constant as the size of the structuring element varies. However, the vHGW CUDA implementation by Thurley and Danell [5] showed that only the horizontal structuring element exhibits the constant-time behaviour. For the vertical structuring element, the computation time increases as the size of the structuring element becomes larger. In order to overcome this problem, Thurley et al. replaced the vertical morphological operation with another algorithm known as transpose-horizontal-transpose (THT). For the THT method, the input image is transposed first before a horizontal structuring element is applied on the image. The processed image is then transposed again to obtain the final output.

3 Implementation of Morphological Operations Using CUDA

Instead of using the data structure of a grayscale image to store a binary image, Van Den Boomgaard and R. van Balen represented the binary image in the form of a bitmap, so that only one bit of memory is required to store the value of a particular binary pixel [6]. As a result of this, a 1-byte memory space is able to store the value of eight binary pixels and the required memory space is reduced by eight times. As most of the modern processors are 32-bit or 64-bit processors, if the binary image is represented in the form of a bitmap, all 32 or 64 binary pixels can be processed in parallel using one single processor. By using the bitmap optimisation technique, the speedup attained can be more than 30 times. Thus, the bitmap representation is used in the proposed implementation.

The morphological operations on bitmapped binary images are implemented by modifying the grayscale implementation. For the implementation of the erosion operation, the ‘min’ operator in the grayscale implementation is replaced with the logical ‘and’ operator. In a similar way, the ‘max’ operator is changed to the logical ‘or’ operator for the implementation of the dilation operation.

The serial implementation of the morphological operations is time consuming, even by using the vHGW algorithm and bitmap representation. Hence, there is a necessity to speed up the implementation of the morphological operations by parallelising it on CUDA. A CUDA-enabled GPU operates based on the parallel execution model of single instruction, multiple threads (SIMT). In this architecture, multiple CUDA threads run concurrently in the GPU. For the implementation of the morphological operations on grayscale images, each CUDA thread is assigned the task of calculating the value of an output pixel. Hence, the number of CUDA threads is set according to the resolution of the image. With the introduction of bitmap representation, one single CUDA thread is able to process 32 binary pixels. Hence, the number of CUDA threads needed is reduced by 32 times. The parallelisation strategy for processing a bitmapped binary image is shown in Fig. 2, where N is the number of CUDA threads per block while M is the number of blocks needed to process the whole image.

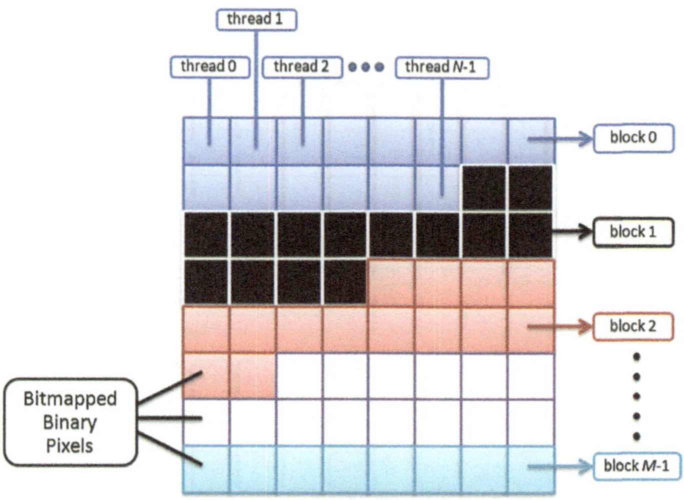

Fig. 2 Parallelisation strategy for processing a bitmapped binary image in the GPU using CUDA

During the morphological operations, the same pixel location is read multiple times. Since the data access to the shared memory is faster than the global memory of the GPU, the pixel data that are accessed repeatedly should be stored in the shared memory. Nevertheless, the size of the shared memory is only limited to 48 kB for GPUs with compute capability 3.5 [1]. Thus, for large binary images with Full High Definition (FHD) resolution, the image is not able to fit into the shared memory. Hence, it is very important to schedule the usage of the shared memory in such a way that at any point of time, only a small segment of the binary image is stored in the shared memory.

Although the bitmap representation reduces the computational complexity of the CUDA implementation, the computation time increases when the size of the structuring element grows larger. In order to resolve this problem, the existing vHGW CUDA implementation for grayscale images is modified and applied to the case of bitmapped binary images. However, for bitmapped binary images, the vHGW algorithm is not suitable for implementing the horizontal morphological operation. Hence, the vHGW algorithm is only implemented for the vertical morphological operation.

For the implementation of the vertical vHGW algorithm, the image columns are divided into multiple vertical segments and each CUDA thread is responsible for the processing of one single segment. Since the limited memory space of the shared memory poses a restriction on the size of the structuring element, both the suffix and prefix arrays are stored in the global memory of the GPU. Figure 3 illustrates the implementation of the vertical vHGW algorithm using CUDA.

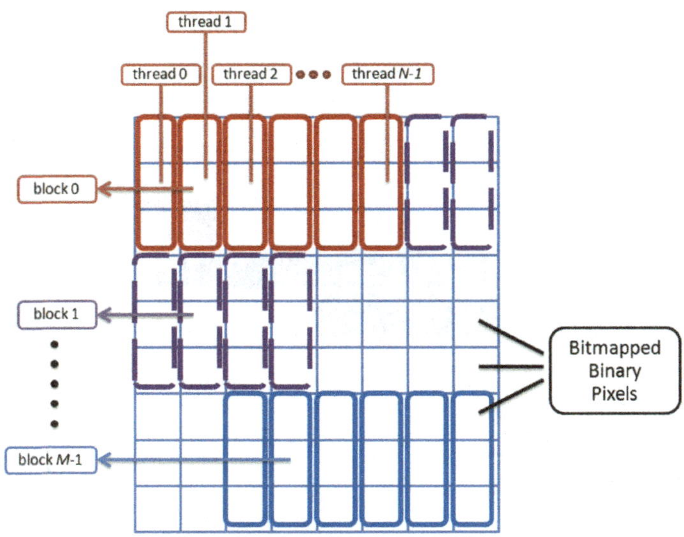

Fig. 3 Parallel implementation of the vertical vHGW algorithm in the GPU using CUDA

4 Results and Discussions

Since the erosion and dilation operations have similar execution steps, the computation time of the erosion operation is similar to the dilation operation. Hence, only the CUDA implementations of the erosion operation are benchmarked. All of the CUDA implementations are benchmarked on a server, which has two Intel Xeon E5-2650 v5 processors and one single Nvidia Tesla K20m graphic card.

4.1 Horizontal Erosion Operation

In this research work, the bitmap representation is incorporated into the implementation of the horizontal erosion operation on binary images. By using the bitmap representation, a 32-bit GPU core is able to process 32 binary pixels in parallel. Furthermore, the number of memory transfer operations required is also reduced substantially. For a horizontal structuring element with a size less than 65 pixels, only 0.125 memory access is needed for each output pixel in the image (four memory accesses for each 32-bit chunk), which is significantly lower than the $7-(4/p)$ memory accesses per pixel for the vHGW algorithm. Since memory access is a very expensive process in the context of the GPU [1], reducing the number of memory transfer operations will definitely improve the performance of the implementation.

The performance of the proposed implementation is compared with the CUDA 6.5 NPP library, OpenCV 3.0 CUDA library and vHGW CUDA implementation [5]. Although the CUDA NPP library, OpenCV library and vHGW implementation are designed to handle grayscale images, it is also possible for these implementations to process binary images by using only two grayscale values (0 and 255) to represent a binary pixel. The performance comparison results for processing a FHD binary image are shown in Fig. 4. For a better representation, the results obtained are illustrated in a log graph.

From Fig. 4, it is observed that the computation time of the proposed CUDA implementation is significantly shorter than the other implementations. Hence, it can process binary images at a faster rate.

For the horizontal erosion operation on bitmapped binary images, each pixel location is read multiple times. Hence, it makes sense to cache the data in the shared memory of the GPU to reduce the memory accesses to the global memory. The results in Fig. 5 compare the performance of the CUDA implementation that uses the shared memory of the GPU to the one that does not use the shared memory. From the graph, it can be observed that both implementations offer similar performance. This means that the usage of shared memory does not offer any further improvement to the proposed implementation. This behaviour is due to the fact that the memory access latency to the global memory is hidden by the L1 and L2 caches of the GPU. Hence, the bottleneck of the algorithm is not on the memory transfer

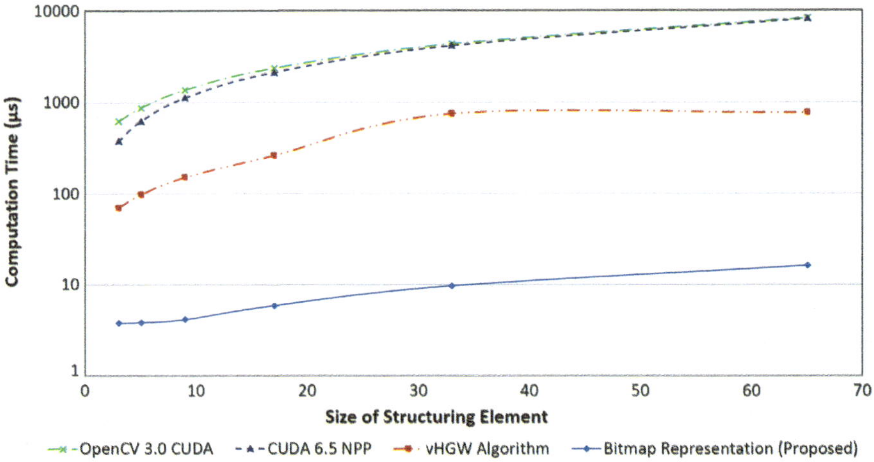

Fig. 4 Performance comparison of different algorithms for the horizontal erosion operation on a FHD binary image

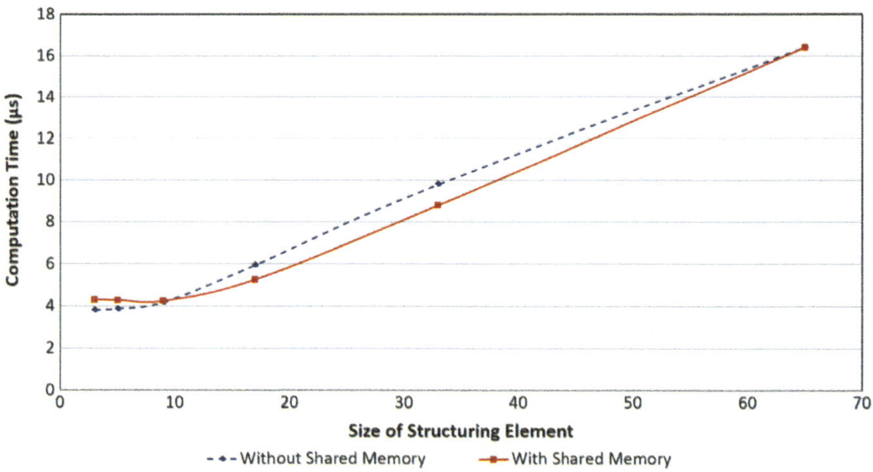

Fig. 5 Computation time of the CUDA implementation with shared memory versus without shared memory

part. In fact, the analysis results using Nvidia Visual Profiler show that the utilisation level of the processing cores is higher than the utilisation level of the memory. Thus, the performance of the implementation is limited by the workload of the processing cores in the GPU rather than the bandwidth of the memory.

4.2 Vertical Erosion Operation

Since most of the binary images are stored in the row-major format in the memory, different memory access patterns have different impacts on the performance of the algorithm. As a result, the vertical erosion operation exhibits different characteristics compared to the horizontal erosion operation. For the vertical erosion operation, the performance of the proposed CUDA implementation that incorporates the bitmap representation is compared with the existing techniques and the results are shown in Fig. 6. The results obtained are illustrated in a log graph to have a better visualisation of the data.

From Fig. 6, the computation time of the proposed technique is significantly faster than the external libraries and vHGW implementation, because the bitmap representation enables one single GPU core to process 32 binary pixels in one shot. At the same time, it is observed that as the size of the structuring element becomes larger, the computation time of the proposed implementation increases at a faster rate compared to the vHGW algorithm. Hence, it is reasonable to modify the vHGW algorithm for grayscale images and apply it to the case of bitmapped binary images. Figure 7 compares the performance of the implementation that incorporates the combination of the bitmap representation and vHGW algorithm to the implementation that utilises only the bitmap representation. From Fig. 7, it is concluded that the implementation that incorporates both the bitmap representation and vHGW algorithm produces a better result.

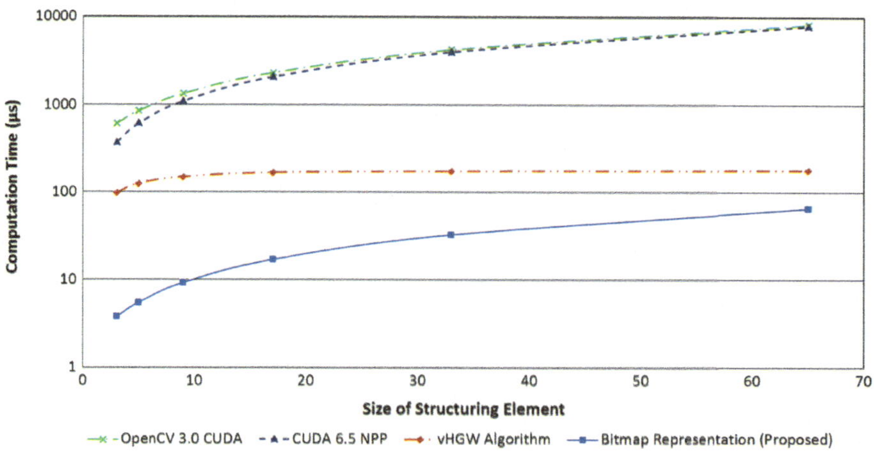

Fig. 6 Performance comparison of different algorithms for the vertical erosion operation on a FHD binary image

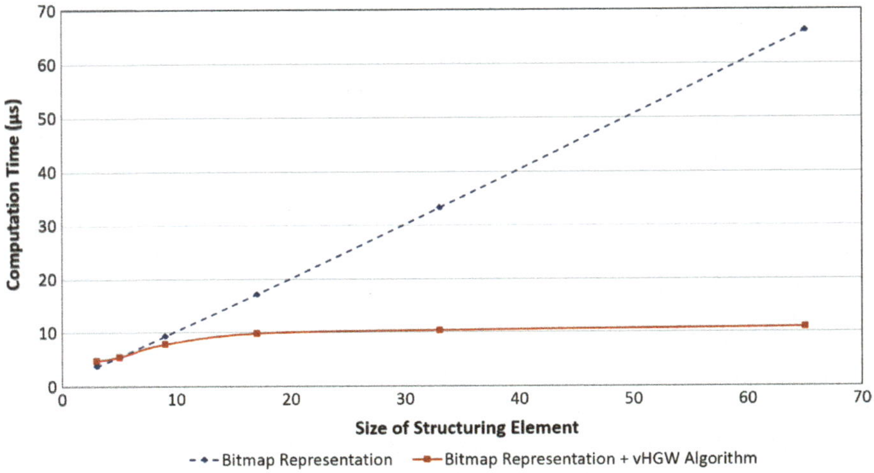

Fig. 7 Performance comparison of the CUDA implementation that incorporates the bitmap representation only with the implementation that incorporates both the bitmap representation and vHGW algorithm

5 Conclusion

Although there are quite a few CUDA implementations for morphology, most of the implementations were designed to process only grayscale images. Hence, we propose two optimisation techniques that make use of the bi-level nature of a binary pixel to further optimise the performance of the CUDA implementation for binary images. One of the optimisation techniques is achieved by representing the binary image in the form of a bitmap. With this optimisation technique, the one-dimensional horizontal erosion operation achieves a speedup up to 70 times, as compared to the existing vHGW CUDA implementation. In addition to the bitmapping of binary images, the existing vHGW algorithm for grayscale images is modified and applied to the case of binary images. With the combination of the bitmap representation and vHGW algorithm, the optimised version of the vertical erosion operation is approximately 20 times faster than the existing vHGW CUDA implementation.

Acknowledgements This project is supported by MOSTI, Malaysia under the e-science funding with a grant number of 010304SF0062.

References

1. Cook S (2012) CUDA programming: A developer's guide to parallel computing with GPUs. Newnes
2. Domanski L, Vallotton P, Wang D (2009) Parallel van herk/gil-werman image morphology on GPUs using cuda. In: GTC 2009 conference posters
3. Gil J, Werman M (1993) Computing 2-d min, median, and max filters. IEEE Trans Pattern Anal Mach Intell 15(5):504–507
4. Solomon C, Breckon T (2011) Fundamentals of digital image processing: A practical approach with examples in Matlab. John Wiley & Sons
5. Thurley MJ, Danell V (2012) Fast morphological image processing open-source extensions for GPU processing with cuda. IEEE J Sel Top Sign Proces 6(7):849–855
6. Van Den Boomgaard R, Van Balen R (1992) Methods for fast morphological image transforms using bitmapped binary images CVGIP. Graph Models Image Process 54(3):252–258
7. Van Herk M (1992) A fast algorithm for local minimum and maximum filters on rectangular and octagonal kernels. Pattern Recogn Lett 13(7):517–521

Non Negative Matrix Factorization for Music Emotion Classification

Nurlaila Rosli, Nordiana Rajaee and David Bong

Abstract Classification of emotion is a fundamental problem in music information retrieval where it addresses the query and retrieval of desirable types of music from large music data set. Until recently, there are only few works on music emotion classification that are carried out by incorporating instrumental and vocal timbre. Generally, vocal timbre alone can be used in distinguishing emotion in music but it became less effective when mixed with the instrumental part. Thus, a new research interest has led to identifying instrumental and vocal timbre as features capable of influencing human affect and analysis of sounds in regards to their emotional content. In this research, non-negative matrix factorization (NMF) is applied to separate music into both instrumental and vocal components. Extracted timbre features from audio using signal processing technique will be used to train and test artificial neural network (ANN) classifier. The ANN learn from supervised and unsupervised training to classify the emotional contents in music data as sad, happy anger or calm. The efficiency of the ANN classifier is verified by a subjective test including inputs from annotators by manual categorization of the audio data. The efficiency of this method reached up to 90 %.

Keywords Classification · Machine learning · Artificial neural network · Signal processing

N. Rosli (✉) · N. Rajaee · D. Bong
Department of Electronics Engineering, Faculty of Engineering,
University Malaysia, Kuala Lumpur, Sarawak, Malaysia
e-mail: laila8805@gmail.com

N. Rajaee
e-mail: rnordiana@feng.unimas.my

D. Bong
e-mail: bbldavid@feng.unimas.my

P.J. Soh et al. (eds.), *Advances in Machine Learning and Signal Processing*,
Lecture Notes in Electrical Engineering 387, DOI 10.1007/978-3-319-32213-1_16

1 Introduction

Basic emotions such as anger, disgust, fear, happiness, sadness and surprise have long existed in human evolution and influence mankind in stimulating rapid judgments and behavior [1]. Given the massive amount of audio features in information and multimedia, available to millions of users in various applications, classification of emotions in music is an interesting topic considering its appeal in influencing music affective of users. The detection and identification of emotion in information is crucial in affective computing which studies on human affects by technological systems and devices particularly in the fields of biomedical engineering, applied mathematics, psychology, consumerism and service industries as interpretations of music are multifaceted and abstractly perceived by human listeners [2].

1.1 Related Works

Music emotion classification (MEC) has been widely studied to accurately tag, segment and categorize music similarities for the purpose of automatic classification of emotive genres [3]. In music emotion classification (MEC), two critical components for music emotion tags are; feature extraction component and classifier learning component [4]. Feature extraction addresses the problem of how to represent the examples to be classified in terms of feature vectors and similarities while classifier learning is to find mapping from feature space to output labels to minimize prediction error [4]. Several methods have been proposed to classify the emotions in music by utilizing audio analysis framework (Sun et al. 2009), supervised and unsupervised data mining algorithm (Chu 2009), regression approach (Yang et al. 2007), content based analysis (Laurier 2011), neural networks (Vempala et al. 2012) and hierarchical SVM (Chiang et al. 2014) [5–10]. The detection of emotion via machine learning approach by Dual Tree Complex Packet Transform (DT_CWPT) has also been proposed to classify emotion using support vector machine (Daimi 2014) and using Gaussian processes (Markov 2014) [11, 12]. The audio features in music can be grouped based on three criterias: timbre, rhythm and pitch. Timbre features capture the tonal quality related to different instrumentation and can be further defined as low-level features. These low level features can be obtained directly from various signal processing techniques such as Fourier Transform, Spectral Analysis and Auto Regressive Modeling. Timbre features have unique attributes in distinguishing emotion in music [4].

However there are several issues on the utilization of instrumentation and vocal timbre features for classification. Firstly, there are only few existing works that incorporating vocal and instrumental timbre in MEC. Vocal timbre alone can be used to classify and differentiate emotion in music but it became less effective when mixed with the instrumental part [13]. Secondly, the local windows used in timbre feature extraction are usually taken from fixed size intervals. Some important timbre feature could be missed by using the fixed size intervals. These issues influence the classification performance in a data-dependent way. Therefore, further investigation on the instrumental and vocal timbre features for classification is proposed to study the classification performance in MEC especially for Malay music. We present the Non-negative matrix factorization (NMF) for separating vocal and instrumental components in music clips. Attributes from the timbre features such as spectral centroid, spectral rolloff, entropy, low energy, irregularity and zero-cross will be extracted for emotion classification in the first stage. Whereas in the second stage, a classifier component for pattern learning of the music emotion is constructed using Artificial Neural Network (ANN). We choose ANNs classifier for its well recognized performance especially in the area of machine learning and pattern recognition as justified in [14]. Finally, the performance of the proposed approach is measured by comparing the automatic classification with the ground truth of human observers.

The rest of this paper is organized as follows; in Sect. 2, methods including isolation process, timbral features extraction, and ANN modelling will be explained thoroughly. Section 3, presents the results and relative of overall findings. We conclude at Sect. 4, with an overview of the systems and suggestion for future work.

2 Classification Methods

Approximately 500 vocal audio data and instrumental data will be pre-processed as input and will be used to train the classifier. The proposed methodology will be based primarily on the extraction of instrumental and timbre features for Malay music for its emotion classification. Six attributes of the instrumental and timbre features such as spectral centroid, spectral rolloff, entropy, low energy, irregularity and zero cross is extracted. Audio pre-processing and timbre extraction will be implemented using audio software converter, audacity, MIR Toolbox, and MIREX. Whereas, ANN classifier will be built, train and testing in MATLAB.

2.1 Audio Pre-processing

Audio pre-processing can be regards as the most significant phase in the area of music emotion classification (MEC). At this stage, music samples are converted to a standard format and split into 20–30 s segment as to reduce the emotion disparity within segment by selecting the most representative part from the whole music clips [15]. However, both instrumental and vocal can contains different emotion and thus it is crucial to separate the original music clips into instrumental and vocal part before it can be split into numbers of frame or segment.

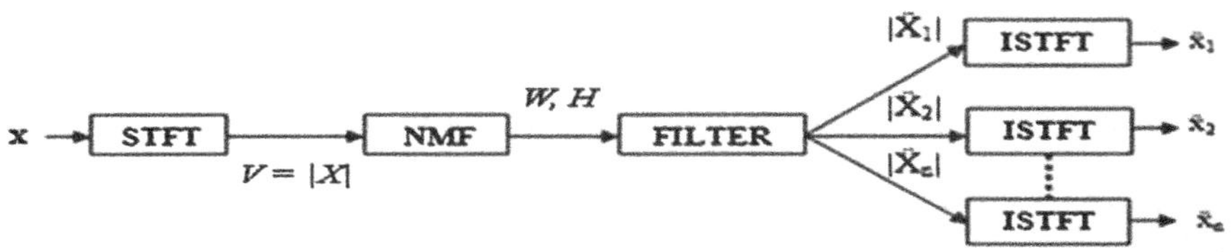

Fig. 1 Separation framework using NMF based on [16]

2.1.1 Isolation Process

Music is the combination or a mixture of instrumental sound and singing voice/vocal. An established source separation method namely, nonnegative matrix factorization (NMF) is used in this research to isolate signal mixtures into instrumental and vocal part. The energy spectrogram from each separated signal modeled using spectral component into time-frequency visualization, based on NMF distributions (Fig. 1).

For the purposed of this research, the information spectrogram constructed using short-time Fourier transform (STFT) to obtain a magnitude spectrogram before decomposed into set of NMF spectral components with inputs $|X|$ and outputs W and H. In NMF, mixture signals V is decomposed into separated source signals, W and H in the form of matrices as shown in Eq. (1).

$$V \approx WH \tag{1}$$

The cost function between W and H is minimized using Eq. (2), where all elements in the matrices are non-negative, and D is a measure of Euclidean divergence as shown in Eq. (3). Further information on NMF can be found in [17].

$$\underset{W,H \geq 0}{\overset{MINIMIZE}{}} \; D(V \| WH) \tag{2}$$

$$D(V\|\hat{V}) = \sum_{i,j} (V_{ij} - \hat{V}_{ij})^2 \tag{3}$$

NMF separation performance is measure using source-to-distortion ratio (SDR) in *BSS-Evaluation Toolbox* [18]. The evaluation modelling is shown in Eq. (4). We have exploits 256 and 512 ms window length of music samples with different spectral component as to compare which window length obtain the best separation performance using NMF based music separation.

$$\mathrm{SDR} = 10\log 10 \frac{\|s\mathrm{dist}\|^2}{\|e\mathrm{interf} + e\mathrm{noise} + e\mathrm{artif}\|^2} \tag{4}$$

2.1.2 File Splitting

500 samples of vocal audio data and instrumental data will be split into many local frames in the first step to facilitate subsequent frame level timbre feature extraction. This will lead to more efficient feature extraction process where spectral analysis can be directly applied to the short-time signals. By applying frame-level feature analysis, the underlying timbre information can be more reliably captured. The proposed methodology is shown in Fig. 2.

2.2 Extraction

Features extraction is a crucial stage whereby accurate extraction determines the accuracy of data generation in the database. After framing, spectral analysis

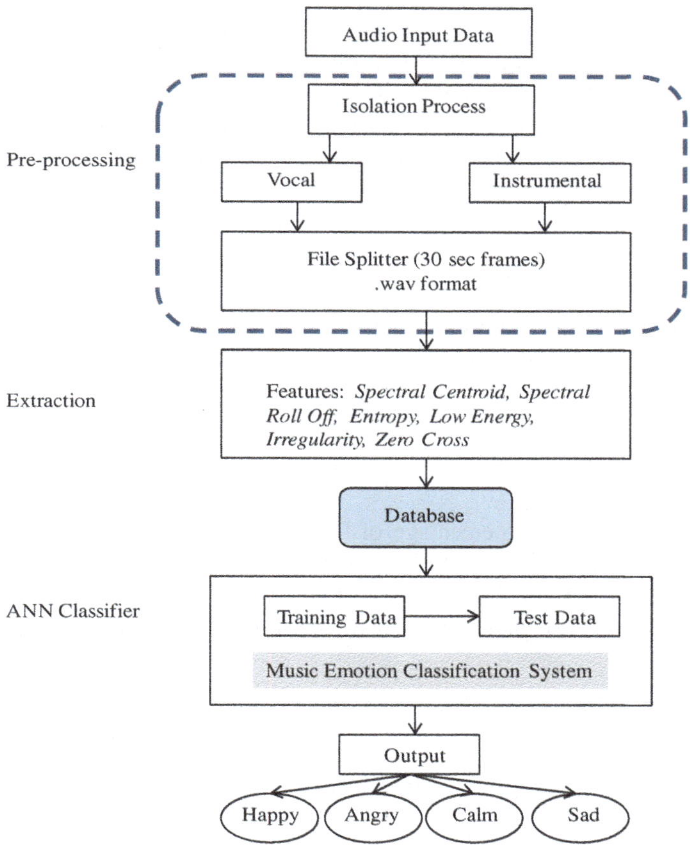

Fig. 2 Proposed flow for ANN music emotion classifier

techniques such as Fast Fourier Transform (FFT) and Discrete Wavelet Transform (DWT) are applied to the windowed signal in the local frames. From the output magnitude spectra, summary of features such as *spectral centroid, spectral rolloff, entropy, low energy, irregularity* and *zero cross* can be derived.

Spectral centroid: refers to the mean of the short time Fourier amplitude spectrum which indicates how bright the music. It is used to measure the center of mass for a particular spectrum and described as the average frequency weighted by amplitudes, divided by sum of amplitudes as given in Eq. (5):

$$\text{Spectral centroid} = \frac{\sum_{k=1}^{N} kF[k]}{\sum_{k=1}^{N} F[k]} \tag{5}$$

Spectral roll off: refers to the point where frequencies getting smaller in amplitude and gives shape of spectrum. 95 % of total spectrum is within this range. The roll off point is the frequency below which N% of the magnitude distribution is determined to distinguish voiced speech from unvoiced. This is because the unvoiced speech has a high proportion of energy contained in the high-frequency range of the spectrum, where most of the energy for voiced speech and music is contained in lower bands. If K is the bin that fulfills, then the Spectral Rolloff frequency is f(K), where x(n)represents the magnitude of bin number n and f(n) represents the center frequency of that bin as given in Eq. (6):

$$\sum_{n=0}^{k} x(n) = 0.85 \sum_{n=0}^{N-1} x(n) \tag{6}$$

Entropy: refers to properties to determine energy not available for work indicating measure of disorder. An estimation of entropy is given in Eq. (7):

$$H(x) = -\int_{x} f(x)\log f(x)\,dx \tag{7}$$

Low energy: refers to percentage of frames with less than average energy. In audio signal processing and speech classifications, the energy curve is used as an assessment of the temporal distribution of energy in order to determine whether it remains constant throughout the signal. Spectral energy density is given as in Eq. (8):

$$E_s(f) = |X(f)|^2 \tag{8}$$

Irregularity: refers to irregularity of the spectrums. Irregularity is given as in Eq. (9):

$$I = \sum_{k=2}^{N-1} \left| a_k - \frac{a_{k-1} + a_k + a_{k+1}}{3} \right| \tag{9}$$

Zero cross: refers to the number of times the signal crosses the zero line indicating the amount of noise. Zero crossing rates are useful to exemplify the divergent audio signals and is widely used in music and speech classification problems as it represents the rate of the signals going from positive to negative and back. Zero cross rate is given in Eq. (10):

$$Z_t = \frac{1}{2} \sum_{n=1}^{N} |\mathrm{sign}(x[n]) - \mathrm{sign}(s[n-1]) \tag{10}$$

2.3 Classifier Model Using Artificial Neural Network

Artificial Neural Network is widely utilized in pattern recognitions including in the detecting of emotions in facial expressions and speech recognition. For this research, we employ a supervised feed-forward with backpropagation network consisting of six input neurons, four hidden neurons and single output neurons to train the ANN classifier as shown in Fig. 3.

Feedforward ANN involves input vector x and response vector y with such a pair (x, y) belongs to a joint probability distribution. The goal is to induce a function $f(x)$ from a set *of* $(x1, y1)$. (xn, yn) of training examples to retrieve the approximation of y. Backpropagation algorithm is used to compute the error derivative for the weight (*EW*) by computing the rate at which the error changes (*EA*). To compute

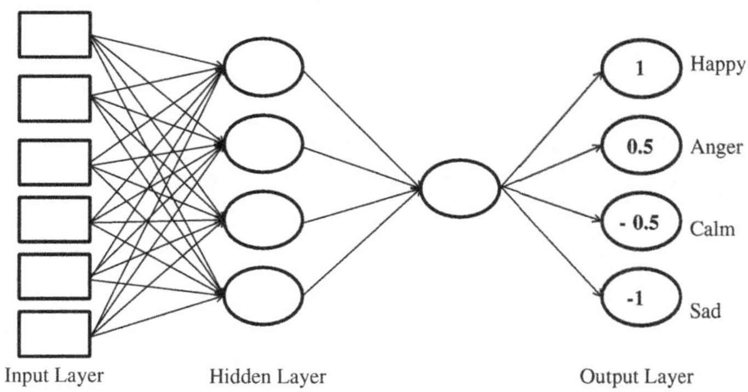

Fig. 3 ANN architecture for classifier

the *EA* for a hidden unit in the layer, all the weights between the hidden unit and output units to which it is connected must be identified [14].

After determining all the *EA* in the hidden layer, repeat computation for other layers in the opposite direction of activities propagating throughout the network. During the classification process, ANN classifier will get the information from the database or (memorized value of musical features) from previous training process and classify emotion from the song by scheming the music features vector as to produced results. Songs with an output ranging from $0 < x \leq 0.5$ were considered as happy songs, songs with output ranging from $0.5 < x \leq 1$ were considered as anger songs, songs with output ranging from $0 > x \geq -0.5$ were considered as calm songs, and songs with output ranging from $-0.5 > x \geq -1$ were considered as sad songs. Testing will be carried out to determine the features of the songs whether they are close to 0.5 (happy), close to 1 (anger), close to −0.5 (calm) or close to −1 (sad) which will verify the efficiency of the ANN classifier. Further testing will also be conducted with (i) only vocal features, (ii) only instrumental features and (iii) vocal and instrumental features, to compare the differences in the classification rates.

3 Results and Discussion

3.1 Music Isolation Results

In this work, music clips is separated by using the source separation method, namely nonnegative matrix factorization. NMF has the ability to decomposing the spectrum of original mixtures signal. Source-to-distortion ratio (SDR) is used to evaluate NMF performance using the *BSS-Evaluation Toolbox*. Higher SDR values suggest better instrumental and vocal separation.

As explained in (Sects. 1 and 2) above, the local windows used in timbre feature extraction are usually taken from fixed size intervals. Some important timbre feature could be missed by using the fixed size intervals. These issues influence the classification performance in a data-dependent way. Thus, to find the best local window length for timbre extraction, we have exploits 256 and 512 ms window length. From the isolation process, NMF has successfully separate music into instrumental and vocal parts and its performance is measured by calculating the mean SDRs of spectrogram factorization for 256 and 512 ms window lengths that has different spectral component. The best performances with the highest mean SDR achieved were obtain using the window length of 256 ms. Though the mean SDR achieved for 512 ms window length are less than when using 256 ms window length, the spectral discontinuity threshold at 0.4 seems to be suitable for both 256 and 512 window length music (Fig. 4).

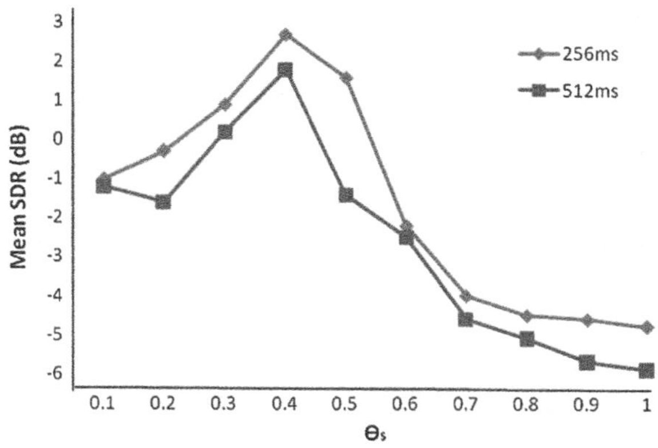

Fig. 4 NMF separation performance evaluation based on source to distortion ratio (SDR) using 256 and 512 ms window length frames

3.2 Classification Results

The accuracy of the classification result can be measured by dividing number of correctly classified songs with the total number of songs (Refer to Eq. 11). A standard method for comparing the performance of different classifier performance measurement is based on the evaluation taken from [19] (Fig. 5).

$$Accuracy = \frac{number\ of\ correctly\ classified\ songs}{total\ number\ of\ songs} \times 100 \quad (11)$$

We have separated 500 music data into both instrumental and vocal components, resulting in 1000 music clips in total. The dataset are then randomly split into two parts, 80 % for training and 20 % for testing. It can be seen from the results that by incorporating ANN with both music components parameters improve the

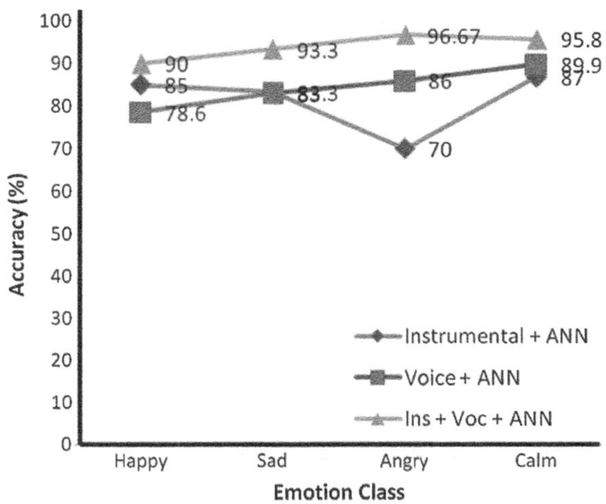

Fig. 5 Music emotion classification comparison using ANN with (i) only vocal features, (ii) only instrumental features and (iii) vocal and instrumental features

classification rate. The classification accuracy when using parameters from both instrumental and vocal components can reach up to 95 %, compare to when using only instrumental features (80 %) or vocal features (86 %) Separation of the music clips into instrumental and vocal part has lessened the burden for automatic emotion annotation in music data representative. Both components contains different spectral information that might leads to different emotive genre within the music frames, thus the isolation process helps in selecting the best timbre information for MEC in both separated instrumental and vocal part.

4 Conclusion and Future Works

This work aim to incorporate the timbre features extracted from the separated music components i.e., instrumental and vocal. We proposed music separation methods in MEC that employs Non-negative Matrix Factorization (NMF) algorithms. Results from the music isolation process shows that the best separation performance can be obtained using windows length 256 ms and spectral discontinuity threshold at 0.4. This work also aim to use ANN with feedforward backpropagation network learning, to classify emotion in selected music clips by conducting different testing algorithm, using (i) only vocal features, (ii) only instrumental features and (iii) vocal and instrumental features as to compare the differences in the classification rates. Results from this testing have proved, that by incorporating instrumental and vocal features in MEC, classification accuracy can be achieved up to 95 % compare to when using only instrumental and only vocal features. Overall, there are two important improvements could be added to this study. First, NMF has proven to be very convincing in providing natural and meaningful music segment representative for MEC though the separation process a bit slow. Thus in the future, we should considered to incorporate NMF to the constraints that related to the music samples, for instance, musical instrument like drums, piano, guitar and etc. as to achieve better and faster convergence. Secondly, exploits other source separation techniques through decomposition of a time-frequency representation as to improve the music separation performance in MEC.

Acknowledgments This work is supported by a grant from University Malaysia Sarawak (UNIMAS).

References

1. LeDoux J (2000) Emotion circuits in the Brain. Annu Rev Neurosci 23:155–184
2. Laurier C, Lartillot O, Eerola T, Toiviainen P (2009) Exploring relationships between audio features and emotion in music
3. Nikalaou N (2011) Music emotion classification, Dissertation for the Diploma of Electronic and Computer, Technical University of Crete

4. Fu Z, Lu G, Ting KM, Zhang D (2011) A survey of audio based music classification and annotation. IEEE Trans Multimedia 13(2):303–319
5. Sun X, Tang Y (2009) Automatic music emotion classification using a new classification algorithm. In: Second international symposium on computational intelligence and design
6. Chu ML (2009) Automatic music genre classification, National Taiwan University Taipei
7. Yang YH, Lin YC, Su YF, Chen HH (2007) A regression approach to music emotion recognition
8. Laurier C (2011) Automatic classification of musical mood by content based analysis, Doctoral Thesis, Universitat Pompeu Fabra
9. Vempala NN, Russo F (2012) Predicting emotion from music audio features using neural networks. In: 9th international symposium on computer music modelling and retrieval
10. Chiang WC, Wang JS, Hsu YL (2014) A music emotion recognition algorithm with hierarchical SVM based classifiers. In: International symposium on computer, consumer and control, pp 1249–1252
11. Daimi Saha (2014) Classification of emotions induced by music videos and correlation with participant's rating. Exp Syst Appl 41(13):6057–6065
12. Markov K, Matsui T (2014) Music genre and emotion recognition using gaussian processes. IEEE J Mag 2:688–697
13. Xu J, Li X, Hao Y, Yang G (2014) Source separation improves music emotion recognition, pp 1–4
14. Basu JK, Bhattacharyya D, Kim T (2010) Use of artificial neural network in pattern recognition. Int J Softw Eng Appl 4(2):23–34
15. Yang Y, Chen HH (2012) Machine recognition of music emotion: A review, ACM Trans Intell Syst Technol 3(3):40. doi:10.1145/2168752.2168754
16. Lee DD, Seung HS (2001) Algorithms for non-negative matrix factorization. In: Advances in neural information processing systems (NIPS), MIT Press, pp 556–562
17. Fevotte C, Bertin N, Durrieu J-L (2009) Nonnegative matrix factorization with the itakura-saito divergence: With application to music analysis, Neural Comput
18. Rafii Z, Pardo B (2013) Repeating pattern extraction technique (REPET): A simple method for music/voice separation, IEEE Trans Audio Speech Lang Process 21(1):73–84
19. Beveridge S, Knox D (2012) A feature survey for emotion classification of western popular music. In: Proceedings of the 9th international symposium on computer music modeling and retrieval, (CMMR): Music and emotions, June, pp 19–22

Speech Enhancement Based on Noise Type and Wavelet Thresholding the Multitaper Spectrum

E.P. Jayakumar and P.S. Sathidevi

Abstract A speech enhancement system for noisy speech signals under different environments, namely airport, babble, car, exhibition, restaurant, station, street and train, is designed and implemented in this paper. Spectral subtraction is the most popular method for speech enhancement which suffers from perceptually annoying musical noise. Musical noise is reduced here by using low variance spectrum estimators and enhancement is achieved by wavelet thresholding the multitaper spectrum of the speech. The main feature of the proposed system is a novel switching mechanism to an optimal wavelet filter bank or wavelet packet filter bank matching to the critical bands of the human ear, based on the type of noise present in the input speech. The system is evaluated using noisy speech signals under different environment noises at different signal to noise ratios. The subjective and objective test results show that the proposed method improves the quality and intelligibility of speech signals.

Keywords Speech enhancement · Musical noise · Multitaper spectrum · Noise classification · Wavelet thresholding

1 Introduction

Performance of most of the speech processing systems such as voice communication, speech recognition, speaker recognition, hearing aids, etc., are affected by the noisy environment where they are used. Therefore speech enhancement is done to neutralise the effect of such environment noises on these speech processing systems. Speech enhancement is a popular topic of research over many years and many techniques were

E.P. Jayakumar (✉) · P.S. Sathidevi
Department of Electronics and Communication Engineering,
National Institute of Technology Calicut, Kozhikode 673601, Kerala, India
e-mail: jay@nitc.ac.in

P.S. Sathidevi
e-mail: sathi@nitc.ac.in

P.J. Soh et al. (eds.), *Advances in Machine Learning and Signal Processing*,
Lecture Notes in Electrical Engineering 387, DOI 10.1007/978-3-319-32213-1_17

proposed. Among the different methods to do the speech enhancement, the spectral subtraction method is the most popular method because it is easy to implement and requires less computation. But the drawback of this method is the introduction of perceptually annoying musical noise. Musical noise is due to the spectral subtraction process which creates isolated peaks in the spectrum occurring at random frequency locations in each frame. The randomly spaced peaks are due to the inaccurate and large-variance estimates of the spectra of the noise and noisy signals [8].

Many methods have been proposed to reduce the musical noise in the enhanced speech. Boll [2] introduced the basic spectral subtraction method and proposed magnitude averaging to reduce the effect of musical noise. Berouti et al. [1] defined spectral floor to make the musical noise inaudible. Ephraim and Malah [7] is based on minimum-mean square error short-time spectral amplitude estimator where a priori signal to noise ratio (SNR) $\gamma_{prio}(\omega_k)$ is smoothened to eliminate musical noise. Virag [18] used masking threshold based on human auditory system to attenuate the musical noise. Hu and Loizou [10] incorporated a psychoacoustical model in frequency domain for enhancing the speech.

Hu and Loizou [11] used multitaper spectrum estimators and wavelet thresholded the log multitaper spectra to get spectral estimates with lower variance.In this method, they used a 5-level Discrete Wavelet Transform (DWT) filter bank with Daubechie db4 wavelet basis for wavelet denoising. This structure is found to be suitable for enhancing the speech under two environments namely, car and babble noises. Therefore, a wavelet packet filter bank matching to the critical bands of the human ear [16] is implemented in this work. Performances of DWT filter bank and wavelet packet filter bank with different wavelet bases are evaluated under different environment noises. Depending upon the type of noise present in the input speech, the proposed system will automatically select the optimum filter bank.

2 Background

2.1 Multitaper Spectrum Estimator

Variance of the spectral estimates has to be reduced since spectral estimates with large variance may result in musical noise. Multitaper spectrum estimator [14], [17] can be used to reduce the variance. To achieve this, a small, say L, number of direct spectrum estimators with different tapers are computed and the mean of L spectral estimates are calculated. Pair wise orthogonal tapers are selected such that the bias and variance of the multitaper spectrum estimate is less. The variance is reduced by a factor of $1/L$. The multitaper spectrum estimate is obtained by the following equation

$$\hat{S}^{mt}(\omega) = \frac{1}{L}\sum_{k=0}^{L-1} \hat{S}_k^{mt}(\omega) \tag{1}$$

where, the spectral estimate $\hat{S}_k^{mt}(\omega)$, which is also called the kth eigen spectrum is given by

$$\hat{S}_k^{mt}(\omega) = \left| \sum_{m=0}^{N-1} a_k(m)x(m)e^{-j\omega m} \right|^2 \tag{2}$$

where, N is the data length. a_k represents the kth data taper used for the computation of spectral estimate $\hat{S}_k^{mt}(.)$. The sine tapers which is proposed by Riedel and Sidorenko [15], have smaller local bias and is given by:

$$a_k(m) = \sqrt{\frac{2}{N+1}} \sin \frac{\pi k(m+1)}{N+1}, \quad m = 0, 1, \ldots.., N-1 \tag{3}$$

The above sine tapers are used in the proposed method.

2.2 Wavelet Thresholding for the Refinement of the Spectrum Estimate

The refinement of the spectral estimate is carried out by wavelet thresholding. It produces a smooth estimate of the logarithm of the spectrum [8, 13, 19]. Improved multitaper spectrum estimates were proposed in [3, 19]. In these methods, the log periodogram is represented as signal added with the noise. If the noise is Gaussian, it can be removed by employing wavelet shrinkage techniques. Better spectral estimates can be obtained using level-independent universal thresholds [6]. If $\hat{S}^{mt}(\omega)$ represents the estimated multitaper spectrum and $S(\omega)$ represents the true power spectrum and if the eigenspectra defined in Eq. 2 are uncorrelated, their ratio $v(\omega)$ should satisfy the following equation.

$$v(\omega) = \frac{\hat{S}^{mt}(\omega)}{S(\omega)} \sim \frac{\chi_{2L}^2}{2L}, \quad 0 < \omega < \pi \tag{4}$$

where, χ_{2L}^2 is a chi-square distribution with $2L$ degrees of freedom. Taking logarithm of both sides in Eq. 4,

$$\log \hat{S}^{mt}(\omega) = \log S(\omega) + \log v(\omega) \tag{5}$$

where, $\log v(\omega)$ is the noise term and is $\log \chi^2$ distributed. Distribution of $\log v(\omega)$ will be very similar to a normal distribution with mean $\phi(L) - \log L$ and variance $\phi'(L)$, where $\phi(L)$ is digamma function and $\phi'(L)$ is trigamma function. The random variable $\eta(\omega)$ is given by

$$\eta(\omega) = \log v(\omega) - \phi(L) + \log L \tag{6}$$

$\eta(\omega)$ will be approximately Gaussian with zero mean and variance $\sigma_\eta^2 = \phi'(L)$ for all ω except near $\omega = 0$ and π. If $Z(\omega)$ is defined as

$$Z(\omega) = \log \hat{S}^{mt}(\omega) - \phi(L) + \log L \tag{7}$$

Using Eq. 5 and since $\log L - \phi(L)$ is a constant, $Z(\omega)$ can be expressed as

$$Z(\omega) = \log S(\omega) + \eta(\omega) \tag{8}$$

Wavelet denoising techniques can be applied to the model in Eq. 8 for removing the noise $\eta(\omega)$ which is Gaussian in nature [4–6].

3 Speech Enhancement by Wavelet Thresholding the Multitaper Spectrum

3.1 Short Time Spectral Amplitude (STSA) Estimator

STSA estimator proposed by Hu and Loizou [11] is used in this paper. The noise signal is assumed to be additive and uncorrelated with the speech signal, i.e., $\mathbf{y} = \mathbf{x} + \mathbf{n}$, where, $\mathbf{y}$ is the the N-dimensional noisy speech, $\mathbf{x}$ is the clean speech and $\mathbf{n}$ is the noise vector. If the N-point discrete Fourier Transform matrix is denoted by F, then, the Fourier transform of $\mathbf{y}$ can be expressed as $Y(\omega) = F^H.\mathbf{y} = F^H.\mathbf{x} + F^H.\mathbf{n} = X(\omega) + N(\omega)$, where, $X(\omega)$ and $N(\omega)$ are the $N \times 1$ vectors containing the spectral components of the clean speech vector and the noise vector, respectively.

The linear estimator of $X(\omega)$ is given by $\hat{X}(\omega) = G.Y(\omega)$, where, G is a $N \times N$ matrix. The error signal in this estimation, $\varepsilon(\omega)$ is expressed as $\varepsilon(\omega) = \hat{X}(\omega) - X(\omega) = \varepsilon_{\mathbf{x}}(\omega) + \varepsilon_{\mathbf{n}}(\omega)$, where, $\varepsilon_{\mathbf{x}}(\omega)$ is the frequency domain representation of speech distortion, which given by $\varepsilon_{\mathbf{x}}(\omega) = (G-1).X(\omega)$ and $\varepsilon_{\mathbf{n}}(\omega)$ is the frequency domain representation of the residual noise, which is given by $\varepsilon_{\mathbf{n}}(\omega) = G.N(\omega)$. The following constrained optimization problem is to be solved to get the optimal linear estimator.

$$\begin{gathered}\min_{G} \varepsilon_{\mathbf{x}}^2(\omega) \\ \text{subject to}: \tfrac{1}{N}\varepsilon_{\mathbf{n}}^2(\omega) \le c\end{gathered} \tag{9}$$

where, c is a positive number, $\varepsilon_{\mathbf{x}}^2(\omega)$ and $\varepsilon_{\mathbf{n}}^2(\omega)$ are the energy of the frequency domain speech distortion and residual noise respectively. The optimum value of G satisfies the following equation [9]:

$$G(F^H.R_{\mathbf{x}}.F + \mu.F^H.R_{\mathbf{n}}.F) = F^H.R_{\mathbf{x}}.F \tag{10}$$

where, μ is the Lagrange multiplier. If G is assumed to be a diagonal matrix, then Eq. 10 can be simplified. If $R_{\mathbf{x}}$ and $R_{\mathbf{n}}$ are assumed to be Toeplitz, then the matrices $F^H.R_{\mathbf{x}}.F$ and $F^H.R_{\mathbf{n}}.F$ are asymptotically diagonal. The diagonal elements of $F^H.R_{\mathbf{x}}.F$ are the power spectrum components $S_{\mathbf{x}}(\omega)$ of the clean speech vector and $F^H.R_{\mathbf{n}}.F$ are the power spectrum components, $S_{\mathbf{n}}(\omega)$ of the noise vector. Let the kth diagonal element of G is denoted as $g(k)$, then for large values of N, Eq. 10 can be modified as

$$g(k) = \frac{S_{\mathbf{x}}(k)}{S_{\mathbf{x}}(k) + \mu.S_{\mathbf{n}}(k)} = \frac{\gamma_{prio}(k)}{\gamma_{prio}(k) + \mu} \tag{11}$$

where, $\gamma_{prio}(k) = S_{\mathbf{x}}(k)/S_{\mathbf{n}}(k)$ is the *a priori* SNR at frequency ω_k and μ is the Lagrange multiplier. The trade-off between speech distortion and residual noise is controlled by μ, where, $\mu \geq 0$. It is decided based on the estimated *a priori* SNR as given by the following equation [9]

$$\mu = \mu_0 - \frac{\text{SNR}_{dB}}{s} \tag{12}$$

where, μ_0 and s are constants, which are experimentally determined and $\text{SNR}_{dB} = 10\log_{10}\text{SNR}$. The power spectrum of the clean speech signal $S_x(\omega)$ is determined using the following expression:

$$\hat{S}_{\mathbf{x}}(\omega) = S_{\mathbf{y}}(\omega) - \hat{S}_{\mathbf{n}}(\omega) \tag{13}$$

where, $\hat{S}_{\mathbf{n}}(\omega)$ is the noise spectrum estimate, obtained during speech-absent frames. A good estimate of *a priori* SNR $\gamma_{prio}(k)$ is very much essential for removing musical noise from the noisy speech. This is done by first calculating the multitaper spectral estimates followed by its refinement done by wavelet thresholding the log of the estimated spectrum.

4 Proposed Architecture of the Speech Enhancement System

A speech enhancement technique based on the wavelet thresholding of the multitaper spectra is proposed by Hu and Loizou in [11]. This method uses 5-level DWT filter bank structure with db4 wavelet basis. The authors have considered only two types of noise signals namely speech shaped noise and car noise. But the speech

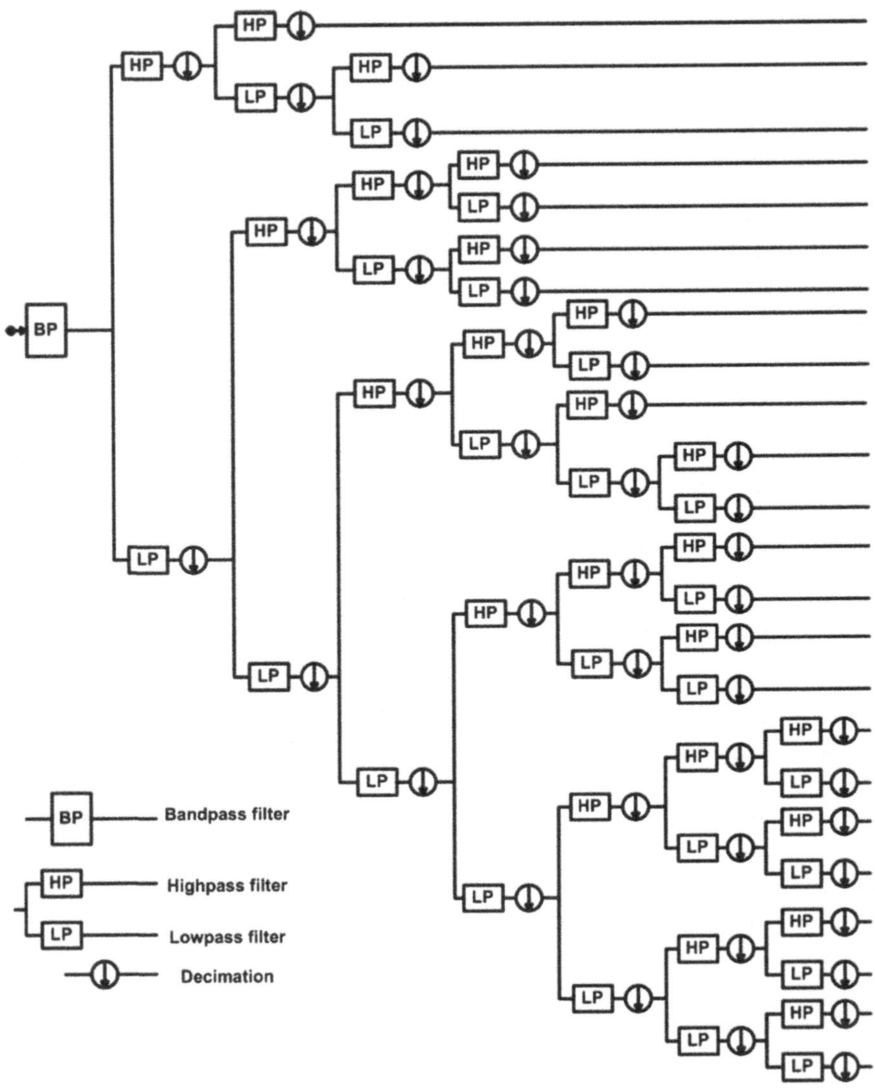

Fig. 1 Wavelet packet filter bank structure matching to the critical bands of human ear [16]

enhancement system proposed in this paper, works well for speech signals corrupted by various types of noise signals namely airport, babble, car, exhibition, restaurant, station, street and train. Experiments were conducted using 5-level DWT filter bank and wavelet packet filter bank matching to the critical bands of human ear as shown in Fig. 1 with different bases (db2 to db40, Haar, coiflets and symlets) and the following observations were made:

1. 5-level DWT filter bank with db4 wavelet gives highest SNR for speech signals corrupted by street and train noise.
2. Wavelet packet filter bank matching to the critical bands of human ear with db2 wavelet gives highest SNR for speech signals corrupted by airport, babble, car, station and restaurant noises.
3. Wavelet packet filter bank matching to the critical bands of human ear with db10 wavelet gives highest SNR for speech signals corrupted by exhibition noise.

Above observations motivated us to design an improved speech enhancement system which adaptively selects DWT filter bank or wavelet packet filter bank with a suitable wavelet basis, depending on the type of noise present in the input speech signal.

4.1 Determination of Noise Type

The features of noise energy distribution in the Bark domain is used for effective noise determination [20]. Let $D(k,l)$ be the Short Time Fourier Transform (STFT) of a noise signal, and $P(k,l)$ be the smoothed noise power spectrum then the smoothing is performed as

$$P(k,l) = \alpha_p P(k,l-1) + (1-\alpha_p)|D(k,l)|^2 \tag{14}$$

where, $\alpha_p = 0.5$ is the smoothing parameter, k denotes the index and l denotes the frame number. Noise energy in each Bark band is calculated as

$$\mathbb{S}(j,l) = \sum_{k=lb(j)}^{ub(j)} P(k,l) \tag{15}$$

where, $ub(j)$ and $lb(j)$ are the upper STFT bin and lower STFT bin respectively and j is the Bark band number, and $j = 1, 2, \ldots .18$. The total noise energy of the lth frame is calculated as

$$\mathbb{S}_t(l) = \sum_{k=1}^{N} P(k,l) \tag{16}$$

where, $N = 128$ is the total number of STFT bins as shown in Table 1.Then the ratio of the noise energy in the jth Bark band of the lth frame to the entire noise energy in the lth frame is calculated as

Table 1 Mapping from 256-point STFT bins to bark bands at a sampling frequency of 8 kHz

Bark band number (*j*)	STFT Bins		Frequencies (Hz)
	Interval (*lb*(*j*)-*ub*(*j*))	Number of bins	
1	1–3	3	0–94
2	4–6	3	94–187
3	7–10	4	187–312
4	11–13	3	312–406
5	14–16	3	406–500
6	17–20	4	500–625
7	21–25	5	625–781
8	26–29	4	781–906
9	30–35	6	906–1094
10	36–41	6	1094–1281
11	42–47	6	1281–1469
12	48–55	8	1469–1719
13	56–64	9	1719–2000
14	65–74	10	2000–2312
15	75–86	12	2312–2687
16	87–100	14	2687–3125
17	101–118	18	3125–3687
18	119–128	10	3687–4000

$$R_{\mathbb{S}}(j,l) = \frac{\mathbb{S}(j,l)}{\mathbb{S}_t(l)} \tag{17}$$

The 18 dimensional feature vector for the lth frame is given by

$$r = (R^{\eta}_{\mathbb{S}}(1,l), R^{\eta}_{\mathbb{S}}(2,l), \ldots, R^{\eta}_{\mathbb{S}}(18,l)) \tag{18}$$

where, the value of η is found out experimentally as 0.25 for better performance. Feature extraction is done on a set of noise frames of a particular noise type. Centroid of a set of such feature vectors is taken as the reference feature vector of that particular noise type. Similarly, the reference feature vectors corresponding to each noise type is obtained and stored in an array. To determine the noise type present in the input noisy speech, the feature vectors of the noise frames are extracted and a measure of closeness is obtained by calculating the Euclidian distance between the feature vector and all the reference feature vectors. If a reference feature vector of a particular noise type is at a smallest distance to the extracted feature vector, then it is determined that the noise frame belongs to that particular noise type.

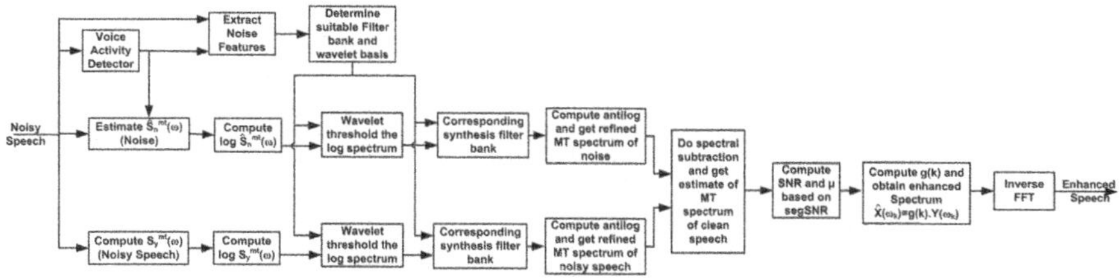

Fig. 2 Block diagram of the proposed speech enhancement system

4.2 Proposed System

The block diagram of the proposed system is shown in Fig. 2. Based on the type of noise present, system will automatically choose DWT filter bank or wavelet packet filter bank with the optimal wavelet basis.

4.3 Implementation Details

For evaluating the proposed method, the noisy speech signal is divided into frames of duration 32 ms and the overlap between the successive frames is chosen as 50 %. The steps involved in the proposed method are described below:

1. Use voice activity detector and identify the presence and absence of speech in the noisy speech $\mathbf{y}$. Estimate the multitaper power spectrum of the noisy speech $S_{\mathbf{y}}^{mt}(\omega)$ and the noise $\hat{S}_{\mathbf{n}}^{mt}(\omega)$ during absence of speech using Eq. 1.
2. Compute the logarithm of the multitaper spectrum found in Step 1.
3. Extract the noise features as explained in Sect. 4.1 and determine the suitable filter bank and wavelet basis based on the noise type. Use this and wavelet threshold [4, 5] the log spectrum of noise and noisy speech individually.
4. Obtain a refined log spectrum of noise and noisy speech by using the corresponding synthesis filter bank and compute their respective antilogarithms.
5. Compute the multitaper power spectrum of the clean speech, $\hat{S}_{\mathbf{x}}^{mt}(\omega)$ by doing spectral subtraction using the following expression.

$$\hat{S}_{\mathbf{x}}^{mt}(\omega) = \begin{cases} S_{\mathbf{y}}^{mt}(\omega) - \hat{S}_{\mathbf{n}}^{mt}(\omega), & \text{if } S_{\mathbf{y}}^{mt}(\omega) > \hat{S}_{\mathbf{n}}^{mt}(\omega) \\ \beta\hat{S}_{\mathbf{n}}^{mt}(\omega), & \text{if } S_{\mathbf{y}}^{mt}(\omega) \leq \hat{S}_{\mathbf{n}}^{mt}(\omega) \end{cases} \tag{19}$$

where, the spectral floor $\beta = 0.002$ [11].

6. Calculate the *a priori* SNR $\gamma_{prio}(k)$ using the following expression

$$\text{SNR} = \frac{\sum_{i=0}^{N-1} \hat{S}_{\mathbf{x}}^{mt}(i)}{\sum_{i=0}^{N-1} \hat{S}_{\mathbf{n}}^{mt}(i)}, \qquad \text{SNR}_{dB} = 10\log_{10}\text{SNR} \tag{20}$$

7. Compute the μ value in Eq. 11 using the following equation:

$$\mu = \begin{cases} \mu_0 - \frac{\text{SNR}_{dB}}{s}, & -5 < SNR_{dB} < 20 \\ 1, & \text{SNR}_{dB} \geq 20 \\ \mu_{\max}, & \text{SNR}_{dB} \leq -5 \end{cases} \tag{21}$$

where, $\mu_{\max} = 10, \mu_0 = (1 + 4\mu_{\max})/5, s = 25/(\mu_{\max} - 1)$ [11].

8. Obtain $g(k)$ using Eq. 11. Compute the enhanced spectrum $\hat{X}(\omega_k) = g(k).Y(\omega_k)$ and obtain the inverse FFT and use the overlap and add method to get the enhanced speech signal.

5 Experimental Results

The performance of the proposed method is evaluated using 10 speech signals from Noizeus [12] database. Clean speech signals corrupted by different noises namely airport noise, babble noise, car noise, exhibition noise, restaurant noise, street noise, station noise and train noise were included. Noisy speech at different SNR values (0, 5, 10 and 15 dB) were used. The proposed method is compared with the MT_SURE method [11] in terms of the SNR and Mean Opinion Score (MOS). Table 2 shows the performance comparison.

Subjective evaluation is done to test the quality of the enhanced speech through listening tests. In subjective evaluation, the speech enhanced by both enhancement methods at various SNR levels were mixed randomly. It is then presented to 10 subjects in the age group of 22 to 26 years. A Sennheiser HD 203 headphone is used for this purpose. Subjects were then asked to rate the quality of the perceived speech in a five point scale (1: bad; 2: poor; 3: fair; 4: good; 5: excellent). The average score from all subjects is referred to as the mean opinion score (MOS). Table 3 compares the mean opinion score (MOS), which is the average score from all subjects obtained with both the methods. It is observed that the MOS of the proposed method is better than that of the MT_SURE method.

Table 2 Performance comparison in terms of SNR for 10 sentences corrupted by different noises at various SNR levels

	Airport				Babble				Car				Station			
(dB)	0	5	10	15	0	5	10	15	0	5	10	15	0	5	10	15
MT_SURE[11]	3.33	7.54	11.53	15.51	3.96	7.42	11.69	15.33	4.15	8.03	11.99	16.04	3.47	7.44	11.72	15.59
Proposed method	3.58	7.76	11.90	15.69	3.97	7.45	11.74	15.43	4.30	8.09	12.26	16.06	3.88	7.51	12.29	15.80
	Restaurant				**Exhibition**				**Street**				**Train**			
(dB)	0	5	10	15	0	5	10	15	0	5	10	15	0	5	10	15
MT_SURE[11]	2.77	6.54	11.28	15.44	3.16	7.63	12.09	15.88	3.05	6.12	11.16	15.43	4.11	8.23	12.16	16.05
Proposed method	3.49	7.08	11.43	15.87	3.37	7.83	12.18	15.99	3.05	6.12	11.16	15.43	4.11	8.23	12.16	16.05

Table 3 Comparison of MOS values obtained for speeches corrupted by different noises at various SNR levels

	Airport				Babble				Car				Station			
(dB)	0	5	10	15	0	5	10	15	0	5	10	15	0	5	10	15
MT_SURE[11]	2.5	2.7	3.1	3.2	2.4	2.6	2.7	3.0	2.9	3.0	3.2	3.5	3.0	3.2	3.3	3.5
Proposed method	2.7	3.0	3.5	3.7	2.6	2.7	3.0	3.5	3.0	3.2	3.5	4.0	3.1	3.2	3.5	3.9
	Restaurant				**Exhibition**				**Street**				**Train**			
(dB)	0	5	10	15	0	5	10	15	0	5	10	15	0	5	10	15
MT_SURE[11]	2.4	2.8	2.8	3.5	2.5	2.5	2.6	3.0	2.5	2.8	2.9	4.0	2.5	2.5	2.7	3.0
Proposed method	2.5	2.9	3.0	3.9	2.7	2.8	3.0	3.5	2.5	2.9	2.9	4.1	2.5	2.5	2.8	3.2

6 Conclusion

In this paper, an improved speech enhancement method is presented, which is well suited for different noise environments. The low variance spectrum estimators based on wavelet thresholding the multitaper spectrum of speech are employed here to reduce the musical noise. Wavelet denoising is effected using either 5-level DWT filter bank or wavelet packet filter bank using optimum wavelet basis according to the type of noise present in the speech. This is an important step in the proposed speech enhancement system leading to the removal of musical noise and is evident from the results obtained through subjective and objective tests. The proposed method gives better SNR and MOS values when compared to the related work [11] available in the literature.

References

1. Berouti M, Schwartz R, Makhoul J (1979) Enhancement of speech corrupted by acoustic noise. In: IEEE international conference on ICASSP'79 acoustics, speech, and signal processing, vol 4, pp 208–211. IEEE
2. Boll SF (1979) Suppression of acoustic noise in speech using spectral subtraction. IEEE Trans Acoust Speech and Signal Process 27(2):113–120
3. Cristán AC, Walden AT (2002) Multitaper power spectrum estimation and thresholding: wavelet packets versus wavelets. IEEE Tran Signal Process 50(12):2976–2986
4. Donoho DL (1995) De-noising by soft-thresholding. IEEE Trans Inf Theory 41(3):613–627
5. Donoho DL, Johnstone IM (1995) Adapting to unknown smoothness via wavelet shrinkage. J Am Stat Assoc 90(432):1200–1224
6. Donoho DL, Johnstone JM (1994) Ideal spatial adaptation by wavelet shrinkage. Biometrika 81(3):425–455
7. Ephraim Y, Malah D (1984) Speech enhancement using a minimum-mean square error short-time spectral amplitude estimator. IEEE Trans Acoust Speech and Signal Process 32 (6):1109–1121
8. Gao HY (1997) Choice of thresholds for wavelet shrinkage estimate of the spectrum. J Time Ser Anal 18(3):231–251
9. Hu Y, Loizou PC (2003) A generalized subspace approach for enhancing speech corrupted by colored noise. IEEE Trans Speech Audio Process 11(4):334–341
10. Hu Y, Loizou PC (2004) Incorporating a psychoacoustical model in frequency domain speech enhancement. IEEE Signal Process Lett 11(2):270–273
11. Hu Y, Loizou PC (2004) Speech enhancement based on wavelet thresholding the multitaper spectrum. IEEE Trans Speech Audio Process 12(1):59–67
12. Hu Y, Loizou PC (2007) Subjective comparison and evaluation of speech enhancement algorithms. Speech Commun 49(7):588–601
13. Moulin P (1994) Wavelet thresholding techniques for power spectrum estimation. IEEE Trans Signal Process 42(11):3126–3136
14. Percival D, Walden A (1993) Spectral analysis for physical applications: multitaper and conventional univariate techniques. Cambridge University Press
15. Riedel KS, Sidorenko A (1995) Minimum bias multiple taper spectral estimation. IEEE Trans Signal Process 43(1):188–195

16. Shao Y, Chang CH (2007) A generalized time–frequency subtraction method for robust speech enhancement based on wavelet filter banks modeling of human auditory system. IEEE Trans Syst Man Cybern 37(4):877–889
17. Thomson DJ (1982) Spectrum estimation and harmonic analysis. Proc IEEE 70(9):1055–1096
18. Virag N (1999) Single channel speech enhancement based on masking properties of the human auditory system. IEEE Trans Speech Audio Process 7(2):126–137
19. Walden AT, Percival DB, McCoy EJ (1998) Spectrum estimation by wavelet thresholding of multitaper estimators. IEEE Trans Signal Process 46(12):3153–3165
20. Yuan W, Xia B (2015) A speech enhancement approach based on noise classification. Appl Acoust 96:11–19

Balinese Character Recognition Using Bidirectional LSTM Classifier

Saad Bin Ahmed, Saeeda Naz, Muhammad Imran Razzak, Rubiyah Yusof and Thomas M. Breuel

Abstract The character recognition of cursive scripts always be provocative. The inherent challenges exists in cursive scripts captured researcher's interest to crop up the issues that surface in building a reliable OCR. There exists many ancient languages that require state of the art techniques to be applied on them. Every such language has its own inherent complex structure. We proposed Balinese character recognition system by Recurrent Neural Network (RNN) approach, so that their characteristics may get substantial attention from research community. The Balinese has Brahmic Indic ancestor having cursive writing style nearest to Devangri, Sinhala and Tamil. We employed BLSTM networks on Balinese character recognition.

1 Introduction

Balinese is considered as one of the ancient languages currently resides in Indonesia. This language is derived from Brahmi Indic in ancient India. Currently, it is speaking by 3.3 million people in Indonesia's Bali Island. Apparently it seems an ornamental in writing closer to Indic scripts of devanagri style as shown in Fig. 1. Balinese in its

S.B. Ahmed (✉) · M.I. Razzak
King Saud Bin Abdulaziz University for Health Sciences, Riyadh, Saudi Arabia
e-mail: isaadahmad@gmail.com

M.I. Razzak
e-mail: mirpak@gmail.com

S. Naz
Department of Information Technology, Hazara University, Mansehra, Pakistan
e-mail: saeedanaz292@gmail.com

R. Yusof
University of Technology, Johor Bahru, Malaysia

T.M. Breuel
University of Technology, Kaiserslautern, Germany
e-mail: tmb@iupr.com

P.J. Soh et al. (eds.), *Advances in Machine Learning and Signal Processing*,
Lecture Notes in Electrical Engineering 387, DOI 10.1007/978-3-319-32213-1_18

Fig. 1 Balinese handwritten text

true form is gradually fading specially in written mode. Most native speakers use it with the combination of other Indonesian languages. Balinese is not been used extensively in it's native ground rather it is extensively being used at transmigration areas outside its native region. Because of its capability of serving as communication medium outside Bali to make it believe that it may play a vital role in its survival. The Balinese main words borrowed from Javanese. The Kawi is considered as official language of Bali island.

The inherent ability of Balinese as a cursive language make it closer to other cursive languages like Chinese, Japanese, Arabic, Indic, Jawi etc. The challenges face by each cursive language is almost same but nature of script expose explicit struggle to that. The Balinese text is written from left to right and words are not separated by spaces as represented in Fig. 1.

In Balinese, there are basic characters and some supporting characters which must appear with base character. The association of such character is really a very challenging task to perform. There are various vowel symbols that may appear with a character to sound it as meaningful. The determination of such patterns with base character is a subtle task. The diacritical marks in Balinese script is not only dots instead, it has different shape characters known as diacritical marks. The complete

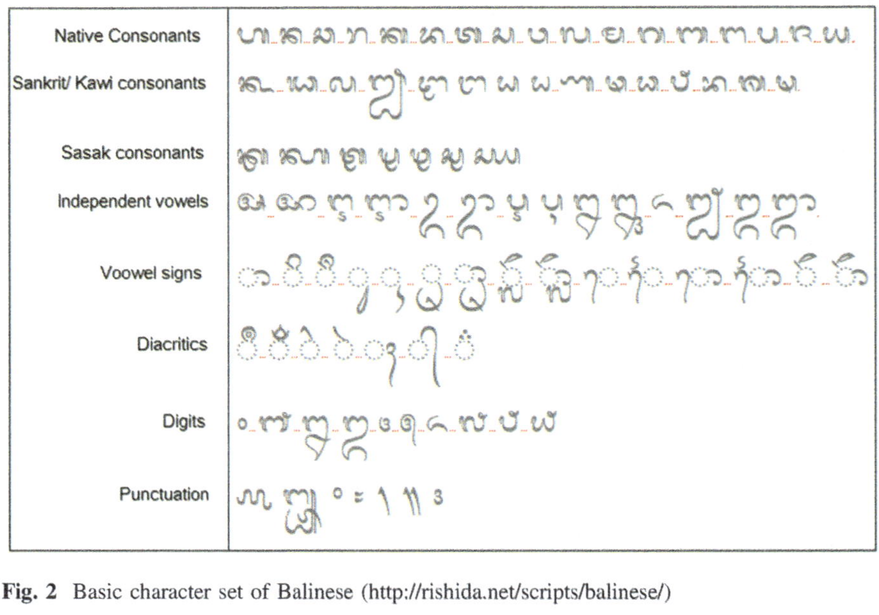

Fig. 2 Basic character set of Balinese (http://rishida.net/scripts/balinese/)

character set of Balinese is represented in Fig. 2. The basic motive of this research work is to explore challenges that have been identified in Balinese character recognition. The next section compiles the literature review. The description about dataset has been explained in Sect. 3. The preprocessing phase and feature extraction have been outlined in Sect. 4. Section 5 represents the methodology that has been followed during recognition and the experimental evaluation and results have been presented in Sect. 6.

2 Literature Review

Although very limited work has been done on Balinese script. But recently, very less research efforts have been made to bring ancient languages into light so that the interest shift of research community towards introduction of state of the art techniques on ancient languages would be possible. The manuscript on recognition of Balinese character represented by [1]. They gather dataset locally and extract features by principle component analysis technique. They computed Eigen vector and Eigen values on covariance matrix of training patterns and kept the highest Eigen vector value. According to Eigen values the known pattern is mapped into image space and weights were stored. They applied backpropagation algorithm as classifier. The evaluation has been performed on variation of different number of dataset at train, validation and test set as mentioned in their manuscript. They evaluate their proposed method by giving different number of hidden layer neurons and momentum value. They reported 96.7 % accuracy on Balinese character recognition.

Another manuscript by [2] proposed semantic feature and K-nearest neighbor algorithm was used for recognition of Balinese script. They extracted features of Balinese characters from KNN. The nearest neighbor value is used to compare with test patterns for the purpose to recognize the given pattern. They take height and width of a character into consideration. Moreover, they divided the characters in three parts such as upper, middle and lower part. They make histogram of each part and evaluate their proposed method on 54 number of test characters which itself considers less samples for evaluation. They reported 88.89 % character recognition accuracy.

Other than Balinese, there are various cursive languages that posed challenges to researchers due to their complex structure and nature. Arabic script is considered as one of such complex cursive language due to its character variability [3, 11]. There have been performed an intensive research in Arabic script recognition since few years. One such work is proposed by [4] for recognition of printed Urdu text. They used window based approach for feature extraction and employed bidirectional long short term memory networks as classifier. They reported 97.63 % accuracy on character recognition of Urdu language. Another work on offline handwritten Urdu recognition by BLSTM is proposed by [5]. They had taken samples from different users on plain paper and applied horizontal profile for text line segmentation. After extracting text line separately, they employed Otsu binarization technique for the purpose to remove noise from scanned images. The skew was corrected by determining the coordinate values. The window of *30 x 1* was used to traverse the image and train the given patterns. As BLSTM is segmentation free technique it implicitly segmented the given characters from text line. They reported significant accuracy on character level.

3 Data Set

It is obvious that data set plays a vital role to investigate the potential of state of the art classifiers. There are many recent dataset have been developed of cursive scripts [4, 6]. We proposed dataset of handwritten Balinese script that was gathered from 20 native Balinese speakers. All writers have written 15 text lines with same text given to them. The detail statistics about dataset is summarized in Table 1. The sample text were taken from booklet named Buku Penunjang Materi bahasa Bali—Widyasari, Sekolah Dasar Kelas written by Tri Agung.

The ground truth is defined by assigning labels to each pattern. Each symbol represents utf-8 encoded Balinese character. We identified 92 different classes of Balinese script. This was a tedious task to identify distinguished classes because same character may appear with vowel sign, with diacritical symbol or may appear as an independent entity. We consider all shapes of same character that may change its appearance, as different classes. The complex structure of Balinese consonants is depicted in Fig. 3. The determination of each shape is a complicated issue to address, as mentioned in figure below.

The dataset was divided into train, validation and test set as depicted in Table 2.

4 Preprocessing and Feature Extraction

The base lines were provided to authors so that less effort may serve for correcting the skew of a given text. The text was taken from blue pen. After taking samples, the pages were scanned. The yielded text was taken in blue color; therefore all other colors were considered as noise and be removed.

Table 1 Balinese dataset characteristics

Balinese dataset detail	Statistics
No. of writers	20
Number of text lines per page	15
Total number of text lines	300
No. of characters per page	~435
Total number of characters	~8700

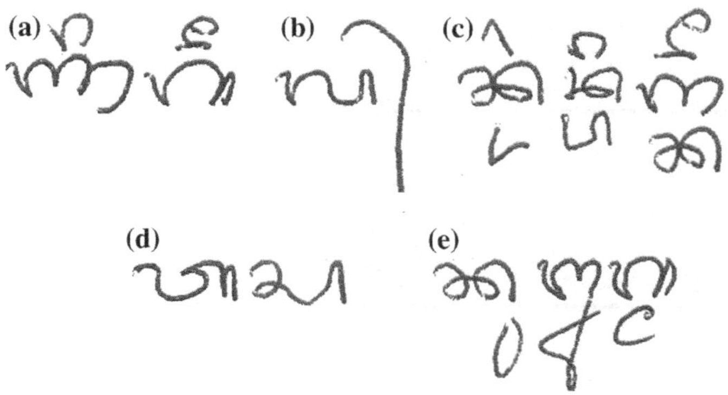

Fig. 3 Different variations of Balinese characters. **a** The native consonants occurred with the vowel signs having base and upper character, **b** The native consonant with diacritical mark, **c** This is so complex structure of Balinese characters. It has upper, base and lower parts. The upper part is a diacritical mark whereas base and lower parts are native consonant symbols, **d** Simple native consonant **e** native consonants with vowel signs at lower part

Table 2 Balinese dataset characteristics

Dataset distribution	No. of transcriptions (text lines)
Train set	150
Validation set	90
Test set	60

The taken samples were segmented into text lines by projection profile method. Algorithm 1 shows the steps that were performed to segment the text page into text lines.

Algorithm 1

1. The Grayscale image (Ig) is converted into Binary Image (Ib).
2. Horizontal projection (IHp) of Ib is computed, which is the sum of each row.
3. IHp is scanned if a non-zero number is found that position is marked as X (line upper row no) and then continue search for zero value if a zero value is found that position is marked as Y (line lower row no).
   Here X and Y indicate the text line positions.
4. The text line positioned at X and Y is cropped from Grayscale image (Ig).
5. The step 3 and 4 is repeated for the whole image.

We employed Otsu thresholding to remove the noise and convert it into gray scale. The skew of image was corrected by calculating variance of horizontal projection method. In this method the image is projected by different angles and only computes the variance of horizontal projection. The windowed approach was used to tune up gray scale pixel values corresponding to each transcription. The feature matrix was formed and passed to BLSTM classifier for training purpose. We used 30×1 window size to get features of given transcriptions as represented in Fig. 4.

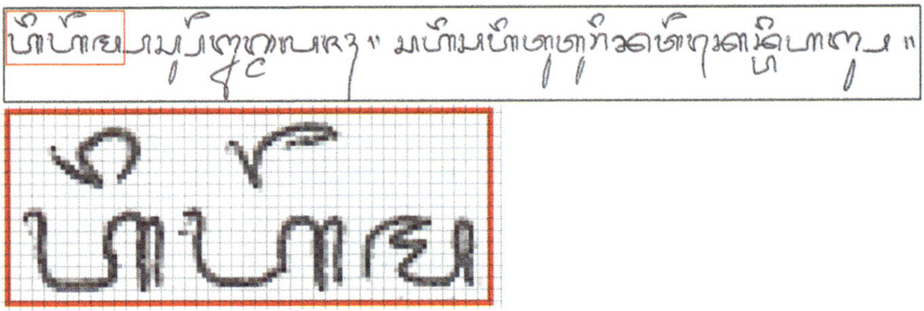

Fig. 4 Highlighted pixel values of corresponding transcription

5 Methodology

This section provides schematic method followed for evaluating Balinese script. The same text was taken from all individuals for the purpose to take variations of same characters that may occur due to individual's writing style. We tried to cover almost all Balinese characters, special symbols and numbers. We have provided text line image and ground truth of every transcription to BLSTM classifier as depicted in Fig. 5. As BLSTM is a RNN technique specialized for learning sequences in forward as well as backward direction. The simple RNN has a capacity to hold data for shorter period of time. When problem grows exponentially and becoming complex then simple RNN cannot handle to manage history of previous computations and resulting in a loss of sequence. To address this vanishing gradient problem [7] Long Short Term Memory (LSTM) was introduced [9]. Because of LSTM we may contain history as long as we need it and discard it later. The LSTM employ both in forward and backward direction.

The tuning parameters are number of hidden LSTM cells, momentum and learning curve values. As learned from recent research on BLSTM as evaluating

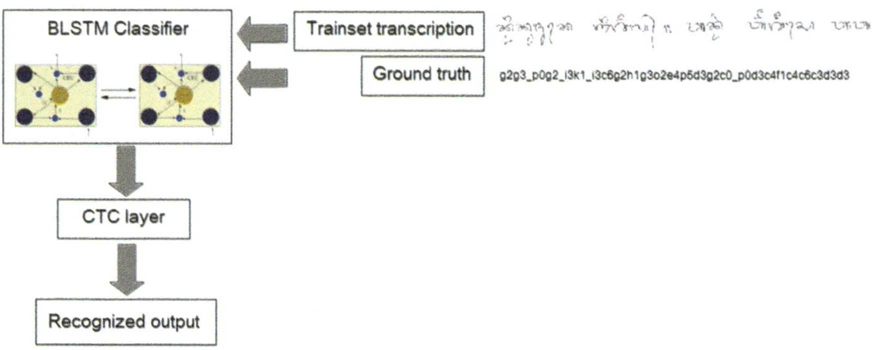

Fig. 5 Proposed methodology

technique [4, 5, 8, 11]. We noticed that every time best result was calculated on 100 LSTM cells. We have decided not to play with hidden layer LSTM cells instead we tried to get better result by tuning the value of learning curve while keeping fixed 100 LSTM cells and momentum 0.9.

By Connectionist Temporal Classification (CTC) output layer [10], the alignment of a sequence is no more difficult to maintain because of the prediction it makes at any point in time against input label until overall sequence of label become correct. Furthermore, the complete sequence of probability is estimated which remove the need for post-processing. The CTC has same number of units as number of input classes with additional blank symbol [7]. The CTC output defines all possible probabilities of occurrence of character symbol against input individual character symbol. The final probability is calculated by considering all probability of individual character symbol.

6 Evaluation

As discussed in previous section that we take learning rate instead of hidden number of neurons as a measure to evaluate the performance of given classifier. We performed experiments by giving various learning rates. By learning rate we can control the size of weight and bias value that changes during learning of training patterns. Learning is dependent on number of training patterns. If there would be large number of training patterns then it might produce desire results in comparison to provide less number of training samples having less variation. We divided our data set as 50 % training, 30 % validation and 20 % testing. With this division we got our desired results. But in our experiments, we tried to inquire, how learning rate can impact the recognition error. Table 3 represents training and validation error that has been reported during training process when provided different learning rates. The CTC error measure was used to evaluate the performance of training because CTC has fastest convergence capability than other character error measures [10]. The graph edit distance measure was used to count number of insertions, deletions and substitutions with respect to each test set transcription. The testing shows a promising results on this ancient cursive script.

Figure 6(a–d), shows the trend of learning curve that has been observed during training phase. As represented in Fig. 6a, the training curve maintains considerable distance from 0 but validation overlapped it at 60th epoch. It means a network is getting over-trained which incurs problem of over-fitting. In Fig. 6b, represents the ideal case of over-training or over fitting one can observe the difference between training and validation curves. It shows network gets trained on same patterns rather than on other variation. This is a reason that two curves are opposite to each other. Figure 6c represents ideal curves that have been made during training phase. The validation maintains reasonable distance from training curve. The best result is recorded on 0.01 learning rate. Figure 6d represents the same pattern of learning as Fig. 6a.

Table 3 Training and validation error with learning rate

No. of epochs	Training error	Validation error
Learn rate = 0.0001		
1	7.22	5.95
20	4.53	4.2
40	3.95	3.72
60	3.46	3.36
Learn rate = 0.001		
1	5.76	4.86
20	2.17	4.33
40	0.635	5.73
60	0.16	6.52
Learn rate = 0.01		
1	12.9	14.4
20	2.5	7.02
40	1.73	7.33
60	1.52	6.42
Learn rate = 0.1		
1	23.2	10.6
20	5.19	4.63
40	5.11	4.53
60	5.05	4.47

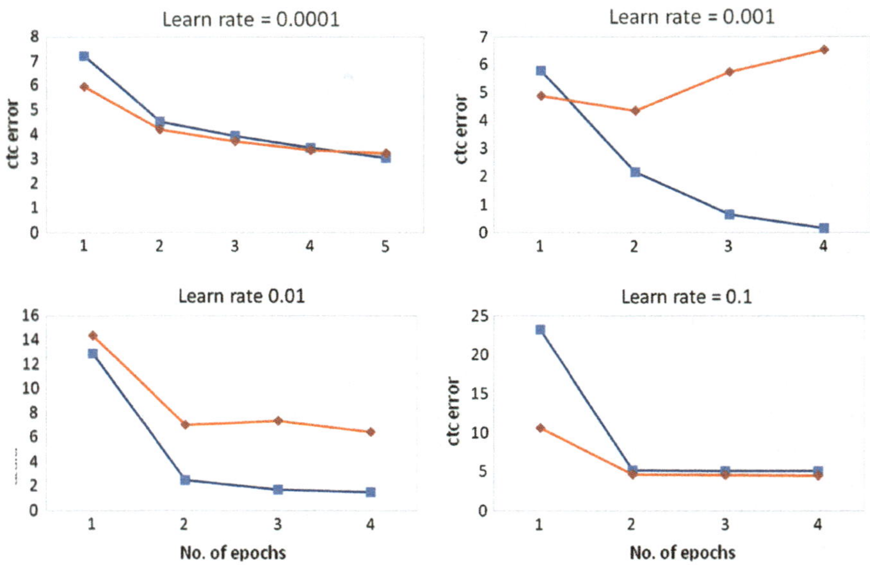

Fig. 6 Performance curve at different learning rates

6.1 Benchmarking of Balinese OCR

As we have observed in previous section that by changing in learning rate we can have varied recognition rate at character level as summarized in Table 4. In this section, we assumed Luh Prapitasar [1] work as a benchmark and evaluate BLSTM classifier on same setting and on same database distribution as mentioned in their manuscript. As momentum and learning rate also contribute in network learning. In our experiments we fixed momentum as 0.9 and learning rate as 0.01. The dataset divided into 40 % as training, 30 % as validation and 30 % as testing. The acquired result by this distribution is mentioned in Table 5.

The experimentation results with the dataset distribution ratio of 50 % as training, 25 % as validation and 25 % as testing are depicted in Table 6. Whereas, the acquired results on 60 % as training, 20 % as validation and 20 % as testing has been summarized in Table 7.

The BLSTM is a context learning approach that may apply on sequential data like in cursive scripts, to get maximum accuracy on intricate data. If more variation

Table 4 Learning rate statistics

Learning rate	Time (s)	Recognition rate	No. of epochs
0.0001	1620	58.01	94
0.001	1427	80.14	87
0.01	1274	98.75	110

Table 5 Experiment result on 40:30:30 dataset distribution on BLSTM classifier and its comparison with BackPropagation Algorithm

Hidden layer neuron/cell	No. of epochs		Accuracy	
BP/BLSTM	BP	BLSTM	BP	BLSTM
10	247	93	73.45	56.72
25	240	102	92.86	73.24
50	325	88	93.42	86.31
75	238	116	93.47	93.58
100	229	123	93.49	96.17

Table 6 Experiment result on 50:25:25 dataset distribution on BLSTM classifier and its comparison with BackPropagation Algorithm

Hidden layer neuron/cell	No. of epochs		Accuracy	
BP/BLSTM	BP	BLSTM	BP	BLSTM
10	259	89	88.24	61.38
25	239	94	94.03	79.02
50	260	105	95.71	89.05
75	235	113	94.67	95.21
100	246	117	94.93	97.89

Table 7 Experiment result on 60:20:20 dataset distribution on BLSTM classifier and its comparison with BackPropagation Algorithm

Hidden layer neuron/cell	No. of epochs		Accuracy	
BP/BLSTM	BP	BLSTM	BP	BLSTM
10	276	73	87.14	73.92
25	251	97	95.78	81.20·
50	245	101	96.00	93.97
75	271	113	96.63	97.91
100	257	110	96.23	98.75

of data is provided during training then the accuracy may stretch to its maximum as observed from performed experiments result. While the BLSTM accuracy on less number of hidden layer cells means that the network does not have enough hidden cells to learn various contexts. Initially, BLSTM classifier has less accuracy due to scarcity in context learning. The maximum accuracy is been reported on 100 hidden layer cells which is greater than the accuracy reported by [1].

7 Conclusion

To recognize ancient characters always been a enthralling endeavor for research community. Balinese script has such a potential that motivate pattern recognition researchers to dig-out solutions for segmentation of such cursive scripts. As explained earlier in manuscript that Balinese has very complex structure and nature of representation. For such complicated scripts the use of segmentation free approaches make the task easier. The powerful structure of BLSTM makes it possible to learn the complex sequences like we have in Balinese without segmentation of characters. Numerous feature extraction approaches may apply to get more accurate results on this intricate script. There exist other classifiers that may help to get desired results other than RNN. But as mentioned earlier that key point of RNN is its sequential learning and for recognition of such subtle scripts, context is more important to learn.

Acknowledgment We would like to say thanks to Mr. Luh Prapitasar and all contributors for providing us Balinese handwritten text samples.

References

1. Prapitasari LPA (2010) Off-line balinese character recognition based on backpropagation neural network. In: Universitas Gunadarma, Proceeding seminar Ilmiah Nasional KOMMIT Nov 2010. ISSN: 1411–6286
2. Sudarma M, Darma WAS (2014) The identification of Balinese scripts' characters based on semantic feature and K nearest neighbor. Int J Comp Appl 91(1):0975–8887

3. Graves A (2012) Offline Arabic handwriting recognition with multidimensional recurrent neural networks. In: Margner V, Abed HE (eds) Guide to OCR for Arabic scripts, ch. 12. Springer, pp 297–231
4. Hasan A, Ahmed SB, Rashid SF, Shafait F, Breuel TM (2013) Offline printed Urdu Nastaleeq script recognition with bidirectional LSTM networks. In: 12th ICDAR, Aug 2013, pp 1061–1065
5. Ahmed S, Naz S, Razzak MI, Rashid SF, Afzal MZ, Breuel TM (2015) Evaluation of cursive and non-cursive scripts using recurrent neural networks. Neural Computing Appl 26(3)
6. Ahmed SB, Naz S, Salahuddin, Razzak MI, Khan AA, Umar AI (2015) UCOM offline dataset—an Urdu handwritten dataset generation. In: International Arab Journal of Information Technology (IAJIT), (in press)
7. Graves A (2008) Supervised sequence labelling with recurrent neural networks. Ph.D. dissertation, Technical University Munich
8. Graves A (2012) Supervised sequence labelling with recurrent neural networks ser. In: Studies in comput intelligence. vol 385, Springer
9. Hochreiter S, Schmidhuber J (1997) Long short term memory. In: Neural computation
10. Graves A, Fern S, Gomez F, Schmidhuber J (2006) Connectionist temporal classification: labelling unsegmented sequence data with recurrent neural nets. In: ICML
11. Naz S, Hayat K, Razzak MI, Khan SU, Anwar MW, Madani SA (2014) The optical character recognition of Urdu-like cursive scripts. Pattern Recognit 47(3)
12. Naz SA, Umar AI, Shirazi SH, Ahmed SB, Razzak MI, Siddiqi I (2015) A review of segmentation techniques for recognition of Arabic-like scripts. Education and Information Technologies 20(2), Springer

An Automated Multiple Choice Grader for Paper-Based Exams

Abrar H. Abdul Nabi and Inad A. Aljarrah

Abstract In this paper an automated multiple choice grader for paper-based exams is implemented. The system consists of two main parts, a software program and a document feeder scanner. The exam papers are fed to the scanner which scans them one by one and send them as an input to the software. The software program recognizes the student Identification Number (ID) and the answers for each exam paper and reports the final results in an Excel sheet. The system starts by applying an aligning procedure and segmenting the scanned image in order to extract form number, student ID, and answers boxes, then a pre-processing step that handles all irregular cases of input is implemented; where in this step a best possible shape that results in the highest recognition accuracy is gained. After getting a proper separated characters and numbers, a feature extraction process is applied on each character/number to calculate its feature vector. The feature vector is then compared with templates of feature vectors for each of the answers choices and numbers with their variations, where both characters and numbers are in English language. After recognizing all the answers and all ID number digits; the system starts grading the student paper and comparing student answer with the pre-entered key answers. A recognition rate of 95.58 % is attained.

1 Introduction

The procedure that recognizes handwritten characters and numbers in scanned images is referred to as Optical Character Recognition (OCR) [1, 2]. The OCR of our system is implemented to meet system domain, where the system is designated for grading the multiple choice exam papers. Any OCR system consists of many

A.H. Abdul Nabi (✉) · I.A. Aljarrah
Faculty of Computer and Information Technology,
Jordan University of Science and Technology, Irbid, Jordan
e-mail: ahabed09@cit.just.edu.jo

I.A. Aljarrah
e-mail: inad@just.edu.jo

P.J. Soh et al. (eds.), *Advances in Machine Learning and Signal Processing*,
Lecture Notes in Electrical Engineering 387, DOI 10.1007/978-3-319-32213-1_19

phases or layers, a specific function is performed in each layer, and the output is forwarded as an input to the next stage.

The system OCR phases are: preprocessing, segmentation, feature extraction and classification. First, the input to the system is preprocessed by converting the gray-level image to a binary image; since binary images are much easier to handle than gray-level images through representing each pixel by only one bit either zero or one. The next step in the preprocessing involves applying several filters on the binary image to eliminate different noise such as scars and isolated pixels. Then the image will be inverted so that the pixels that represent the object/character have values of ones while the background has values of zeroes.

The system has two parts according to its domain, the first one is the student ID recognizer, and the other one is the answers recognizer. The segmentation stage will be applied only on student ID, where digits are split from each other. The feature extraction phase includes calculating the features vector, which consists of four features: Polar Shape vector, Euler Number, Eccentricity and Solidity properties. The last phase is the classification; in which the resultant feature vector is matched to pre prepared database of feature vectors that includes samples of each of the English numbers for student ID and the first five English letters that represent the options of answers in all their forms. A resizing factor is applied to the input to guarantee a certain scale; the same as the template. Two classification methods are performed in parallel, one is to directly perform Correlation Coefficient formula to detect the closest template pattern that highly matches the input depending on the resultant values; which is the lowest hence the closest. This formula does not need any feature extraction; it is applied directly after the preprocessing phase finished. Meanwhile we perform in parallel the features extraction techniques to further more classify them later on by calculating the nearest neighbor distance. A voting procedure between the results of both paths is applied to reach the final decision of the classification. In other words, the classification procedure we perform is the nearest neighbor between each extracted vector of the input and the calculated one for each template simply by calculating the absolute distance. This problem is one-to-many classification; meaning that we calculate the difference for each input against all stored templates, then we compare these differences against previously selected thresholds to reach the final decision. Our feature vector is formed as the following; we extract shape properties such as eccentricity, solidity, Euler number, polar shape vector and/or detection of corners, edges and lines, then we concatenate these values into one feature vector which is the final form that is ready for classification. We conduct two experiments where the first one we use absolute difference, and in the other experiment we perform the Euclidean distance [3, 4], the results from both experiments were almost the same, so we announce the results for the first one only due to the space limitation.

The threshold were selected carefully according to the first order statistics such as the mean; where for each class type (number or letter) we collect many templates covering almost all the possible variations that may occur in the real time of writing, then we extract the feature vector for all these templates and calculate the mean vector per each category. For example, for the letter A class, we average all the features related to templates that are from this class; this mean vector form the final threshold value that is ready to be used during classification. For a given input image; we can compare its feature vector against all the mean vectors for all classes; the minimum result indicates the nearest neighbor which is the winner to join this input to it. The results of our system are performed on test images that are selected from real exam experiments, noting that there is no any intersection between our templates dataset and the tested images.

To improve the accuracy of the system, the number of the samples in the database should increase in order to reduce the rate of misclassifying characters.

The rest of this paper is organized as follows: Sect. 2 presents the related work. Section 3 discusses the proposed methodology. Section 4 talks about the experiments and gives some experimental results. Finally, the conclusion and the future work are presented in Sect. 5.

2 Related Works

The development of Optical Character Recognition (OCR) can be traced back to the work performed on two main applications which are: expanding telegraphy and creating reading devices for the blind [5]. Then the OCR engine has been developed and widely used in many applications; such as: data entry for business documents, e.g. check, passport, invoice, bank statement and receipt, automatic number plate recognition, making textual versions of printed documents like book scanning, making digital or electronic images of printed documents searchable like Google books, and converting handwriting in real time to control a computer which is known as pen computing, and many other applications that operate mainly on the OCR engine [1, 2, 6–11]. Most of the recent and previous work that is done in developing any OCR software usually contain a common similar pre-processing step before recognizing the characters or words in a given input image; this step begins with image alignment to make the lines of the text perfectly horizontal and vertical, then applying some filters that smoothen the edges and reduce or remove the noisy pixels, then a binarization step is applied on the image to convert it into a binary image; still this step is not needed in some algorithms [12], after that the non-glyph lines and boxes should be removed and cleaned up, then identifying the major objects such as paragraphs, columns, captions, etc; this is often called "Zoning".

Then detection methods are used to detect words and characters; in case of the input being an image with multiple lines; then words should be segmented to get separated, after that the processing should handle character after character; by applying some segmentation techniques, but still the problem of processing connected characters due to any artifacts or noises is a quite challenging in this step; since these characters should be isolated correctly, as the same problem of having a single character that is broken into multiple segments; in this case the character should be connected correctly. Finally, a scaling factor should be applied to unify the size of the characters in order to ease the recognition process against the stored templates in later phases. Mainly two approaches are used in character recognition [13]; Matrix matching and Feature extractions. The first method tends to compare the input image to a stored glyph or to use pixel-by-pixel basis as in the pattern matching or recognition algorithms; but this technique is still constrained to the variety and capacity of the database that contains the stored templates [14, 15].

The other technique tend to decompose the glyph or the input character image into set of features; like lines, lines intersections and directions, closed loops, etc., then these features are concatenated to represent one Feature Vector; which can be used in many classifiers to decide the similarity of the input feature vector and the feature of the stored templates. The most commonly used classifiers in the domain of OCR's are the Nearest neighbor classifiers such as the k-nearest neighbors algorithm [16], which compares the image features with stored glyph features and choose the nearest match [17]. Many software systems have been developed and released; the most well-known are Tesseract and Cuneiform. Both of these software use a two-pass approach; where the additional second pass uses the letter shape that is recognized with high confidence on the first pass; in order to better recognize the remaining letters on the second pass; this can give a very good performance in case of low-quality scans where the font is faded or blurred [18–20].

Many other types of software were developed using different algorithms in extracting features or even in the similarity measure phase. Our software is customized and dedicated for the purpose of recognizing certain characters in the process of grading a multiple choice exam scanned forms.

Different than the previous works, in this paper we propose to design and implement our system following simple classification procedures and selecting various types of features that can work well in our target domain; resulting in gaining compromised speed and accuracy and in the same time satisfying the target need of such software. The types of features that we mainly extract are of shape-related types [21], which can perform better against Tesseract software.

3 Implementation Details and Methods

After the scanner finish scanning all papers, the system then starts processing them image by image, so the input of the system is a gray level scanned image representing the student exam paper, and the final output after grading all students' papers is an Excel file sheet, as shown in Fig. 1.

A detailed view for three subsystems; the Parser and both ID and Answers OCR systems is illustrated separately in Figs. 2 and 3. Firstly, the image goes through the Parser five steps, and then the resultant is forwarder to either ID recognizer or Answers recognizer. The output of the Parser is belong to three kind of block image: ID block, Form number block and Answer block, both student ID and form number blocks are forwarded to ID recognizer, and the answer block is forwarded to Answer recognizer. The recognizers almost have the same basic steps

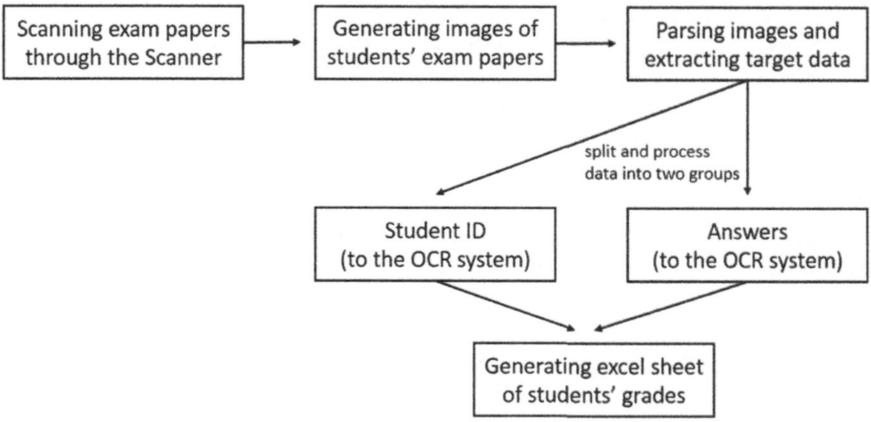

Fig. 1 General view of the system

Fig. 2 Parser system

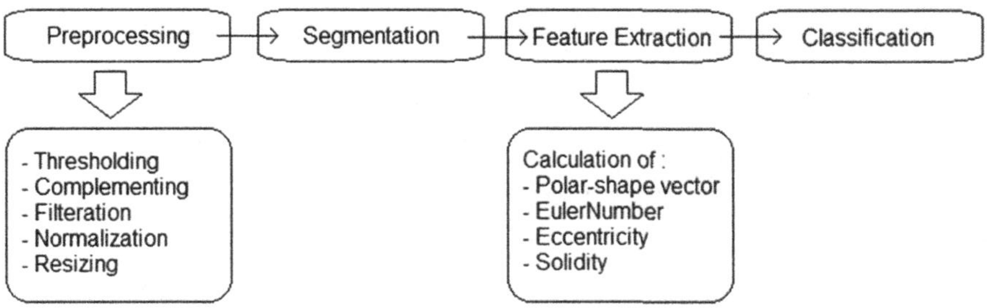

Fig. 3 Proposed OCR system

functionality but with difference of that the Segmentation step is eliminated from the Answer OCR, also some minor changes in some steps implementations are applied. The system keeps tracking and maintaining every student ID and the answers that belong to it in order to grade it later depending on the answer key which is entered previously by the user of the system. Each of any steps will be discussed independently later on.

In order to make sure that a small misalignment in the scanning process will not result in a wrong location of the required fields (form number, student ID, and answers); an aligning procedure is applied on the input image. Some calculations are made on a stored template exam paper to figure out the distances between a predetermined reference point and the answer boxes and the ID box. This reference point is located in the center of the upper line of the form number block. Since these distances are fixes relatively to the reference point, all the difficulty is stated in how to locate properly this reference point, upon this step the remaining part of locating answers blocks and ID block will be very easy.

This reference point is located in the center of the upper line of the form number block. Since these distances are fixed relatively to the reference point, all the difficulty is stated in how to locate properly this reference point, such trick can ease the locating of the remaining part of answers blocks and ID block; however the key challenge is how accurately detect this point, and this solved by enforcing small constrain on the exam paper format.

The format of the exam paper should have some properties to ease the process of parsing; the most important one is that the boxes (answers, ID and form #) should be surrounded by a border with thickness of 2px minimum, such a thickness can

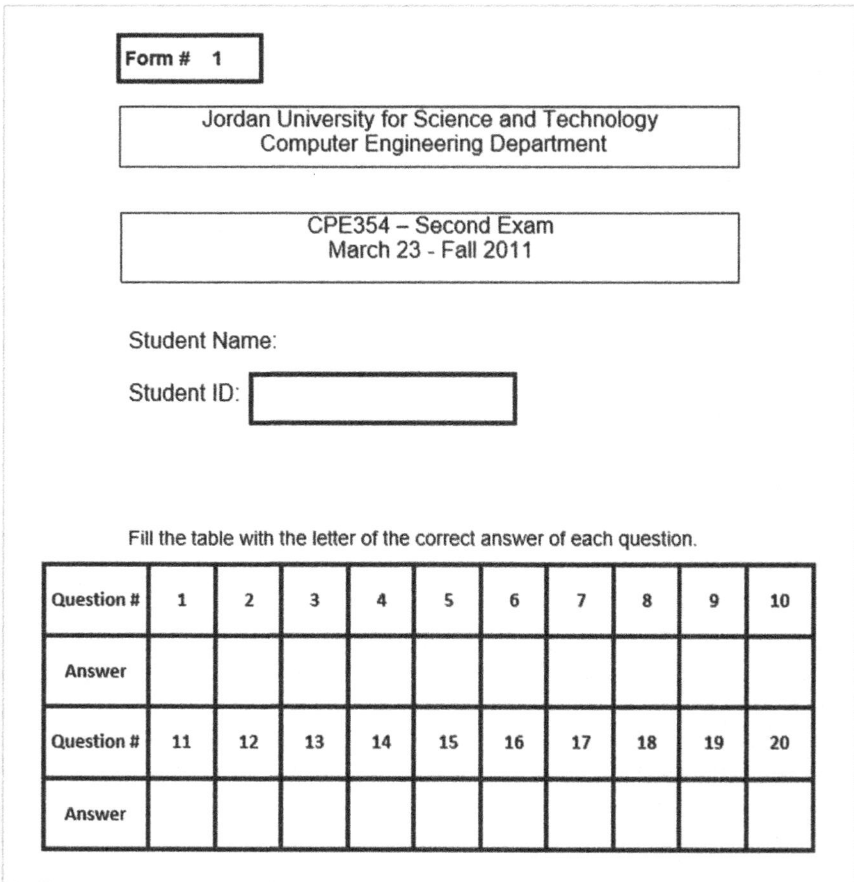
Form # 1

Jordan University for Science and Technology
Computer Engineering Department

CPE354 – Second Exam
March 23 - Fall 2011

Student Name:

Student ID:

Fill the table with the letter of the correct answer of each question.

Question #	1	2	3	4	5	6	7	8	9	10
Answer										
Question #	11	12	13	14	15	16	17	18	19	20
Answer										

Fig. 4 A sample of the multiple choice exam paper

ease the line detection procedure which further locating the reference point comfortably.

The classification is performed by calculating the correlation coefficient between the input sample and the stored templates. Ten template samples are used in this work for each symbol (letter or digit). Figure 4 shows an example sample that is used throughout our experiments in the proposed software system.

4 Experimental Results

The proposed approach has been implemented and tested using Matlab alongside C# environment. The system is operated on different images taken from 27 exam papers belong to students from a random class, where these exam papers have been scanned and saved as images for testing purposes. For each sample (exam paper), the number of correctly recognized student answers and ID digits are counted and the recognition rate is computed. The recognition rate for the form number is 100 %; this rate is reached because the form number is written in the computer English language and printed out in the exam paper, so it is not from human writings.

The total time of processing each input image (starting with scanning the exam paper and ending with grading it) is 1–4 s, for the first couple of exam papers, the scanner initially spend around 3 s due to mechanical issue, then the time difference is decreased as the scanner begin to be faster for scanning the rest of exam papers. Our processing time for recognition is only 0.5–1 s on normal office desktop computer [6 GB RAM, i7 2.20 GHz per core]. Many other OCR engines spend more time to process such input image, and may generate higher recognition accuracy, knowing that some use higher hardware capacity, or even too much complicated fancy preprocessing or post processing techniques to reach such very high accuracy. However, the nature of our problem is quite different from the usual character recognition problem; since recognizing handwriting of exam papers are extremely hard, and du to the fact that the nature of human in the exam conditions tends to have huge pressure and stress which consequently affect his writings, which leads to have higher variations and noises, that means more difficult processing in such scanned images from these exam papers is needed. However, our main contribution is to design a holistic recognition system application that can potentially release as mature product to toggle the problem of grading exam papers with automated version software, trading off two issues; accuracy and performance.

Recently, some organizations conduct student exams using online-examination servers as an alternative solution to handwritten ones. However, there are many accompanying limitations to such alternative, mainly: (1) online exams usually suffers from limited resources as the number of students in each exam session is huge; this limits the capacity of resources that the organization can offer during that session; (2) typical sudden power failures which thus requires re-conducting that exam session and thus affecting mainly the students psychology state and adding more pressure and stress on them, in addition to the waste of time and resources that are needed to be re-allocated to re-conduct that exam session. Such limitations rise the need to develop our proposed automated system which can be replaced such online alternative system limitations and in the same time retaining the benefits of the handwritten multiple-choice exams.

Even though our system accuracy results are not 100 % and can generate some erroneous grades, this is usually can be solved by simply depending on the students to flag any erroneous grades; as the original papers well be distributed back to the students alongside their grades. Thus, any erroneous result will be identified right away by the student and get returned back to the teacher who can shortly revise the erroneous paper by comparing the exam paper results with the key answers and grade it to give it back to the student. However, this automated software system prediction faults is similar to the typical human faults; as the teacher can always have a potential erroneous grade by mistaking the answers by either reading it wrongly or unintentionally mistaken the key answer due to high volume of papers to be graded that can add huge pressure and stress on the teacher during the grading process.

But this is the first release of this software, still there are many enhancements can be integrated In this work such as increasing the size of templates dataset or performing many experiments on more diverse testing sets and etc., however many possible enhancements can be conducted to reach better performance and more reliability in the future.

Table 1 Results of the answer's OCR according to the 27 exam papers test set

Symbol	All samples	Wrong	Correct
A	114	10	104
B	115	13	102
C	102	2	100
D	101	3	98
E	108	1	107
Total	540	29	511
Average	100 %	5.37 %	94.63 %

Table 2 Results of ID digits according to the 27 exam papers test set

Digit	All	Wrong	Correct
0	102	2	100
1	69	0	69
2	36	1	35
3	6	1	5
4	3	1	2
5	10	0	10
6	6	0	6
7	35	0	35
8	7	0	7
9	23	3	20
Total	297	8	289
Average	100 %	2.69 %	97.31 %

Tables 1 and 2 report the results for some of these samples. As can be figured out from Table 1, the estimated overall recognition rate for answers samples is 94.63 %. The results reported in Table 2 show a recognition rate of 97.31 % for ID digits samples. The total Error rate is 4.42 % mean while the total Accuracy rate is 95.58 %.

5 Conclusion and Future Work

In this paper a software program that can be used in conjunction with a scanner is developed to implement a multiple choice exam grading system. The main prospective of the system is to support instructors with new tools to facilitate and speed up the educational process. It is believed that this project can be translated into a product that can deal with a real necessity. Approximately, a recognition accuracy of 94.63 % is accomplished so far for characters group, and about 97.31 % for numbers group, but this is just the start and the accuracy is expected to improve considerably by working more on it, like increasing the number of template samples and by employing more features and classifiers.

References

1. Plamondon R, Srihari SN (2000) On-line and off-line handwriting recognition: a comprehensive survey. IEEE Trans Pattern Anal Mach Intell
2. Dreuw P et al (2012) RWTH OCR: a large vocabulary optical character recognition system for Arabic scripts, guide to OCR for Arabic scripts
3. Hyvärinen A et al (2009) Natural image statistics. Springer
4. Burger W et al (2011) Principles of digital image processing: fundamental techniques. Springer
5. Schantz HF (1982) History of OCR, optical character recognition. Recognition Technologies Users Association
6. Margner V, Abed HE (2007) Arabic handwriting recognition competition. In: ICDAR'07: proceedings of the ninth international conference on document analysis and recognition (ICDAR 2007), vol 2. IEEE Computer Society, Washington, DC, USA, pp 1274–1278
7. Espana-Boquera S et al (2011) Improving offline handwritten text recognition with hybrid HMM/ANN models. IEEE Trans Pattern Anal Mach Intell
8. Jayadevan R et al (2011) Offline recognition of Devanagari script: a survey. IEEE Trans SMC-Part C Appl
9. Romero VN et al (2013) The ESPOSALLES database: an ancient marriage license corpus for off-line handwriting recognition. Pattern Recognit
10. Gimenez A et al (2014) Handwriting word recognition using windowed Bernoulli HMMs. Pattern Recognit Lett
11. Bluche T et al (2013) Feature extraction with convolutional neural networks for handwritten word recognition. In: International conference on document analysis and recognition (ICDAR)
12. How does OCR document scanning work?. Explain that stuff, 2012. http://www.explainthatstuff.com/how-ocr-works.html
13. http://www.dataid.com/aboutocr.htm

14. Aljarrah I, Al-Khaleel O, Mhaidat K, Alrefai M, Alzu'bi A, Rababah M (2012) Automated system for arabic optical character recognition with lookup dictionary. J Emerg Technol Web Intell 4(4):362–370
15. Aljarrah I et al (2012) Automated system for Arabic optical character recognition. In: Proceedings of the 3rd international conference on information and communication systems (ICICS '12), Article 5:6 pages, ACM, New York, USA
16. http://codingplayground.blogspot.sg/2010/01/nearest-neighbour-on-kd-tree-in-c-and.html
17. The basic pattern recognition and classification with openCV|Damiles, 2013. http://blog.damiles.com
18. Smith R (2007) An overview of the tesseract OCR engine. In: International conference on document analysis and recognition (ICDAR), IEEE Computer Society
19. Marosi I (2007) Industrial OCR approaches: architecture, algorithms and adaptation techniques. In: Document recognition and retrieval XIV, the international society of optics and photonics SPIE
20. Kay A (2007) Tesseract: an open-source optical character recognition engine. Linux J
21. Mingqiang Y, Kidiyo K, Joseph R (2008) A survey of shape feature extraction techniques. Pattern Recognition Techniques, Technology and Applications, Vienna
22. Resig J (2009) OCR and neural nets in JavaScript, 2009. http://Ejohn.org
23. Nixon M, Aguado A (2002) Feature extraction and image processing, 1st edn
24. Theodoridis S, Koutroumbas K (2003) Pattern Recognition, 2nd edn. Elsevier, San Diego, Calif, USA
25. http://www.codeproject.com/Articles/1376/NET-TWAIN-image-scanner
26. http://code.google.com/p/tesseract-ocr/
27. http://www.mathworks.com/products/image/
28. Haralick R et al (1992) Computer and robot vision. Addison-Wesley Publishing
29. Duda R et al (2001) Pattern classification. Wiley
30. Pratt W (2001, 2007) Digital image processing. John Wiley and Sons
31. Bow ST (2002) Pattern recognition and image preprocessing. Marcel Dekker
32. Graves A, Schmidhuber J (2009) Offline handwriting recognition with multidimensional recurrent neural networks. In: Neural Information Processing Systems (NIPS)

Nonlinear System Modelling Utilizing Second Order Augmented Statistics Complex Value Algorithm

Chukwuemena Cyprian Amadi, Bukhari Che Ujang and Fazirulhisyam Bin Hashim

Abstract The recently introduced augmented complex nonlinear gradient descent (ACNGD) algorithm for complex domain adaptive filtering which utilises the full second order statistical information is shown to be suitable for the processing of both circular and noncircular signals. By virtue of the underlying widely nonlinear model, the complex nonlinear gradient descent (CNGD) is shown to successfully model conventional system, however, unable to model the widely nonlinear system, and the ACNGD is capable to model both the conventional and widely nonlinear system. Simulations in adaptive modelling context for signals with different probability distributions and degrees of circularity support the analysis.

Keywords Widely nonlinear modelling · Augmented complex statistics · Augmented (CNGD) · Improperness · Nonlinear systems · Circular and noncircular complex signals · System modelling

1 Introduction

Signals processing in terms of circular and noncircular data, plays a vital role in wireless communication development [1], radar and environmental science [2]. In the meantime, machine learning theories and novel signal processing have been developed to address the arising challenges in circular and noncircular signal processing.

C.C. Amadi (✉) · B.C. Ujang · F.B. Hashim
Department of Computer and Communication Systems Engineering,
Universiti Putra Malaysia, Serdang, Malaysia
e-mail: amadi1@live.com

B.C. Ujang
e-mail: bukhari.ujang@gmail.com

F.B. Hashim
e-mail: fazirul@upm.edu.my

P.J. Soh et al. (eds.), *Advances in Machine Learning and Signal Processing*,
Lecture Notes in Electrical Engineering 387, DOI 10.1007/978-3-319-32213-1_20

In the past years, much noticeable research effort was put to develop new efficient approaches for noncircular dilemma, however in practical point, circular models are applied and adapted in a suitable way. A question needs to be raised, to find a definitive answer whether the nonlinear model system would satisfy the condition of both circular and noncircular signals. T. Adali, was able to conclude an important result in that direction, where nonlinear model exhibits optimal performance for the modelling of linear signal, particularly in the presence of noise [3]. Many benefits can be noticed in the usage of nonlinear models for linear noisy dilemma especially in sonar literature and radar systems [4].

In many technologies, the signals are sometimes complex-valued [5]. In order to process such signals, complex least mean square (CLMS) algorithm has been introduced in 1975 to fulfil that need. Since that time complex-valued adaptive filters have attracted various researchers. However, reducing the computational complexity and performing fast convergence are important issues in processing complex valued signals. Hence a prominent gradients based algorithms; complex nonlinear gradient descent (CNGD) and augmented complex nonlinear gradient descent (ACNGD) are selected. Both algorithms proved to be stable and robust, but with enhanced performance for the ACNGD over the CNGD due to the existence of the augmented statistics. The "augmented" [6] has path ways for the efficient possibility of designing adaptive filtering algorithms appropriate for circular and noncircular signal processing.

Both of these algorithms have been established in signal processing. Yet, the analysis on the performance in system modelling background is still a subject of ongoing research [3, 5, 11]. In this paper, the nonlinear system and widely nonlinear system are used for the modelling performance, evaluation of CNGD and ACNGD algorithm.

2 Element of Augmented Complex Statistics

Augmented statistics in the complex domain are estimated to be straight forward expansion of those in real domain. For example, the covariance matrix of a zero mean complex vector, Z is normally considered to be the extension of the real covariance $E\{xx^T\}$, however, when the transpose operator $(.)^T$ is replaced with hermitian operator $(.)^H$ it becomes $E\{xx^H\}$ [7]. This is only true for circular complex data and does not hold true in many application [8].

The augmented complex statistic accounts for both covariance and pseudo-covariance matrix by which for any complex random vector x $\epsilon\mathbb{C}^m, E\{x\} = 0$. The covariance matrix can be define as,

$$C_{xx} = E\{xx^H\}, P_{xx} = E\{xx^T\} \tag{1}$$

where C_{XX} is the covariance matrix and P_{XX} is the pseudo-covariance matrix [9]. To allow the use of all data present in the complex random vector, the augmented complex vector x^a can be define as,

$$x^a = \begin{bmatrix} x \\ x^* \end{bmatrix} \quad (2)$$

where $(\cdot)^*$ represents the complex conjugate operator. Then the covariance matrix C_{XX}^{aa} is given by,

$$C_{x^a x^a} = \begin{bmatrix} C_{xx} & P_{xx} \\ P_{xx}^* & C_{xx}^* \end{bmatrix} \quad (3)$$

This information contains both the covariance and pseudo-covariance matrix of x^a. For circular complex data, $P_{XX} = 0$, but for non-circular data, only the C_{XX} and not the P_{XX} matrix results in under modelling [10]. To design adaptive filter suitable for general complex processes, the second order statistical information present within the signal of the ACNGD [11] will use the augmented complex vector as the input to the filter.

3 Data Driven System Modelling Algorithm

The adaptive filter model structure shown in Fig. 1, operates in a system modelling setting, consists of x(n) which is the random circular or noncircular input to the adaptive filter, y(k) is the filter output signal and d(k) is the desire signal from the system output.

$$e(k) = \frac{1}{2}|e(k)|^2 = \frac{1}{2}|d(k) - y(k)|^2 \quad (4)$$

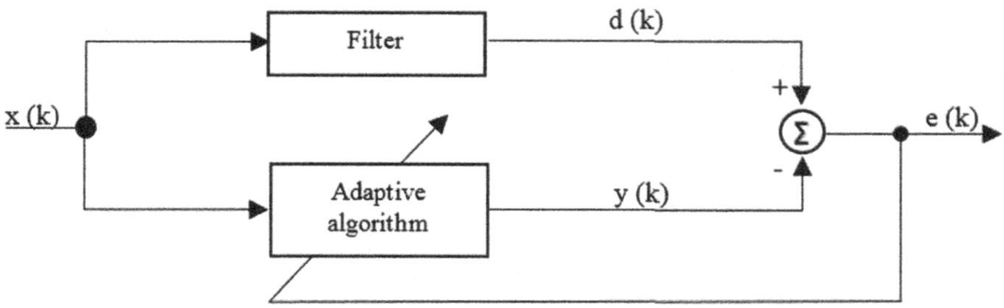

Fig. 1 System modelling block diagram

where e(k) is dissimilar between the desire signal output and the filter output signal used to update the weight vector to reduce cost function. The weight update is shown as,

$$h(k+1) = h(k) + \mu \nabla_W e(k) \tag{5}$$

where μ is the real-valued learning rate, ∇ is the gradient and e(k) is given as (4). The CNGD and ACNGD algorithms operate on the principles of (4) and (5). However, the ACNGD model further enhances the CNGD model to fully capture and utilizes the second order statistics. The algorithm weight update vector uses a stochastic gradient based adaptation which is discussed in the next section.

4 The CNGD and ACNGD Algorithm

The standard output of the conventional nonlinear finite impulse response filter trained by the CNGD algorithm is given as,

$$y(k) = \Phi(x^T(k)w(k)) \tag{6}$$

and the weight update is given as,

$$w(k+1) = w(k) + \mu\{\Phi'(k)\}^* e(k)x^*(k) \tag{7}$$

where Φ is the fully nonlinear function chosen from a class of fully complex nonlinear function [3], e(k) is the output error, μ is the learning rate and y(k) is the output signal, w(k) is the weight update vector at time instant k, while $\Phi'(\cdot)$ is the derivative of the nonlinear activation function. The output of the widely nonlinear version of the CNGD, called the augmented CNGD (ACNGD), proposed in [12], [13] can be summarized as,

$$y(k) = \Phi(\underbrace{x^T(k)h(k)}_{\text{Standard part}} + \underbrace{x^H(k)g(k)}_{\text{Augmented part}}) \tag{8}$$

and the weight update is given as,

$$h(k+1) = h(k) + \mu\{\Phi'(k)\}^* e(k)x^*(k) \tag{9}$$

$$g(k+1) = g(k) + \mu\{\Phi'(k)\}^* e(k)x(k) \tag{10}$$

where h(k) and g(k) are the standard and conjugate weight update vectors of the filter,

5 Simulations

A comprehensive comparison for the performance between the training CNGD and ACNGD algorithms for the block diagram shown in Fig. 1 is provided in Figs. 2, 3, 4, 5. The nonlinear FIR filter is trained with fully nonlinear gradient functions, i.e., CNGD [12] and ACNGD [13], in which the learning algorithms are used for the modelling of various systems, the tanh nonlinear activation function is used for all algorithms. However, simulations were conducted in system identification setting, to generate the circular and noncircular input feed to the system and filter, the independent identical distributed (iid) is chosen to be white Gaussian noise defined as,

$$A_1 : z(k) = \Phi(0.35w(k) + w(k-1) + 0.35w(k-2)) \tag{11}$$

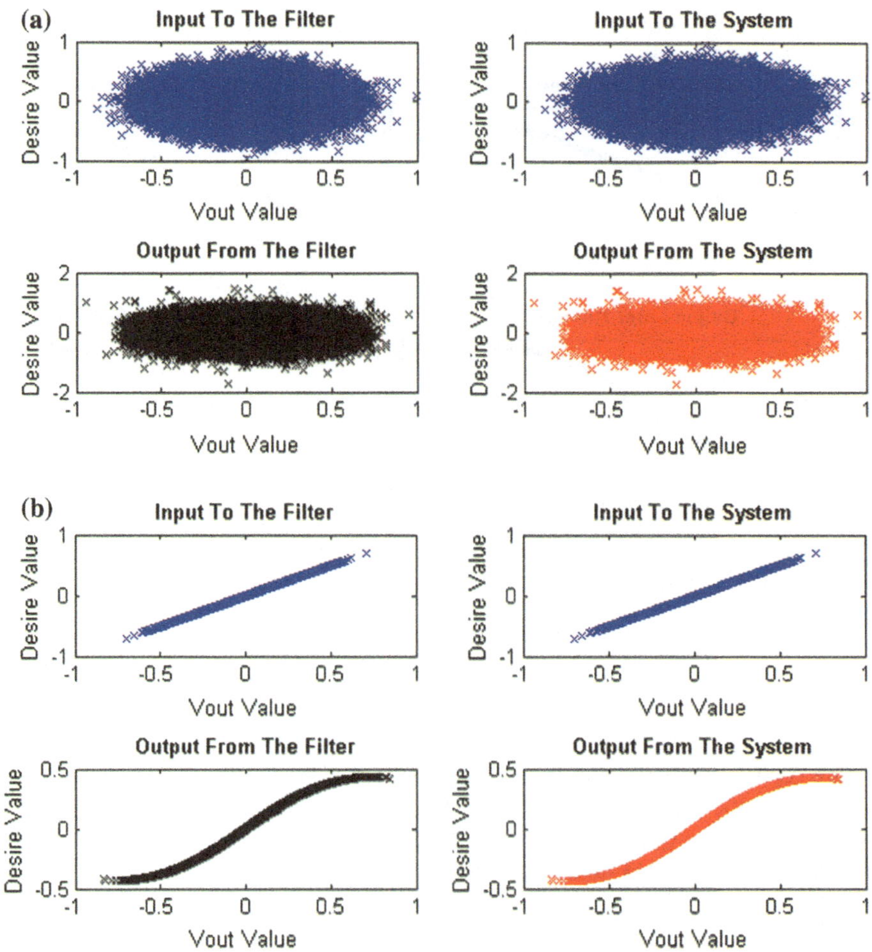

Fig. 2 Circular and noncircular CNGD performance for NMA (3) modelling. **a** Circular signal, **b** Noncircular signal

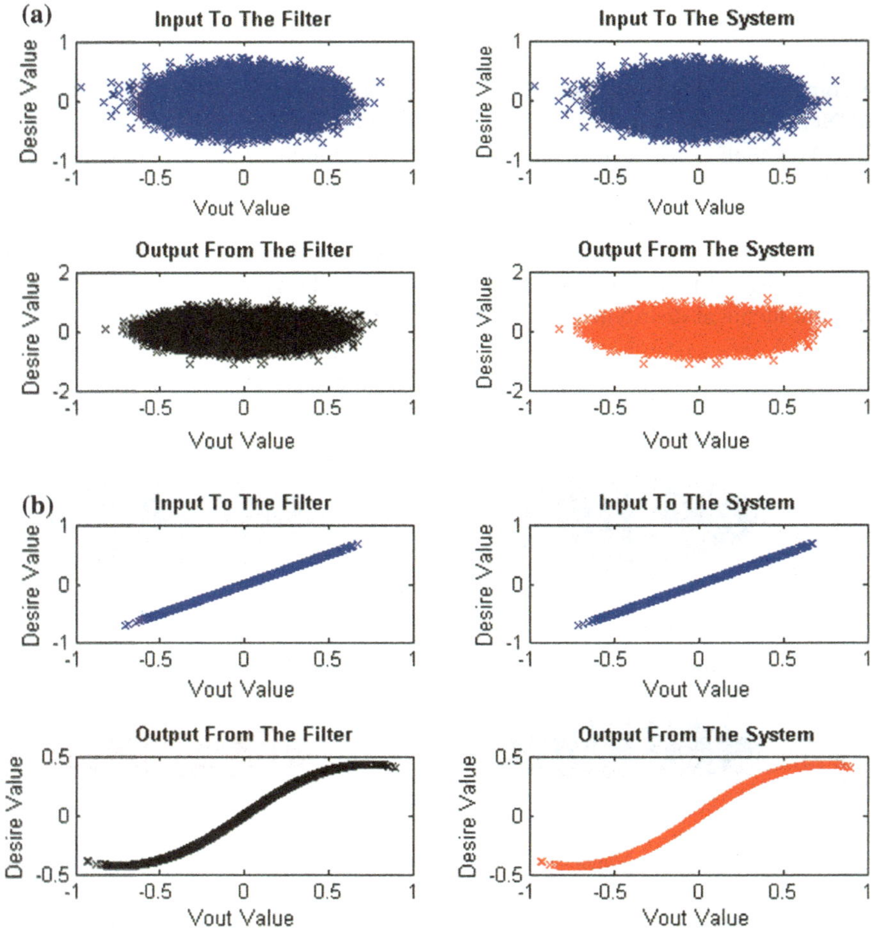

Fig. 3 Circular and noncircular ACNGD performance for NMA (3) modelling. **a** Circular signal, **b** Noncircular signal

$$A_2 : z(k) = \Phi(0.35w(k) + w(k-1) + 0.5(w(k-1)*) + 0.35w(k-2) + 0.25(w(k-2))*); \tag{12}$$

where A_1 is a nonlinear system NMA (3), A_2 is a widely nonlinear system WNMA (3) which is the extension of NMA (3). $\Phi(\cdot)$ is the nonlinearity tanh function which is the same as the CNGD and ACNGD algorithm nonlinearity function used. The simulation is conducted under two cases:

i. Modelling of NMA (3)
ii. Modelling of WNMA (3)

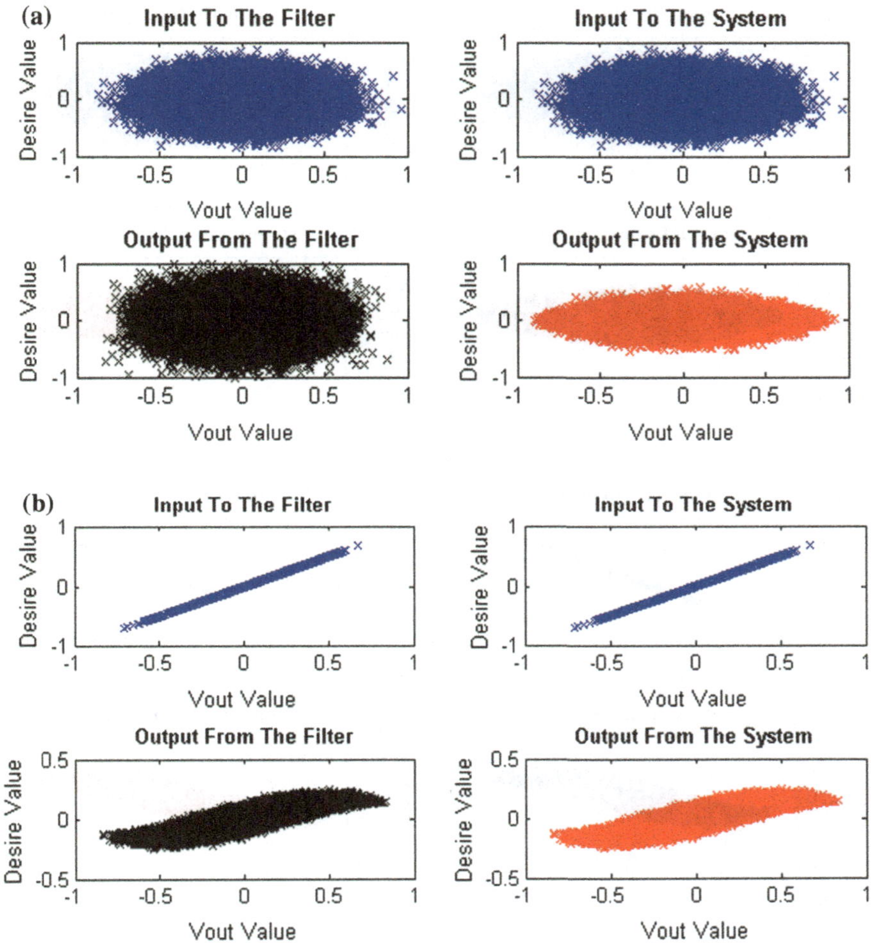

Fig. 4 Circular and noncircular CNGD performance for WNMA (3) modelling. **a** Circular signal, **b** Noncircular signal

This is to illustrate the usefulness of comparison of both algorithms for system order identification of augmented and conventional complex system.

5.1 Nonlinear MA(3)

For this simulation, the signal length L is chosen to be 200,000, learning rate = 3×10^{-1}, and tap length = 3. In the first set of simulation, the performance of CNGD and ACNGD algorithms are analysed for a NMA (3) system.

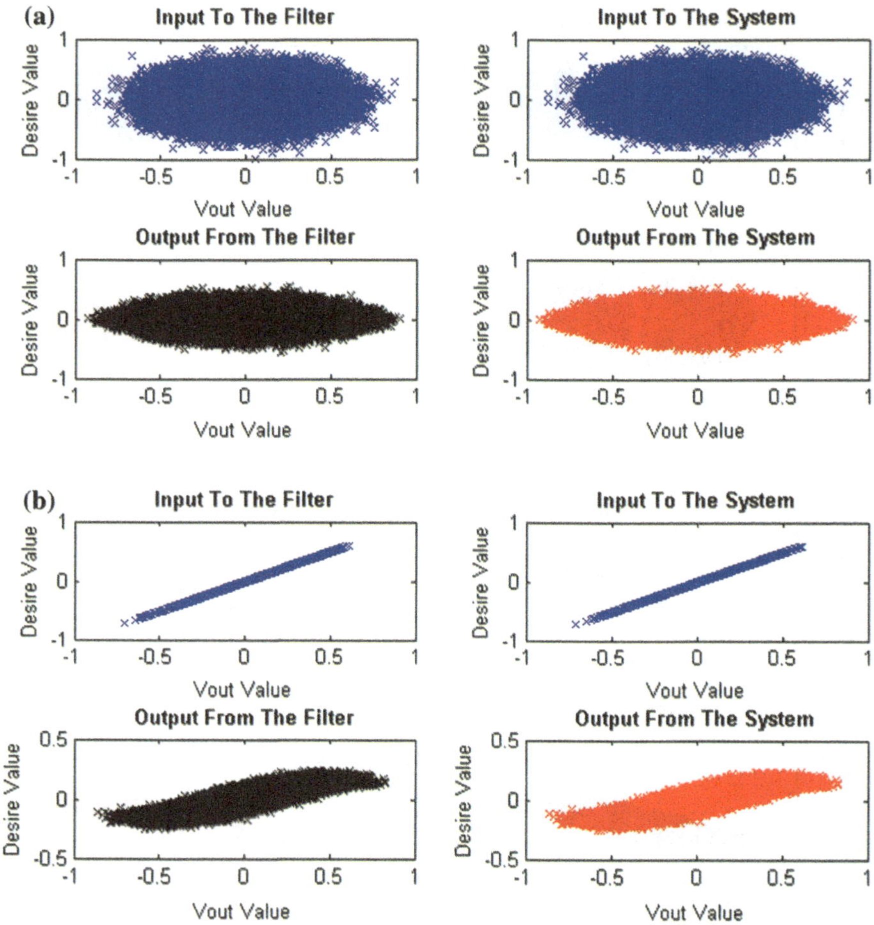

Fig. 5 Circular and noncircular ACNGD performance for WNMA (3) modelling. **a** Circular signal, **b** Noncircular signal

Figure 2 depicts the simulation for the modelling of nonlinear MA (3) system given in (11) using the CNGD algorithm with circular and noncircular random noise signal to the filter and system. The FIR filter was able to model the system output with little changes in the circularity difference. Table I depicts the circularity difference.

Figure 3 shows a similar figure trend using the ACNGD learning algorithm. This proves that both algorithms will perform identically with little changes in the circularity difference due to the present of the standard part (8) in the algorithm.

Table 1 illustrates the circularity difference between the CNGD and ACNGD with nonlinear MA (3) system given in (11), the input to the filter and system will always be the same. Hence, both algorithm performed identically with same input circularity into the filter and system, and same output circularity from the filter and system. This

Table 1 The circularity difference for nonlinear MA (3) system

Algorithm	Signal input	Filter input	System input	Filter output	System output
CNGD	Circular	0.0028	0.0028	0.0045	0.0045
ACNGD	Circular	0.0028	0.0028	0.0060	0.0060
CNGD	Noncircular	1.0000	1.0000	0.9957	0.9957
ACNGD	Noncircular	1.0000	1.0000	0.9972	0.9972

depicts the relationship between the CNGD and ACNGD standard weight present in the algorithms values with the nonlinear MA (3) system given in (11).

5.2 *Widely Nonlinear MA(3)*

In this simulation, the performance of CNGD and ACNGD algorithms are analysed for a widely nonlinear MA (3) system. Same parameters for simulating the nonlinear MA (3) system are used.

Figure 4 illustrates the performance CNGD algorithm for the modelling of widely nonlinear MA (3) system shown in (12) with circular and noncircular signal to the filter and system. It was observed that the CNGD algorithm filter output for the circular signal was unable to model the widely nonlinear MA (3) system output due to the absences of augmented weight. However, it was able to model the nonlinear MA (3) system given in (11) efficiently.

On the other hand, the ACNGD algorithm, Fig. 5 depicts similar modelling curves of widely nonlinear MA (3) system. The ACNGD algorithm filter output was able to model the widely nonlinear MA (3) system output for circular signal, this is due to the presence of the augment weight (10) in the algorithm.

Table 2 shows the circularity difference between the CNGD and ACNGD with the widely nonlinear MA (3), the input to the filter and system will always be the same for each simulation. It was realized that both algorithms performed identically with same output circularity from the filter and system for the noncircular signal. However, on the circular signal of the CNGD algorithm, the CNGD filter output was incapable of modelling the widely nonlinear MA (3) system output. This illustrate the differences between CNGD weight in Eq. (7) and ACNGD weight in Eq. (10) with the widely nonlinear MA (3) system in Eq. (12).

Table 2 The circularity difference for widely nonlinear MA (3) system

Algorithm	Signal input	Filter input	System input	Filter output	System output
CNGD	Circular	0.0028	0.0028	**0.0096**	0.7492
ACNGD	Circular	0.0028	0.0028	0.7516	0.7516
CNGD	Noncircular	1.0000	1.0000	0.9617	0.9617
ACNGD	Noncircular	1.0000	1.0000	0.9612	0.9612

The bold value shows the incapability of the CNGD algorithm to model the widely nonlinear MA(3) system when circular signal input is fed in

5.3 The Performance of the CNGD and ACNGD for the Tracking of NMA(3) and WNMA(3) System

In this simulation, the signal length L was chosen to be 14,000, learning rate = 4×10^{-2}, and tap length = 3. The performance of CNGD and ACNGD are analysed for the NMA (3) system tracking.

Figure 6 shows the weight value for the tracking of nonlinear MA (3) system A_1 using the CNGD algorithms. It can be seen that CNGD algorithm was able to capture the coefficient values of the nonlinear MA (3) system given in A_1. With W_1 on 0.35, W_2 on 1 and W_3 on 0.35. These values are the coefficient of the system give in (11). This proves that the CNGD algorithm is able to model the nonlinear MA (3) system efficiently.

Figure 7 illustrates similar weight value tracking for the modelling of nonlinear MA (3) system A_1 using the ACNGD algorithms. The ACNGD standard part (8) was able to capture the coefficient values of the nonlinear MA (3) system given in A_1. With H_1 on 0.35, H_2 on 1 and H_3 on 0.35. The weight values tracking of the ACNGD has the same directional flow as the CNGD. It shows that both algorithm has identical performance to the modelling and tracking of the nonlinear MA (3) system, The augmented CNGD weight G_1, G_2, G_3 values are zero. However, this shows the relationship with the nonlinear MA (3) system. These weight values are display to realize better clarity. Both algorithms are zero imaginary valued and are removed due to the coefficient of the nonlinear MA (3) system being real-valued.

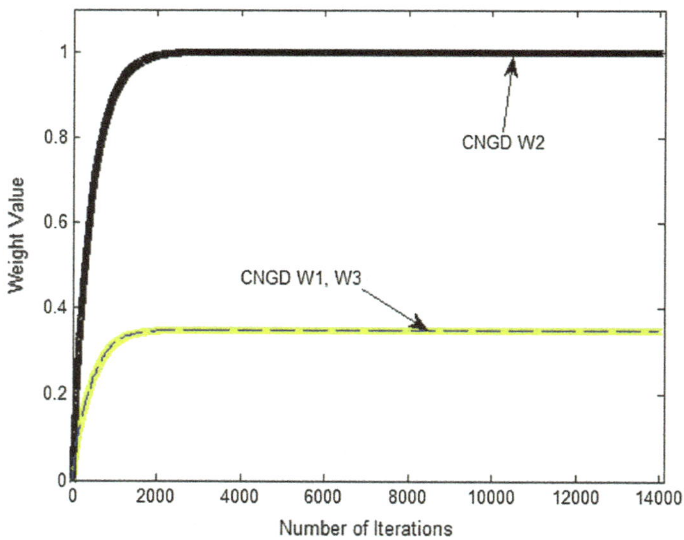

Fig. 6 The performance of CNGD for NMA (3) system tracking

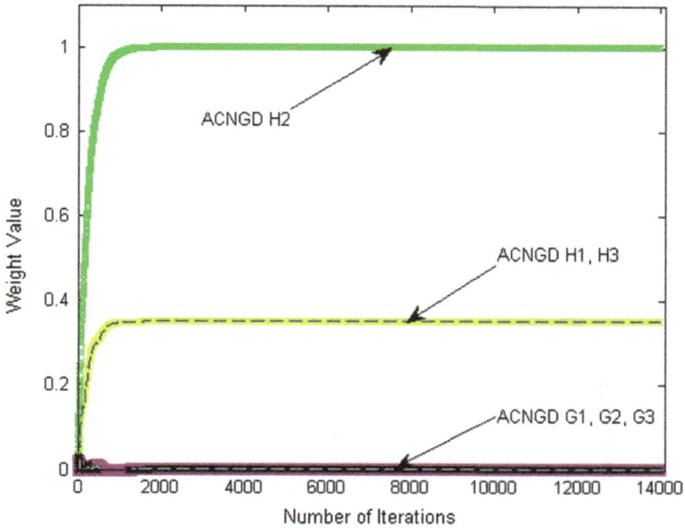

Fig. 7 The performance of ACNGD for NMA (3) system tracking

Figure 8 shows the weight value of CNGD algorithms for the tracking of the widely nonlinear MA (3) system A_2. The CNGD algorithm was unable to capture the coefficient values of the augmented weight G_1 on 0, G_2 on 0.5 and G_3 on 0.25, of the widely nonlinear MA (3) system given in (12). As a result of nonexistence of augmented weight in the CNGD algorithm.

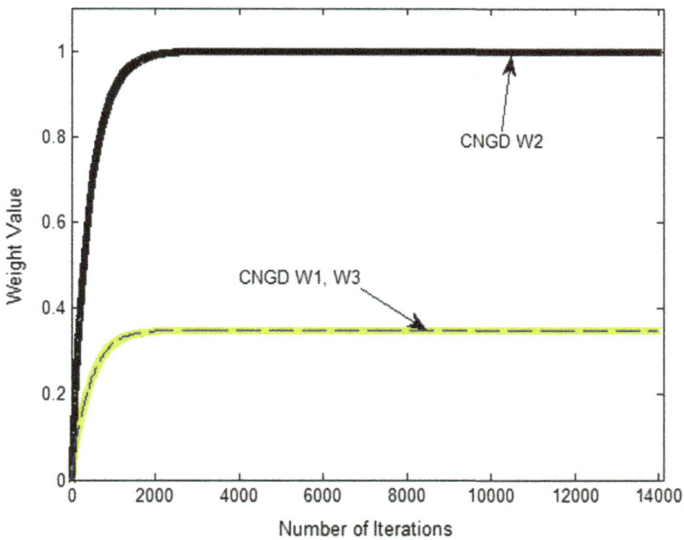

Fig. 8 The performance of CNGD for WNMA (3) system tracking

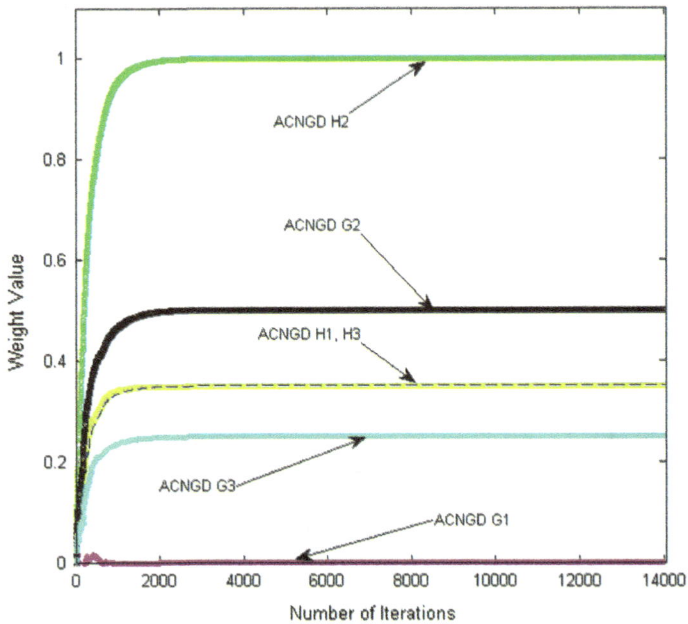

Fig. 9 The performance of ACNGD for WNMA (3) system tracking

Figure 9 depicts the weight value for the ACNGD algorithms for the tracking of widely nonlinear MA (3) system A_2. The ACNGD was capable of capturing the coefficient values of both standard part (9) H_1 on 0.35, H_2 on 1 and H_3 on 0.35 and the augmented part G_1 on 0, G_2 on 0.5 and G_3 on 0.25 of the widely nonlinear MA (3) system given in (12). Due to the existence of augmented weight in the ACNGD algorithm.

6 Conclusion

The performance of a class of complex valued adaptive filter trained by the complex nonlinear gradient decent (CNGD) and augmented (CNGD) for the modelling of various complex valued systems has been analysed. The CNGD algorithm does not perform well for the modelling of widely nonlinear system when the input is circular. However, the ACNGD works for every circumstances. Furthermore, the CNGD can only model conventional system regardless of the circularity, while the ACNGD model all kind of systems due to the presents of augmented weights. Hence, in order to model the widely nonlinear system, augmented statistic is considered. The augmented nonlinear gradient decent (ACNGD) algorithm has depicted a superior modelling performance for both A_1 and A_2 system. This complements the ability of ACNGD to capture the full second order statistics of the complex domain to model both second order circular (proper) and noncircular (improper) signal, and track their weight updates. Simulations support the approach on model selection and identification. This can be extend to the modelling of real world process.

References

1. Mandic DP, Goh SL (2009) Complex valued nonlinear adaptive filters: noncircularity, widely linear and neural models. Wiley
2. Hirose A (2003) Complex-valued neural networks: theories and applications. World Scientific Publishing
3. Kim T, Adali T (2003) Approximation by fully complex multilayer perceptrons. Neural Comput 15(7):1641–1666
4. Pritzker Z, Feuer A (1991) Variable length stochastic gradient algorithm. IEEE Trans Signal Process 39(4):997–1001
5. Widrow B, McCool J, Ball M (1975) The complex LMS algorithm. Proc IEEE 63(4):719–720
6. Xia Y, Jelfs B, Van-Hulle MM, Principe JC, Mandic DP (2009) An augmented echo state network for nonlinear adaptive filtering for complex noncircular signals. IEEE Trans Neural Net
7. Xia Y, Javidi S, Mandic DP (2010) A Regularised Normalised Augmented Complex Least Mean Square Algorithm. In: Proceedings of the IEEE
8. Cheong-Took C, Mandic DP (2010) A quatemion widely linear adaptive filter. IEEE Trans Signal Process 58(8):4427–4431
9. Neeser FD, Massey JL (1993) Proper complex random processes with applications to information theory. IEEE Trans Inf Theory 39(4):1293–1302
10. Kreutz-Delgado K (2006) Complex gradient operator and the ICIR calculus, Course lecture supplement ECE275A, Department of Electrical and Computer Engineering, University of California, San Diego
11. Ujang BC, Took CC, Mandic DP (2014) Adaptive convex combination approach for the Identification of improper quaternion processes. IEEE Trans Neural Networks and learning system 25(1)
12. Goh SL and Mandic DP (2004) A class of low complexity and fast converging algorithm for complex-valued neutral networks. In: IEEE workshop on machine learning for signal processing
13. Xia Y, Jelfs B, Van Hulle MM, Principe JC, Mandic DP (2011) An augmented echo state network for nonlinear adaptive filtering of complex noncircular signals. IEEE Trans on Neural Net 22(1)

Perceived Stress Scale and Brainwave Characteristic of Breastfeeding Women

Najidah Hambali, Noor Izzati Abd Halin, Zunairah Haji Murat and Nur Idora Abdul Razak

Abstract Breastfeeding is a process of nursing a baby with milk straight from the women's breast. It is suggested that a baby should be breastfed within one hour of birth, exclusively breastfed for the first six months, and then breastfed until age two with age-appropriate, nutritionally sufficient and harmless complementary diets. This paper focuses on the examination of breastfeeding women on their stress level and the brainwave characteristics. The brainwave signals were documented using wireless Electroencephalogram (EEG) equipment via Bluetooth while. The stress level was performed using a questionnaire. The EEG signals were derived using an existing system with the electrodes attached to human scalp. These electrodes measure the electrical signals which are produced by the activities of the neurons in the human brain, such as Delta, Theta, Alpha and Beta waves in the brain to determine the Brainwave Balancing Index (BBI) and also the brain hemispheric dominance which is right or left dominance. The outcome presented balanced BBI of the breastfeeding women after the breastfeeding session, although most of them were high in stress. Right brain dominance was also recorded for the majority of them for both sessions, before and after breastfeeding. Statistical investigation leads to no significant correlations between stress and BBI of breastfeeding women after breastfeeding session, in terms of breastfeeding method and in terms of breastfeeding women' categories.

Keywords Breastfeeding · Brainwave balancing index · Brain hemisphere dominance · EEG · Stress

N. Hambali (✉) · N.I.A. Halin · Z.H. Murat · N.I.A. Razak
Faculty of Electrical Engineering, Universiti Teknologi, MARA 40450 Shah Alam, Selangor, Malaysia
e-mail: najidah@salam.uitm.edu.my

N.I.A. Halin
e-mail: izzatihalin1011@gmail.com

Z.H. Murat
e-mail: zunai194@salam.uitm.edu.my

N.I.A. Razak
e-mail: nuridora@salam.uitm.edu.my

P.J. Soh et al. (eds.), *Advances in Machine Learning and Signal Processing*,
Lecture Notes in Electrical Engineering 387, DOI 10.1007/978-3-319-32213-1_21

1 Introduction

Breastfeeding has proved to give so much benefit to the babies and the mothers. Therefore, the ability of women's to produce breast milk was also important. In few places like Norway, Jining City, Australia and Kelantan, one of the influence factor to a problem in breastfeeding is stress that comprises breastfeeding rates, low breast milk production and breastfeeding cessation among the women [1–4]. The career problem, low breastfeeding milk production and financial difficulty are the potential reasons for breastfeeding women.

The decision not to continue breastfeeding may come from the personal problem suffered by the mother. A study found that, there was a significant relationship between life stress and breastfeeding duration that were on financial problem and traumatic stress [5]. These kind of stress had resulted in discontinuation of breastfeeding. A new mother in United States was reported with positive diagnostic criteria of posttraumatic stress disorder after childbirth that may contribute the mother's decision to stop breastfeed their babies [6]. In addition, depressive symptoms that affected to the women is twofold compared to men for the period of the reproductive years. Due to this factor, some women tends to quit breastfeeding in the earliest three months after delivery. Besides that, socioeconomic status, age, parity and academic qualification of mother will affect the depression breastfeeding women [1, 7].

Human physiological features are linked with stress conditions. Some of the changes that has been reported during tense state are increased heart rate, reduced skin resistance and increased blood pressure. Dealing with variations of individual stress response, there is a need to introduce the person specific parameter [8]. The stress and anxiety symptoms are in a direct regression as forecasted at six months postpartum and has been reported by World Health Organization (WHO) [1]. Thus, women with greater stress condition are more likely to gain their stress after stop breastfeeding. Not only that, the women capability to bring out the milk is impacted by the postnatal anxiety and depressive disorder [2]. Hence, to accommodate the individual difference, the stress evaluation method is used.

Research on association between breastfeeding and Sudden Infant Death Syndrome (SIDS) based on meta-analysis methods reported that the longer the duration of breastfeeding up to six months or a baby who is exclusively breastfed had stronger effect in protecting SIDS [9]. In terms of early brain development, breastfeeding is absolutely related with enhanced myelin water fraction in somatosensory, auditory and language which improved language presentation, visual response and motor control presentation. Other than that, initial breastfeeding associated with better expansion in late maturation of white matter area that regularly connected with advanced-order intellect such as social-emotional operative and language [10].

Breastfed kids and prolonged breastfeeding period presented a constructive association in numerous brain regions and neural development [10]. Moreover, investigation on maternal brain activation of the mothers was conducted using MRI in reaction to their baby's cry versus controlled-baby's cry [11]. Electroencephalogram (EEG) was implemented to examine the respondent's brainwave pattern that relates with their stress level [12]. Therefore, a preliminary study has been conducted on the brain wave of the breastfeeding women in correlation to the stress level [13].

This project is specifically analyzing the stress scale, brain dominance and the pattern of BBI of women who breastfed their children. The results and discussion section presents the stress level, BBI and brain hemispheric dominance of breastfeeding women. The correlations of stress and a few categories of BBI is also investigated.

2 Methodology

All of this experimentation was conducted at Biomedical Research and Human Potential, Faculty of Electrical Engineering, Universiti Teknologi MARA. This study was an extension of [13]. The participants was increased to forty volunteering breastfeeding women in Malaysia from Klang Valley zone. Two stages of data collection have been completed, a questionnaire survey and experimental procedure.

2.1 Survey Questionnaire

Firstly, the participants were given a perceived stress scale survey (PSS) [14]. This assessment is used to evaluate the stress perceptual experience. The score is given by the different numerical score based on the responses. This perceived stress scale shows the stress level based on the total score. Table 1 shows the interpretation perceived stress scale score that attributes to the score of the participant based on the survey.

2.2 Experimental Process

Two breastfeeding techniques, either through direct feeding or milk expressing by means of a breast pump were employed for EEG data accumulation for both before

Table 1 Perceived stress scale scores [14]

PSS index	Score level	Health concern level	Description
1	0–7	Very low in stress	Low stress
2	8–11	Low in stress	
3	12–15	Average in stress	
4	16–20	High in stress	High stress
5	21 and over	Very high in stress	

Fig. 1 Electrodes connection

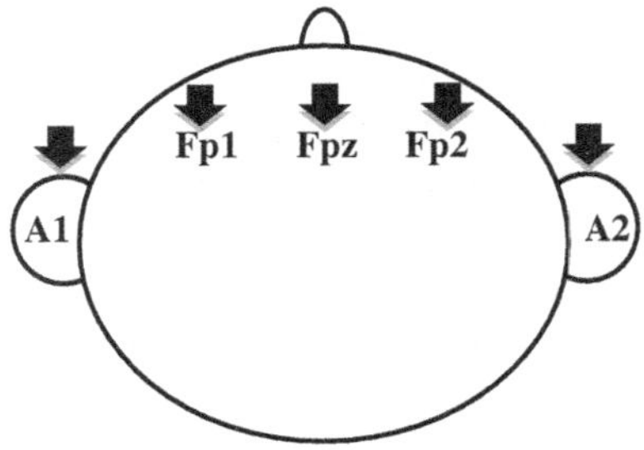

and after breastfeeding. In this study, thirteen through the direct feeding and twenty seven from the milk expressing were documented for EEG.

The EEG data acquisition was executed by using the G-Mobilab equipment. Two channel bipolar connection was employed using five pieces of gold electrodes. The points chosen were Fp1 for the left side of the forehead and Fp2 for the right side of the forehead, connected to Channel 1 and Channel 2 respectively. Fpz was used for the reference at the center of forehead and also both A1 and A2 for the earlobes reference points. The electrodes connection follows the International Standard 10–20 electrode placement system as shown in Fig. 1.

The points chosen were Fp1 for the left side of the forehead and Fp2 for the right side of the forehead, connected to Channel 1 and Channel 2 respectively. Fpz was used for the reference at the center of forehead and also both A1 and A2 for the earlobes reference points. The electrodes connection follows the International Standard 10–20 electrode placement system. Concurrently, Bluetooth is utilized to drive the EEG raw data to the processing device. Then, the logged EEG signal was evaluated via readily obtainable Brainwave Balancing Index System [15, 16] by offering the results of BBI and brainwave dominance. The system concentrates on the frontal region of the brain. The BBI was calculated based on [16] as Eq. (1);

$$\text{Percentage Difference} = 2 \times \left| \frac{\sum \text{left} - \sum \text{right}}{\sum \text{left} + \sum \text{right}} \right| 100\,\% \quad (1)$$

There are five levels of BBI index described from [16] based on the percentage difference between left and right brainwaves as in the mentioned equation, where Table 2 shows the element of the BBI. The lowest score of percentage difference is categorized for highly balanced while the highest score of percentage difference is classified for unbalanced BBI.

Table 2 Brainwave balancing index [15, 16]

BBI index	Stress index (SI)	Description
5	Highly balanced	Balanced
4	Balanced	
3	Moderately balanced	
2	Less balanced	Unbalanced
1	Unbalanced	

3 Results and Discussion

The results are divided into four parts as follows; the survey questionnaire, BBI, brain dominance and statistical analysis. The breastfeeding women volunteers' age is between twenty to forty years old while their babies are between three to fifteen months. From forty participants, thirteen was practiced direct feeding and twenty seven for milk expressing method. Besides that, thirty three is Full Time Working Mothers (FTHM) and seven is Stay At Home Mothers (SAHM) to differentiate the correlation between occupations and stress level of the breastfeeding women.

3.1 *Perceived Stress Scale*

Figure 2 shows the stress assessment for breastfeeding women, according to Table 1. As illustrated, most of the breastfeeding women were high on stress, which was 77.5 % in total. Another 22.5 % was nominated by breastfeeding mothers with low in stress. Most participants were under stress before they went through breastfeeding activities.

3.2 *Brainwave Balancing Index*

Figure 3 shows the recorded BBI results based on Table 2 of breastfeeding women before and after direct feeding or milk expressing method. As can be seen in the figure,

Fig. 2 PSS of breastfeeding women

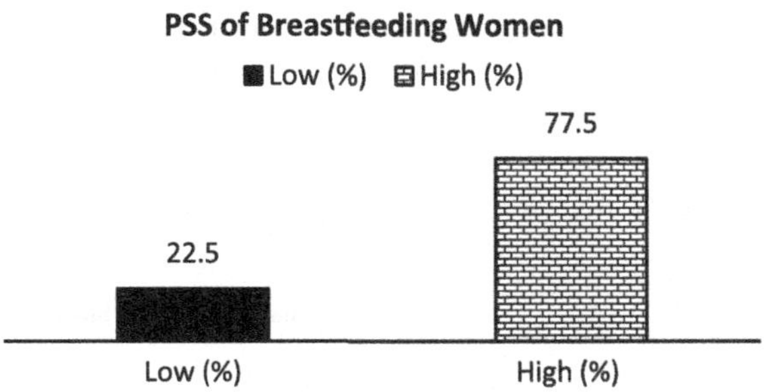

Fig. 3 BBI of breastfeeding women

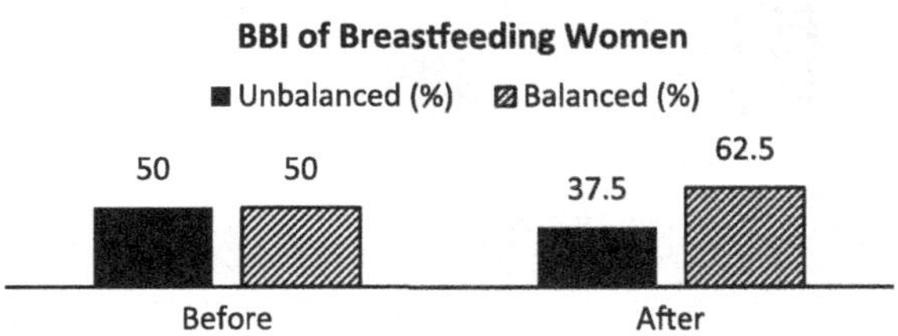

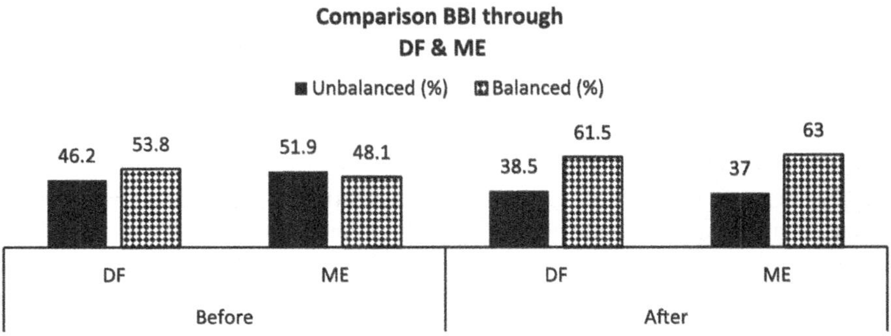

Fig. 4 Comparison BBI between direct feeding & milk expressing method

before they went through any breastfeeding activity, the participant that was recorded with balanced BBI showed a similar percentage with the participant that was recorded with unbalanced BBI which was 50 %. After breastfeeding activity, the percentage of participants with balanced BBI showed a significant increment to 62.5 %, while the percentage of participants with unbalanced BBI had decreased to 37.5 %.

Figure 4 shows the comparison BBI results of breastfeeding activities in terms of breastfeeding method. Based on the results of direct feeding, before breastfeeding activity, 53.8 % was recorded with a balanced BBI and 46.2 % was recorded with unbalanced BBI. Later on, after going through breastfeeding activity, the percentage of participants of balanced BBI had increased to 61.5 % and had pulled down the percentage of participants of unbalanced BBI to 38.5 %. While for milk expressing method, before breastfeeding activity, 48.1 % was recorded with a balanced BBI and 51.9 % was recorded with unbalanced BBI. After going through breastfeeding activity, the percentage of participants of balanced BBI had increased to 63 %, while the percentage of participants of unbalanced BBI decreased to 37 %.

Figure 5 shows the comparison BBI results between FTWM and SAHM. For FTWM, only 45.5 % was recorded with balanced BBI as compared to

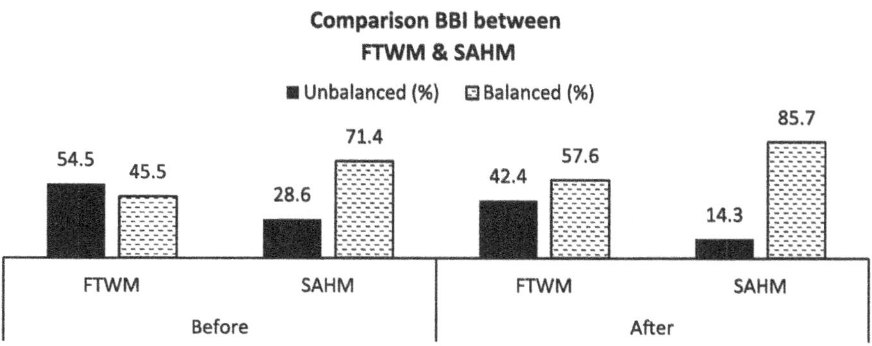

Fig. 5 Comparison BBI between FTWM and SAHM

unbalanced BBI, which was recorded to be 54.5 % before these mothers went through any breastfeeding activity. After breastfeeding activity, the percentage of participants with balanced BBI had increased to 57.6 %, while the percentage of participants with unbalanced BBI had decreased to 42.2 %. As depicted for SAHM, 71.4 % participants were recorded with balanced BBI and left behind by far in percentage of participants with unbalanced BBI, which was recorded to be 28.6 % before these mothers went through any breastfeeding activity. After breastfeeding activity, the percentage of participants with balanced BBI went through a huge increment to 85.7 %, while the percentage of participants with unbalanced BBI had decreased to only 14.3 %.

3.3 *Brain Dominance*

Figure 6 shows the brain hemispheric dominance of breastfeeding women for both sessions, before and after breastfeeding. As depicted, only 25 % participants were in left hemispheric brain dominance while 55 % were in right hemispheric brain dominance.

Besides that, it can be observed in Fig. 6 that there was a brain dominance changes either from left to right (L-R) with 2.5 % or right to left (R-L) with 17.5 % participants.

The brain hemisphere dominance of breastfeeding women before and after breastfeeding is displayed in Fig. 7. Based on the result, it can be seen that most of the breastfeeding mothers were in right hemispheric dominance before and after breastfeeding activity which resulted in 57.5 % and 72.5 % as compared to left hemisphere dominance that was only presented with 42.5 % and 27.5 %

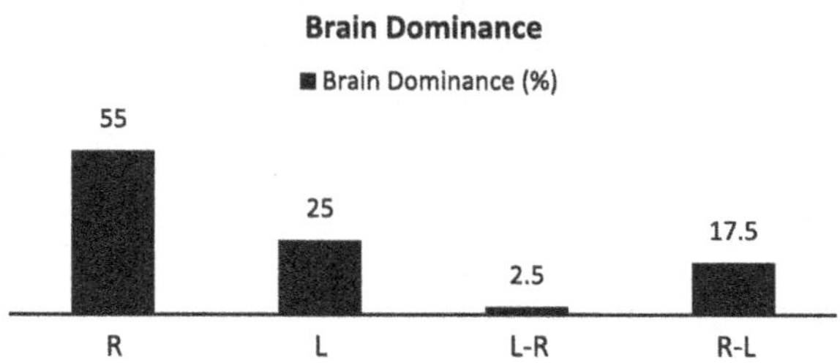

Fig. 6 Brain dominance of breastfeeding women

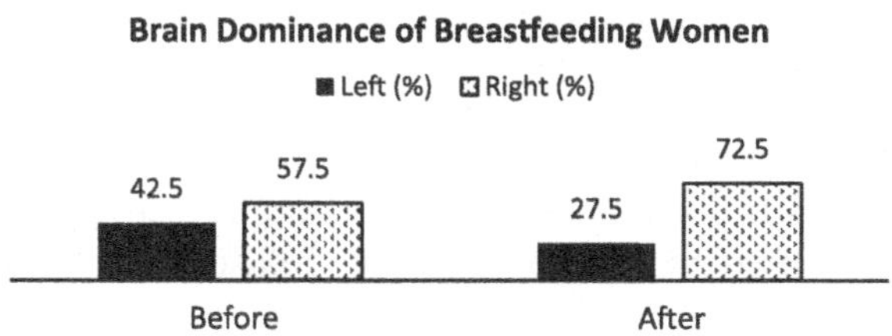

Fig. 7 Brain dominance of breastfeeding women

Table 3 Spearman rank order correlations of PSS and BBI

PSS	BBI after	BBI after (direct feeding)	BBI after (milk expressing)	BBI after (FTWM)	BBI after (SAHM)
Correlation coefficient	0.247	0.378	0.210	0.277	0.168
Sig. (1-tailed)	0.62	0.101	0.147	0.060	0.359
Sample size (N)	40	13	27	33	7

respectively. Right brain dominant people are described as more sensitive to the emotional features. The greater blood oxytocin levels during breastfeeding lead to the women's body to create more receptors, eternally enhancing the emotions of love and also the power to experience loved.

3.4 *Statistical Analysis*

Statistical Analysis was implemented to investigate the correlation between PSS and BBI of breastfeeding women after they went through breastfeeding activity. The data were computed into the Statistical Package for the Social Sciences (SPSS) to analyze the relationship.

Table 3 shows a Spearman Rank Order Correlations summary that was figured to explore the connection between PSS and BBI after breastfeeding for overall, for direct feeding, for milk expressing, for FTWM and for SAHM. The result showed weak positive correlations for all categories of PSS and BBI with the correlation coefficient, r_s between 0.168 and 0.378.

As a conclusion for these two variables, there were no significant correlations since the p value was greater than 0.05 for all categories of BBI; after breastfeeding for overall, direct feeding, milk expressing, FTWM and SAHM. That means, the result of PSS index did not significantly correlated to BBI of the breastfeeding women with $p = 0.062$, 0.101, 0.147, 0.060 and 0.359 respectively.

4 Conclusion

The results and discussion of PSS and brainwave characteristic of breastfeeding women were documented. The PSS of breastfeeding women was examined from the questionnaire. In addition, the BBI and brain hemispheric dominance were determined by utilizing the readily available BBI system. The correlations between PSS and few categories of BBI were also generated.

After going through the breastfeeding activity, it shows that the brainwave of breastfeeding women was more balanced as compared to before breastfeeding activity. Both methods either direct feeding or milk expressing produced a balanced BBI after breastfeeding session. SAHM produced a more balanced BBI before and after breastfeeding as compared to FTWM.

In addition, most of the breastfeeding women maintained brain dominance before and after breastfeeding either left or right and others were changing. Brain dominance analysis showed a majority of breastfeeding women were in right dominance.

For the correlations between PSS and BBI of breastfeeding mothers after breastfeeding session, in terms of breastfeeding method and in terms of FTWM and SAHM, there were no significant evident according to the statistical analysis. Consequently, breastfeeding activity will improve the balancing of the brainwave.

Another factor that can be investigated is to compare the stress level and the pattern of BBI in different surrounding including religious recitation, such as 'Asmaul Husna' and 'zikir', classical music and heavy noise.

Acknowledgments This project supported by a Research Acculturation Grant Scheme (RAGS) 600—RMI/RAGS 5/3 (58/2013) funded by the Ministry of Higher Education, Malaysia. The authors also gratefully acknowledge the use of the facilities of the Biomedical Research and Human Potential, Faculty of Electrical Engineering, Universiti Teknologi MARA, Malaysia.

References

1. Ystrom E (2012) Breastfeeding cessation and symptoms of anxiety and depression: A longitudinal cohort study. BMC Pregnancy Childbirth 12(1):36
2. Yang B-F, Song H-M, Wang S-L, Liu X-H, Zhuang B (2008) Psychological and physiological characteristics and their contributing factors in 505 postpartum women in Jining city. In: Fourth international conference on natural computation. IEEE, vol 2, pp 67–71
3. O'Brien M, Buikstra E, Fallon T, Hegney D (2009) Exploring the influence of psychological factors on breastfeeding duration, phase 1: Perceptions of mothers and clinicians. J Human Lact 25(1):55–63
4. Tengku ATI, Wan AMWM, Zaharah S, Rohana AJ, Nik Normanieza NM (2012) Perceptions and practice of exclusive breastfeeding among Malay women in Kelantan, Malaysia: A qualitative approach. Malaysian J Nutr 18(1):15–25
5. Dozier AM, Nelson A, Brownell E (2010) The relationship between life stress and breastfeeding outcomes among low-income mothers. Adv Prev Med 2012:902487
6. Beck CT, Gable RK, Sakala C, Declercq ER (2011) Posttraumatic stress disorder in new mothers: Results from a two-stage US national survey. Birth 38(3):216–227
7. Aubuchon-Endsley NL, Kennedy TS, Gilchrist M, Thomas DG, Grant S (2015) Relationships among socioeconomic status, dietary intake, and stress in breastfeeding women. J Acad Nutr Diet 115(6):939–946
8. Xu Q, Nwe T, Guan C (2015) Cluster-based analysis for personalized stress evaluation using physiological signals. IEEE J Biomed Health Inform 19(1):275–281
9. Hauck FR, Thompson JMD, Tanabe KO, Moon RY, Vennemann MM (2011) Breastfeeding and reduced risk of sudden infant death syndrome: A meta-analysis. Pediatrics 128(1):103–110
10. Deoni SCL, Dean DC, Piryatinsky I, O'Muircheartaigh J, Waskiewicz N, Lehman K, et al (2013) Breastfeeding and early white matter development: A cross-sectional study. Neuroimage 82:77–86

11. Kim P, Feldman R, Mayes LC, Eicher V, Thompson N, Leckman JF et al (2011) Breastfeeding, brain activation to own infant cry, and maternal sensitivity. J Child Psychol Psychiatry Allied Discip 52:907–915
12. Adnan N, Murat ZH, Kadir A, Yunos NHM (2012) University students' stress level and brainwave balancing index: Comparison between early and end of study semester. In: IEEE student conference on research and development (SCOReD), pp 42–47
13. Hambali N, Hassan HNNA, Murat ZH, Razak NIA (2015) The preliminary study of interrelationship of perceived stress to brainwave characteristic of breastfeeding women. Electr Electron Eng 5(1A):1–6
14. Sheldon C, Kamarck T, Mermelstein R (1983) A global measure of perceived stress. J Health Soc Behav 385–396
15. Murat ZH, Taib MN, Lias S, Kadir RSSA, Sulaiman N, Mustafa M (2010) EEG analysis for brainwave balancing index (BBI). In: Second international conference on computational intelligence, communication systems and networks, pp 389–393
16. Murat ZH, Taib MN, Lias S, Kadir RSSA, Sulaiman N, Hanafiah ZM (2011) Development of brainwave balancing index using EEG. In: Third international conference on computational intelligence, communication systems and networks, pp 373–378

Improved Speech Emotion Classification from Spectral Coefficient Optimization

Inshirah Idris and Md Sah Salam

Abstract In order to improve the performance of speech emotion recognition systems, and to reduce the related computing complexity, this work proposed two approaches of spectral coefficient optimization. The two approaches are (1) optimized based on discrete spectral features and (1) combine spectral features. Experimental studies have been performed through the Berlin Emotional Database, using a support vector machine (SVM) classifier, and five spectral features including MFCC, LPC, LPCC, PLP and RASTA-PLP. The experiment results have shown that speech emotion recognition based on optimized coefficient numbers can effectively improve the performance. There were significant improvements in the accuracy 2 % for the first approach and 4 % for the second approach compared to that using the existing approaches. Moreover the second approach outperformed the first approach in the accuracy. This good accuracy came with reducing the features number.

Keywords Spectral features · Coefficients · MFCC · LPC · LPCC · PLP · RASTA-PLP · SVM

1 Introduction

Speech Emotion Recognition (SER) has become a hot research topic in recent years, due to its ability to identify the mood of a particular person from his or her voice. This makes it an important part of Human-Computer Interaction (HCI), as

I. Idris (✉)
Computer Science Department, Sudan University of Science and Technology, Khartoum, Sudan
e-mail: inshirah15@hotmail.com

M.S. Salam
Software Engineering Department, Universiti Teknologi Malaysia (UTM), Skudai, Johor, Malaysia
e-mail: sah@utm.my

P.J. Soh et al. (eds.), *Advances in Machine Learning and Signal Processing*, Lecture Notes in Electrical Engineering 387, DOI 10.1007/978-3-319-32213-1_22

used for many important applications including e-learning, robotics, healthcare, security, entertainment and so on. In general, SER is a pattern recognition system which uses a vector of extracted speech features from an emotional speech database, in order to recognize a persons emotional state, through the use of a classifier.

Since the feature extraction stage plays an important role in the performance of any pattern recognition system, the first issue in this area involves finding the best features that can help increase SER accuracy. Literature shows that there are four categories of acoustic speech emotion features, which include voice quality, prosodic, spectral and wavelet features. According to Wang et al. [1] the most commonly-used features include prosodic and spectral features.

When working with spectral features, including the Mel-Frequency Cepstral Coefficients (MFCC), Linear Predictive Coefficients (LPC), the Linear Predictive Cepstral Coefficients (LPCC), Perceptual Linear Prediction (PLP), Relative Spectral Transform—Perceptual Linear Prediction (RASTA-PLP), the first and most important question is to determine how many coefficients are suitable for use. However there are no guidelines regarding how to choose the best number of coefficients. The tradeoffs in having large number of coefficients is that it may help to accommodate suitable features in the features vectors but it will also increase the feature dimensionality and possible redundancy which lead in increasing computational cost. On the other hand, small number of coefficients may lead to insufficient suitable features which may result in low recognition.

From the literature, researchers are used several number of coefficients in developing their SER systems (Fig. 1). Pierre-Yves [2] has used 10 MFCC coefficients. Rong et al. [3], Schuller et al. [4] and Lee et al. [5] have used 12 MFCC coefficients. Lee et al. [6], Wang and Guan [7] and Lugger and Yang [8] have used 13 MFCC coefficients. Schuller et al. [9] has used 15 MFCC coefficients. Several authors also have chosen to use the same number of coefficients for different

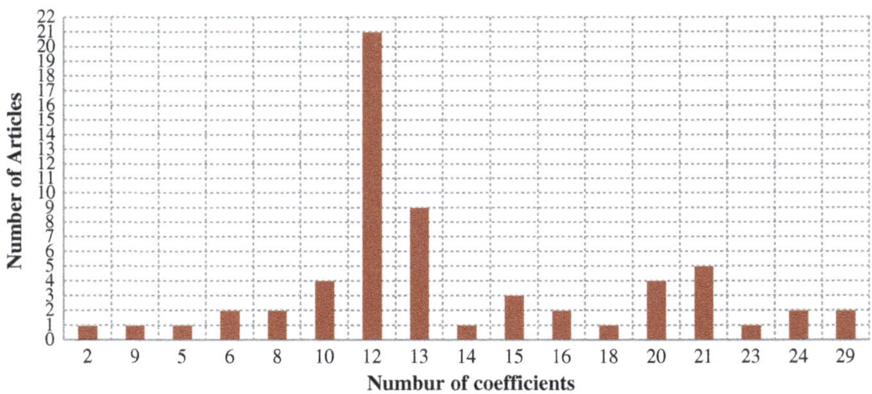

Fig. 1 Shows some coefficient numbers used by researchers in conjunction with spectral features for developing their systems, as obtained in a survey conducted between 2000 and 2015, with 40 papers

spectral features. For example, Kim et al. [10] has used 12 coefficients for both LPC and MFCC. Other researchers chose to use different numbers of coefficients for different spectral features. For example, Nwe et al. [11] chose to use 16 coefficients for LPCC, and 12 coefficients for both MFCC and LFPC. Fu et al. [12] also selected 10 coefficients for LPCC, and 12 coefficients for MFCC.

There are some researchers who also chose to test different numbers of coefficients for the same spectral features. For example, Koolagudi et al. [13] used 6, 8, 13, 21 and 29 coefficients for both LPCC and MFCC, while Murugappan et al. [14] used 13, 15 and 20 coefficients for MFCC, and Milton et al. [15] used MFCC with 10, 15, 24 and 23 coefficients. To reduce the dimensionality and computation of the SER system Hegde et al. [16] used the F-ratio technique to select a subset of 12 MFCC coefficients within the Hidden Markov Model (HMM), and concluded that the selection of 8 MFCC coefficients offers a better classification accuracy than that which could be achieved when selecting all 12 coefficients.

Based on the works mentioned above it is clear that there are no uniform patterns used to choose a suitable number of coefficients. This paper has proposed two approaches of selecting optimized numbers of coefficients, depending on the classifier, that could help to increase SER system accuracy while reducing feature vector dimensionality.

2 The Proposed System

Figure 2 shows the proposed speech emotion recognition system architecture, as based on optimized coefficients. The system process used is as follows:

1. The system starts with the speech records from the emotional database, which are described in Sect. 3.
2. The features step involves spectral features pre-processing and extracting using the selected scope number of coefficients, then the optimization of the number of coefficients for spectral features, the main method and algorithm described in Sect. 4.
3. After the optimizing process the features vectors are fed to the classifier, which provides the classification result (accuracy or class label). The classification method is described in Sect. 5.

3 The Berlin Emotional Database (EMO-DB)

A significant number of emotional speech databases have been developed for use when testing SER systems. Some of these databases are publicly available, while others have been created in order to meet a researchers particular needs. Emotional

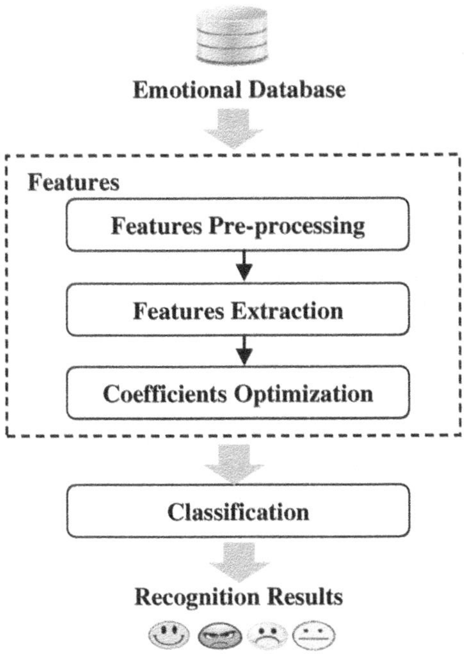

Fig. 2 The proposed system architecture

speech databases can be categorized into three different categories, namely acted, spontaneous and Wizard-of-Oz databases. It is more practical to use a database that has collected samples from real-life situations, and this can serve as a good baseline for creating real-life applications within a specific industry. However the acted database has been consider the easiest one to collect, and different studies have proven that it can offer strong results. It is therefore suitable for theoretical research.

Within this study, Berlin Emotional Database (EMO-DB) was selected as one of the most well-known acted emotional speech databases [17]. It also has been used with spectral features in many studies [18, 19]. The EMO-DB is an acted German emotional speech database recorded at the Department of Acoustic Technology, at TU-Berlin, and is funded by the German research community. It was recorded using a Sennheiser microphone set at a sampling frequency of 16 kHz, with the help of ten professional actors including five males and five females. These actors were asked to simulate seven emotions which included anger, boredom, disgust, fear, happiness, sadness and a neutral emotion, for ten utterances. Following the recording, twenty judges were asked to listen to the utterances in a random order in front of a computer monitor. They were allowed to listen to each sample only once, before they had to decide on the emotional state of the speaker. After the selection process, the database contained a total of 535 speech files.

4 Features

4.1 Features Pre-Processing and Extraction

In this work, we considered five different spectral features namely, MFCC, LPC, LPCC, PLP and RASTA-PLP. MFCC considered being the most used feature of speech [20–22]. It has been widely utilized within speech recognition and speech emotion recognition systems, and Poa et al. [23] reported it as the best and the most frequently acoustic features used in SER. LPC also has been considered one of the most dominant techniques for speech analysis [23]. LPCC is extension of the LPC that has the advantage of less computation, its algorithm is more efficient and it could describe the vowels in better manner [24].

PLP are also an improvement of LPC by using the perceptually based Bark filter bank. PLP analysis is computationally efficient and permits a compact representation [25]. While RASTA-PLP is improvement of the PLP method by adding a special band-pass filter was added to each frequency sub-band in traditional PLP algorithm in order to smooth out short-term noise variations and to remove any constant offset in the speech channel.

MATLAB R2012a was employed in order to compute 30 coefficients of the five features for a frame length of 25 ms every 10 ms, while ten different statistical measurements including minimum, maximum, stander deviation, median, mean, range, skewness, and kurtosis, were utilized for five spectral features from all speech samples.

4.2 Coefficients Optimization

Within this study, two approaches have been proposed for optimizing the number of coefficients for spectral features. The classifier has been used to compare a different number of coefficients, and then to select the coefficients that offer the best accuracy and the lowest number of features for speech emotion recognition. According to literature, the number of coefficients used in the past range from 2 to 29. From this the range of numbers of chosen coefficients was from 0 to 30 for MFCC, PLP and RASTA-PLP. However the first coefficients for LPC and LPCC have the same value for all records, namely 1 for LPC and −1 for LPCC, so the range of numbers of coefficients for both of them are chosen from 1 to 30. The coefficients optimization process as shown in Fig. 3 is as follows:

1. The first coefficient number in the search scope (0 for MFCC, PLP and RASTA-PLP and 1 for LPC and LPCC) has been chosen.
2. Then the features that corresponding to this coefficient number has been choosing from the extracted features vector.

Fig. 3 The coefficients optimization process

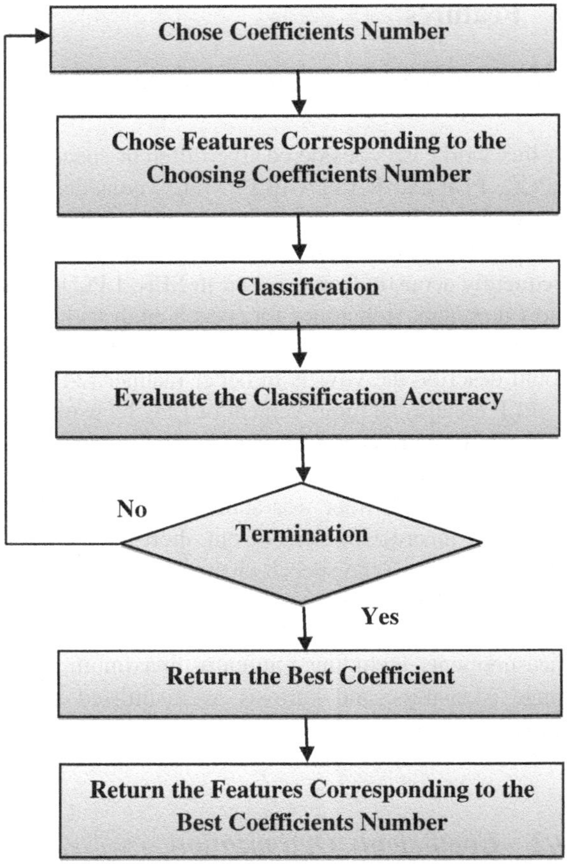

3. Using SVM the accuracy of classification was calculated these steps are repeated until reaching the final number of coefficient number in the search scope (30 for the five features).
4. The coefficient numbers that give the highest accuracy with lowest number of features has been choosing, and the corresponding features have been choosing.
5. Finally the features have been combined in one vector.

The first approach was used to optimize the number of coefficients for the five features separately. The second approach was used to optimize the number of coefficients for the five features in a combination, The selection and evaluating of the coefficient number according to the classification accuracy have been done manually.

5 Classification

Several types of classifiers have been used in SER systems, including the Hidden Markov Model (HMM), the K-Nearest Neighbors (KNN), the Artificial Neural Network (ANN), the Gaussian Mixtures Model (GMM) and the Support Vector Machine (SVM). According to the literature [25] SVM and ANN are the most popular classifiers. Within this paper, SVM was adopted because it shows a strong performance when working with limited training data that has many features. SVM is a binary classifier used for classifications and regression. It can basically handle only two-class problems. SVM classifiers are mainly based on the use of kernel functions to nonlinearly map original features within a high dimensional space, in which data can be effectively classified using a linear classifier.

Classification with all speech utterances and spectral features was performed through the use of MATLAB R2012a. The radial basis kernel function (RBF) was employed with optimized g (in Gaussian function) and C (penalty parameter). The optimization of these classifier parameters was used in order to improve classifier accuracy. The scope of g is he scope of g is 2(−10:1:10) and the scope of C is 2 (−5:1:5). 5-fold. Cross-validation was performed for parameters selection. The performance analysis was undertaken using accuracy, which is the percentage of correctly-classified instances over the total number of instances.

6 Experiments and Analysis of Results

6.1 Optimized Based on Discrete Spectral Features

Within the first approach, the coefficients were separately optimized for the features, and the accuracy of the individual features was calculated. The result is shown in Fig. 4, where the x-axis indicates the number of coefficients, and the y-axis indicates the corresponding accuracy value. From the figure it can be observed that LPC gives the best accuracy of 58 % with 5 coefficients, and LPCC gives the best accuracy of 74 % with 12 coefficients.

For MFCC, as Fig. 5 shows, the best accuracy was 86 % with 20 coefficients. PLP gives the best results with 15 coefficients with an accuracy of 62 %, and finally RASTA-PLP gives the best accuracy of 54 % with 4 coefficients.

The results show that the MFCC feature provides the best accuracy among all features. This good result relates to the largest number of coefficients. LPCC and PLP provide good accuracy, with a reasonable number of coefficients. LPC and RASTA-PLP give the lowest numbers of coefficients and the worst accuracy. After separately determining the best coefficient values for every feature, the five features were combined. This provided an overall accuracy of 84 %, with 437 features.

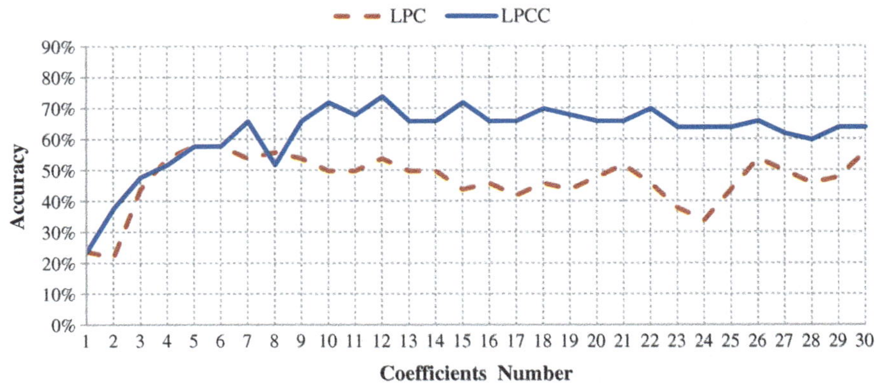

Fig. 4 The accuracy of LPC and LPCC for numbers of coefficients from 1 to 30

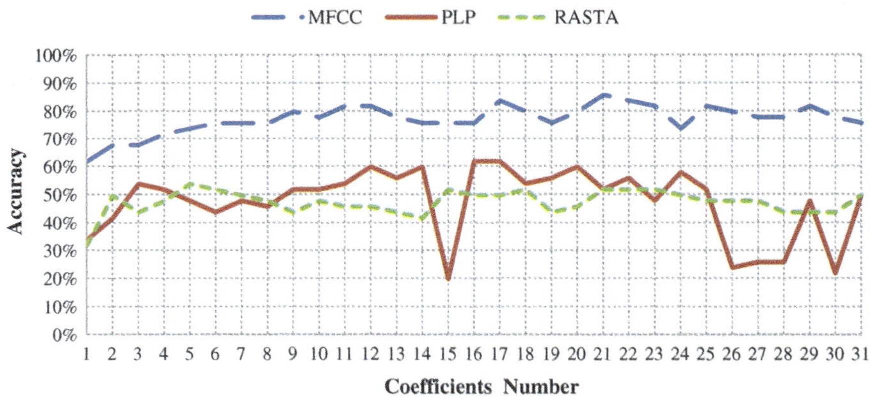

Fig. 5 The accuracy of MFCC, PLP and RASTA-PLP for all numbers of coefficients from 0 to 30

6.2 Optimized Based on Combine Spectral Features

Within the second model, the five features were combined first before coefficients optimization. Figure 6 showed that the best accuracy for the combined features was 88 % with 8 coefficients and 286 features.

The two approaches offered remarkable results as shown in Table 1. However, the second approach offered the highest accuracy with the lowest number of features.

When compared this study method undertaken with the greatest number of coefficients used in the past, namely 12 and 13 coefficients, as shown in Fig. 1, the result in Table 2 has shown that the number of coefficients selected by the two proposed Approaches can offer much greater accuracy than the number of coefficients used in the past. Additionally, the greater accuracy came with fewer features.

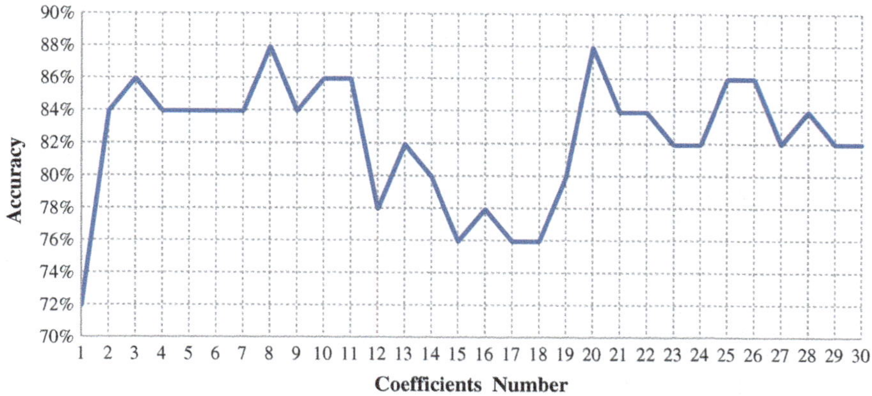

Fig. 6 The accuracy of the combined features for all numbers of coefficients, from 1 to 30

Table 1 Approaches accuracy

Approaches	Number of coefficients	Accuracy (%)	Number of features
Approach (1)	LPC(5), LPCC(12), MFCC(20), PLP(15), RASTA-PLP(4)	84	437
Approach (2)	8	88	286

Table 2 Comparison with the most number of coefficients used in SER

Number of coefficients	Accuracy (%)	Number of features
12	78	414
13	82	446

7 Conclusion

In this paper, two approaches for optimizing the coefficients numbers of spectral features, and for establishing a speech emotion model based on optimized coefficients, were proposed. Experiments have shown that the methods utilized for optimizing coefficients numbers not only increase the accuracy of the system when compared to the most commonly-used coefficients, but also reduces the numbers of features. This also shows that optimizing coefficient numbers for spectral features in combined, results in fewer features and better performance in speech emotion recognition, than when it is optimized separately before combination. Other Approaches used to optimize coefficients numbers will be studied in future works.

References

1. Wang F, Sahli H, Gao J, Jiang D, Verhelst W (2014) Relevance units machine based dimensional and continuous speech emotion prediction. Multimedia Tools Appl 1–18
2. Pierre-Yves O (2003) The production and recognition of emotions in speech: features and algorithms. Int J Hum Comput Stud 59(1):157–183
3. Rong J, Li G, Chen Y-PP (2009) Acoustic feature selection for automatic emotion recognition from speech. Inf Process Manag 45(3):315–328
4. Schuller B, Steidl S, Batliner A (2009) The inter-speech 2009 emotion challenge. In: INTERSPEECH, vol 2009. Citeseer, pp 312–315
5. Lee C-C, Mower E, Busso C, Lee S, Narayanan S (2011) Emotion recognition using a hierarchical binary decision tree approach. Speech Commun 53(9):1162–1171
6. Lee CM, Yildirim S, Bulut M, Kazemzadeh A, Busso C, Deng Z, Lee S, Narayanan S (2004) Emotion recognition based on phoneme classes. In: INTER-SPEECH, pp 205–211
7. Wang Y, Guan L (2005) Recognizing human emotion from audiovisual information. In: Proceedings of the IEEE international conference on acoustics, speech, and signal processing (ICASSP'05), vol 2. IEEE, pp ii–1125
8. Lugger M, Yang B (2008) Psychological motivated multi-stage emotion classification exploiting voice quality features. Speech Recognition, In-Tech, pp 395–410
9. Schuller B, Muller R, Lang MK, Rigoll G (2005) Speaker independent emotion recognition by early fusion of acoustic and linguistic features within ensembles. In: INTERSPEECH, pp 805–808
10. Kim EH, Hyun KH, Kim SH, Kwak YK (2007) Speech emotion recognition using eigen-FFT in clean and noisy environments. In: RO-MAN 2007 the 16th IEEE international symposium on robot and human interactive communication. IEEE, pp 689–694
11. Nwe TL, Foo SW, De Silva LC (2003) Speech emotion recognition using hidden markov models. Speech Commun 41(4):603–623
12. Fu L, Mao X, Chen L (2008) Speaker independent emotion recognition based on SVM/HMMS fusion system. In: International conference on audio, language and image processing, ICALIP 2008. IEEE, pp 61–65
13. Koolagudi SG, Barthwal A, Devliyal S, Rao KS (2012) Real life emotion classification using spectral features and gaussian mixture models. Procedia Eng 38:3892–3899
14. Murugappan M, Baharuddin NQI, Jerritta S (2012) DWT and MFCC based human emotional speech classification using LDA. In: 2012 International conference on biomedical engineering (ICoBE). IEEE, pp 203–206
15. Milton A, Roy SS, Selvi S (2013) SVM scheme for speech emotion recognition using MFCC feature. Int J Comput Appl 69(9)
16. Hegde S, Achary K, Shetty S (2015) Feature selection using fisher's ratio technique for automatic speech recognition. arXiv preprint arXiv:1505.03239
17. Burkhardt F, Paeschke A, Rolfes M, Sendlmeier WF, Weiss B (2005) A database of german emotional speech. In: Interspeech, vol 5, pp 1517–1520
18. Wu S, Falk TH, Chan W-Y (2011) Automatic speech emotion recognition using modulation spectral features. Speech Commun 53(5):768–785
19. Zhang Q, An N, Wang K, Ren F, Li L (2013) Speech emotion recognition using combination of features. In: 2013 fourth international conference on intelligent control and information processing (ICICIP). IEEE, pp 523–528
20. Kockmann M, Burget L et al (2011) Application of speaker-and language identification state-of-the-art techniques for emotion recognition. Speech Commun 53(9):1172–1185
21. Waghmare VB, Deshmukh RR, Shrishrimal PP, Janvale GB (2014) Emotion recognition system from artificial marathi speech using MFCC and LDA techniques. In: Fifth international conference on advances in communication, network, and computing, CNC, 2014

22. Kuchibhotla S, Vankayalapati H, Vaddi R, Anne K (2014) A comparative analysis of classifiers in emotion recognition through acoustic features. Int J Speech Technol 17 (4):401–408
23. Pao T-L, Chen Y-T, Yeh J-H, Liao W-Y (2005) Combining acoustic features for improved emotion recognition in mandarin speech. In: Affective computing and intelligent interaction. Springer, pp 279–285
24. Ingale AB, Chaudhari D (2012) Speech emotion recognition. Int J Soft Comput Eng (IJSCE) 2231–2307. ISSN
25. Pao T-L, Chen Y-T, Yeh J-H, Li P-J (2006) Mandarin emotional speech recognition based on SVM and NN. In: 18th international conference on pattern recognition ICPR 2006, vol 1. IEEE, pp 1096–1100

Classification of EEG Signals Using Single Channel Independent Component Analysis, Power Spectrum, and Linear Discriminant Analysis

Handayani Tjandrasa and Supeno Djanali

Abstract Epilepsy is a neurological disorder of the brain that can generate epileptic seizures when abnormal excessive activity occurs in the brain. The seizure is marked by brief episodes of involuntary movement of the body, and sometimes followed by unconsciousness. In this study, the EEG classification system was performed to predict whether EEG signals belong to normal individuals, epileptic patients in seizure free or seizure condition. The EEG dataset contains 5 sets of 100 EEG segments which is referred to as set A to set E. The classification system consisted of three scenarios. One of the scenarios involved the methods of Single Channel Independent Component Analysis (SCICA), power spectrum, and a neural network. The results were compared to the results without implementing SCICA. The last experiment showed the effect of using Linear Discriminant Analysis (LDA) to reduce the features of power spectrum. The results gave the accuracies for 3, 4, and 5 classes. By applying SCICA, all the accuracies were improved significantly with the maximum accuracy of 94 % for 3 classes.

Keywords Electroencephalogram (EEG) signals · Single channel ICA (SCICA) · Power spectrum · Linear discriminant analysis (LDA) · Multilayer perceptron network (MLP) · Radial basis function network (RBFN)

1 Introduction

Non-invasive EEG signals represent brain activities recorded through electrodes located on the scalp. The magnitude of the amplitude and frequency of EEG depends on the location and time of recording brain activity. A routine EEG is generally a recording of brain activity during a time interval, approximately 20–40 min [1].

H. Tjandrasa (✉) · S. Djanali
Department of Informatics, Sepuluh Nopember Institute of Technology (ITS), Surabaya, Indonesia
e-mail: handatj@its.ac.id

P.J. Soh et al. (eds.), *Advances in Machine Learning and Signal Processing*, Lecture Notes in Electrical Engineering 387, DOI 10.1007/978-3-319-32213-1_23

Brainwaves may appear irregular and have no general pattern, although particular patterns can occur such as spikes related to epilepsy.

EEG signals can be analyzed by extracting alpha (α), beta (β), delta (δ), and theta (Θ) rhythms. The frequency band of the alpha waves is 8–13 Hz, recorded from the occipital area of a relaxed person. The alpha waves become clearly noticeable when the eyes are closed. The beta waves is in the range of 13–30 Hz, recorded from the parietal and frontal lobes. As the brain activity increases, the frequency rises and the amplitude gets lower. The delta waves is in the range of 0.5–4 Hz, and theta waves is in the frequency band of 4–8 Hz.

An epileptic seizure occurs owing to a burst of abnormal electrical activity in the brain, and characterized by brief episodes of involuntary movement of the body, and sometimes followed by unconsciousness. The number of people worldwide suffered from epilepsy was approximately 50 million. Recent studies showed that up to 70 % people with epilepsy could be successfully treated [2].

Analyzing bulk of EEG recordings manually for detecting abnormal EEG data requires precision and a lot of time. Therefore, an automatic detection system helps diminish the burden of clinical works to give good interpretation of the brain waves.

There had been many researches on epileptic EEG classification for 2 classes. Many studies utilized discrete wavelet transform (DWT) to extract the features of EEG data. Daubechies wavelets were used to detect epileptic form discharges of 3-Hz spike in non-seizure epileptic patients [3]. The maximum, minimum, mean, and standard deviation for all DWT subbands of EEG signals were computed and used as the inputs of an ANFIS classifier [4]. Another research applied PCA, ICA, and LDA for DWT feature reduction and used a SVM classifier [5]. DWT coefficients were clustered by K-means clustering method, then the probability distribution of the clustered wavelet coefficients were computed and inputted into a multilayer perceptron network [6]. Also, an EEG classification method extracted the features from the cross correlation and power spectrums of healthy and epileptic EEG [7]. The features extracted from the intrinsic mode functions (IMFs) were used to differentiate the epileptic seizure and the normal EEG signals [8]. The classification method comprised of DWT features such as the mean value, standard deviation, energy, curve length, and skewness; and the combination of genetic programming and a KNN classifier, gave 2 and 3 classes of normal and epileptic EEG [9]. Features extracted from permutation entropy and a SVM classifier were applied to detect epileptic seizure EEG from non-seizure or normal EEG [10].

Feature extraction based on the power spectral analysis and other complexity features was used to classify 2 and 3 classes of EEG records of normal controls, mild cognitive impairment, and early Alzheimer's disease [11]. Power spectral analysis was applied to detect awareness in brain-injured patients [12].

In this study, the EEG classification system was applied to determine whether EEG signals belong to normal individuals, epileptic patients in seizure free or seizure condition based on 3, 4, and 5 classes.

2 Data and Methods

2.1 EEG Dataset

The EEG dataset used in this study is available online [13]. The EEG dataset contains 5 sets of 100 EEG signal segments. The sets are referred to as set A to set E. The sampling rate for data acquisition is 173.61 Hz and each segment has 4097 samples for a duration of 23.6 s. Both set A and set B were recorded non-invasively from five healthy persons in a relaxed and awake condition, with eyes open in set A, and eyes closed in set B. Sets C, D, and E were intracranially recorded from five patients with epilepsy. Set D and C were taken under complete seizure control. Set D was from the epileptogenic area. Whereas set C was taken from the hippocampal formation of the opposite hemisphere of the brain. Set E was recorded from all positions with seizure activity.

2.2 Preprocessing Using SCICA

ICA is a statistical model with the goal to separate mixed signals denoted by the vector $\mathbf{x}$, into the independent sources $\mathbf{s}$, from the mixing equation $\mathbf{x} = \mathbf{A}\mathbf{s}$, such that $\mathbf{s} = \mathbf{W}\ \mathbf{x}$, where the de-mixing matrix $\mathbf{W} = \mathbf{A}^{-1}$. The matrix W is computed using FastICA algorithm by applying negentropy function.

Single channel ICA (SCICA) is the use of the ICA method for a single channel signal by breaking up into several parts of similar length and treating the signal parts as a vector of signals [14, 15]. In this study, single channel ICA was applied for each of 500 signal segments of 5 data sets A-E. A signal segment which consist of 4097 samples was divided into 4 subsegments of 1024 samples, with the last sample removed. The resolved independent components were combined to get the original size of 4096 samples. EEG independent components could be selected to remove artifacts [16].

2.3 EEG Power Spectrum

EEG signals were sampled at discrete time intervals before further processing. The common process to analyze EEG data is to transform the discrete time domain into the discrete frequency domain since many EEG evaluations are related with certain rhythms such as alpha (α), beta (β), delta (δ), and theta (Θ). The magnitude of the discrete Fourier transform computed using FFT algorithm, is squared to get the power spectrum. In this study the number of discrete data was reduced by taking the summation of every 25 or 50 data values of power spectrum.

2.4 Feature Extraction

One straightforward approach, the power spectrum of EEG data can be calculated by taking the square of the FFT magnitude. The frequency range is limited to 500 samples which approximates 21.2 Hz, since beyond that range the magnitude approaches to zero.

Linear Discriminant Analysis (LDA) transforms variables into reduced features that discriminate classes maximally based on the within-class and between-class scatter matrices.

$\mathbf{S}_W$ is the within-class scatter matrix, defined by the following equation,

$$\mathbf{S}_W = \sum_{i=1}^{c} \sum_{x_k \in X_i} (x_k - \mu_i)(x_k - \mu_i)^T \tag{1}$$

and $\mathbf{S}_B$ is the between-class scatter matrix, given by,

$$\mathbf{S}_B = \sum_{i=1}^{c} N_i(\mu_i - \mu)(\mu_i - \mu)^T \tag{2}$$

where N_i is the number of samples in class X_i, c is the class number, and μ_i is the class mean of X_i.

The purpose is to maximize $\mathbf{S}_B$ while minimizing S_W, which can be satisfied by finding $\mathbf{w}$ that maximizing the following objective,

$$J(\mathbf{w}) = \frac{|\mathbf{w}^T \mathbf{S}_B \mathbf{w}|}{|\mathbf{w}^T \mathbf{S}_W \mathbf{w}|} \tag{3}$$

$\mathbf{w}$ are the largest eigenvectors of $\mathbf{S}_W^{-1}\ \mathbf{S}_B$ and the maximum rank is of $c - 1$.

2.5 Classifiers

In this study we applied two kinds of classifiers, the first classifier is a multilayer perceptron (MLP) network using 10-fold cross validation, and the second classifier is a radial basis function network (RBFN) using the same cross validation.

A multilayer perceptron network has an input layer, followed by one or several hidden layers, and an output layer. For the training stage, MPL implements backpropagation algorithm. The algorithm consists of two steps which are the feedforward and the backpropagation steps. In the feed forward step, the computation starts from the input layer up to the output layer, using network weights that are initialized to small random numbers. In the backpropagation step, the errors are propagated backward from the output layer to the input layer.

A basic RBF network comprises of an input layer, a hidden layer, and an output layer. The input layer are the input nodes. Each neuron of the hidden layer computes the distance between the centre of the neuron and the input vector. The output neuron is given by computing the value of the radial basis function for the distance. The output layer uses the supervised learning process similar to MLP.

2.6 Implementation

The features from the power spectrum were calculated by summing every 50 component values to produce one feature. The frequency range of 4097 samples is the same as the sampling rate which is 173.61 Hz. The samples were limited to 500 to get 10 features. The examples of normalized power spectrum features for the 20th segment of each set from A to E, are shown in Fig. 1. The figure shows low intensity in set A (normal, eyes open), alpha waves clearly noticeable in set B (normal, eyes closed), and the peaks of set E (EEG seizure) are closed to theta waves and between alpha to beta waves. The spectrums in set C and set D look more similar than the other sets, such that the class separation of set C and D reduces the classification accuracy. The spectrum characteristics can show some variations for other segments.

The total number of training and testing data from all five sets (A to E) was 500 signal segments. The amplitudes of EEG signals differed significantly especially for

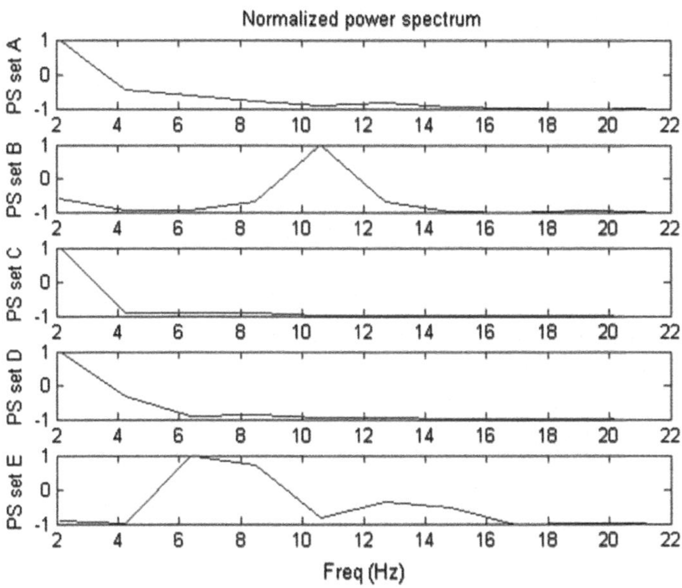

Fig. 1 Examples of power spectrum with 10 features from set A-E

the seizure EEG signals which had the biggest energy. Therefore the power spectrum needed to be normalized in the range −1 to 1 in order to be able to differentiate the EEG signals based on the spectrum characteristics. Using LDA, features reduced to a smaller number which was less than the number of classes. The classes of testing data were determined using MLP and RBF network classifiers.

The process was repeated for obtaining 20 features by summing every 25 components of power spectrum, and both results of different choices of component number were compared.

The classification system was carried out using three scenarios as follows:

1. Combination of PS (power spectrum) and ANN classifier (MLP and RBFN).
2. Combination of SCICA, PS (power spectrum), and ANN classifier.
3. Combination of PS, LDA, and ANN classifier.

In the study, all EEG segments of set A-E were applied and grouped alternatively in 3 classes (AB-CD-E), 4 classes (A-B-CD-E), and 5 classes (A-B-C-D-E). The previous researches mostly worked on 2 classes (normal and ictal stage) [5, 17, 18]. Several researches studied on 3 classes (normal, interictal, ictal stages) [18]. Another review showed that the number of researches using wavelet based processing, was 4 for 4 classes, and 1 for 5 classes [19].

3 Results and Discussion

SCICA was used as preprocessing to improve the performance of the classification system. Each signal was divided into 4 subsignals which were treated as multichannel signals. The computation of ICA were carried out on the subsignals. The results were combined into one segment of 4096 data. Figure 2 shows the example of the process using set E.

The next process was to apply the FFT algorithm and reduced the data into 10 or 20 features. The performance of the classification system was compared with the system without using preprocessing.

The total number of the training dan testing data was 500 signal segments from set A-E. Using MLP and RBFN as classifiers, the results of classification with 10-fold cross validation for 10 and 20 features are shown in Table 1. LDA reduced 10 or 20 features into 2 features. For 5 classes, each set from A to E, are separately classified, and the average accuracies for MLP and RBFN without SCICA were 69.1 and 66.4 % respectively and improved to 76.2 and 75 % by using SCICA. The lower accuracies than other classes were mostly caused by the similarity of set C and D. The accuracies were improved significantly by applying SCICA as preprocessing. By merging set C and D, the accuracies of 4 classes increased significantly to the average of 88.2 % for MLP and 83.5 % for RBFN without SCICA, and the accuracies improved to 91.3 and 88.9 % respectively with SCICA. Further merging set A and B resulted in 3 classes, the first class was for EEG signals of

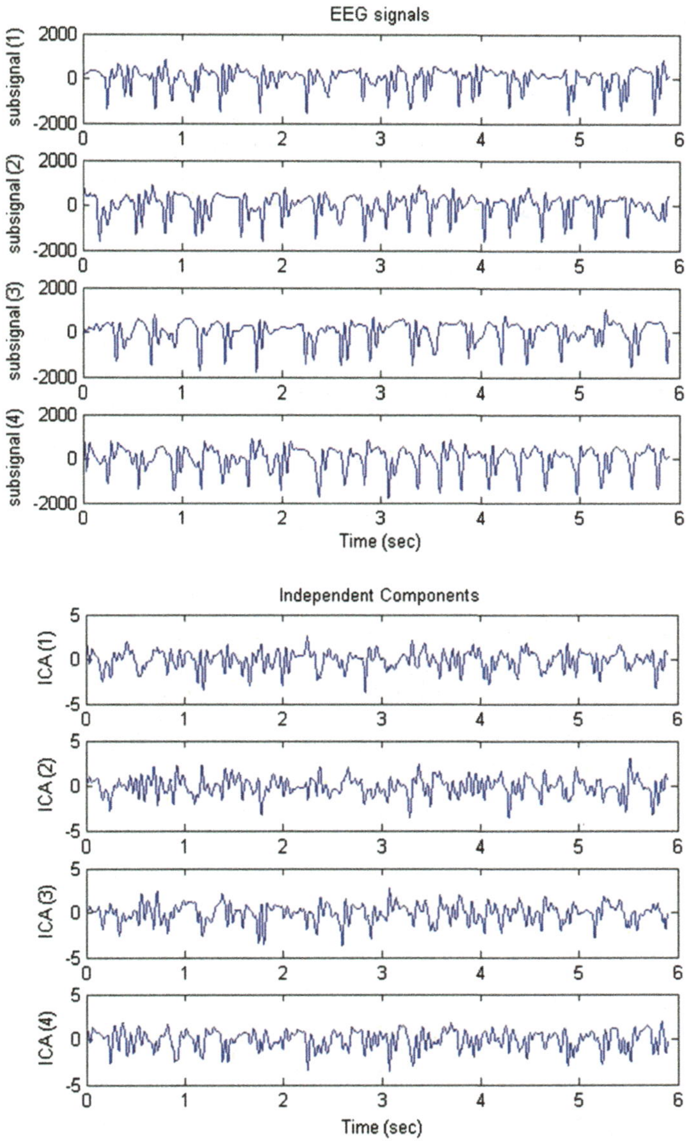

Fig. 2 Examples of SCICA process for set E

healthy volunteers, the second class was for EEG signals of epilepsy patients during the seizure free condition, the third class was for EEG signals during epileptic seizures. The average accuracies for MLP and RBFN without SCICA as preprocessing were 90.5 and 84.5 % respectively, and the accuracies were improved to 92.9 and 90.5 % respectively with SCICA. The results of feature reduction using LDA with 2 features for 4 and 3 classes are illustrated in Figs. 3 and 4 respectively.

Table 1 Accuracy for 3, 4, 5 classes

Grouping dataset	SCICA-PS-ANN		PS-ANN		PS-LDA-ANN	
	MLP (%)	RBFN (%)	MLP (%)	RBFN (%)	MLP (%)	RBFN (%)
AB-CD-E 20 features	94	91.6	91.8	84.6	89	89
AB-CD-E 10 features	91.8	89.4	89.2	84.4	88.4	89.2
A-B-CD-E 20 features	91.4	89.2	89.2	84.2	81.8	82
A-B-CD-E 10 features	91.2	88.6	87.2	82.8	80.6	84
A-B-C-D-E 20 features	78.2	75.2	69.6	66.8	–	–
A-B-C-D-E 10 features	74.2	74.8	68.6	66	–	–

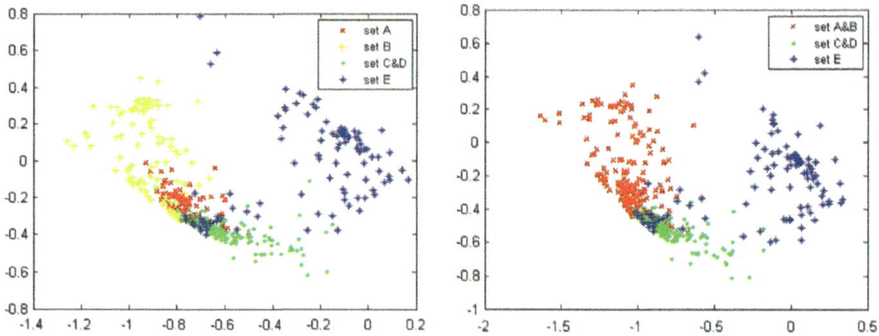

Fig. 3 Plots of EEG data using 2 features reduced from 10 features for 4 and 3 classes respectively

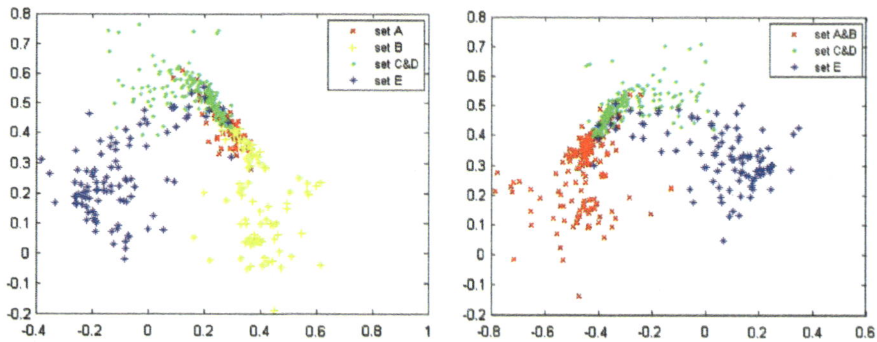

Fig. 4 Plots of EEG data using 2 features reduced from 20 features for 4 and 3 classes respectively

4 Conclusion

The EEG classification system for 5 classes had low accuracy since the high level of similarity between set C and D from epileptic EEG. MLP gave higher accuracy than RBFN for 10 and 20 features which each represents the summation of power spectrum within the frequency interval 2.12 and 1.06 Hz respectively. Applying SCICA as preprocessing improved the accuracy significantly.

Applying LDA to reduce 10 and 20 features to 2 features, gave better accuracy for RBFN compared to MLP. For LDA, the number of features is less than the number of classes. By having 2 features, the EEG data could be observed visually. The computation results showed that the EEG classification system gave very good accuracy results for 4 classes and 3 classes of EEG signals A-E. The maximum accuracy was 94 %.

Acknowledgments This work is supported by the grant from Directorate General of Higher Education, Ministry of Research, Technology and Higher Education, Indonesia.

References

1. Niedermeyer E, Da Silva FL (2004) Electroencephalography: basic principles, clinical applications, and related fields. Lippincot Williams & Wilkins. ISBN 0-7817-5126-8
2. World Health Organization (2015) Epilepsy. Fact sheets. Retrieved 22 Aug 2015
3. Adeli H, Zhou Z, Dadmehr N (2003) Analysis of EEG records in an epileptic patient using wavelet transform. J Neurosci Methods 123:69–87
4. Güler İ, Übeyli ED (2005) Adaptive neuro-fuzzy inference system for classification of EEG signals using wavelet coefficients. J Neurosci Methods 148:113–121
5. Subasi A, Gursoy MI (2010) EEG signal classification using PCA, ICA, LDA and support vector machines. Expert Syst Appl 37:8659–8666
6. Orhan U, Hekim M, Ozer M (2011) EEG signals classification using the K-means clustering and a multilayer perceptron neural network model. Expert Syst Appl 38(10):13475–13481
7. Iscan Z, Dokur Z, Demiralp T (2011) Classification of electroencephalogram signals with combined time and frequency features. Expert Syst Appl 38:10499–10505
8. Pachori RB, Bajaj V (2011) Analysis of normal and epileptic seizure EEG signals using empirical mode decomposition. Comput Methods Programs Biomed 104:373–381
9. Guo L, Rivero D, Dorado J, Munteanu CR, Pazos A (2011) Automatic feature extraction using genetic programming: an application to epileptic EEG classification. Expert Syst Appl 38:10425–10436
10. Nicolaou N, Georgiou J (2012) Detection of epileptic electroencephalogram based on permutation entropy and support vector machines. Expert Syst Appl 39:202–209
11. McBride JC, Zhao X, Munro NB, Smith CD, Jicha GA, Hively L, Broster LS, Schmitt FA, Kryscio RJ, Jiang Y (2014) Spectral and complexity analysis of scalp EEG characteristics for mild cognitive impairment and early Alzheimer's disease. Comput Methods Programs Biomed 114:153–163
12. Goldfine AM, Victor JD, Conte MM, Bardin JC, Schiff ND (2011) Determination of awareness in patients with severe brain injury using EEG power spectral analysis. Clin Neurophysiol 122:2157–2168

13. Andrzejak RG, Lehnertz K, Mormann F, Rieke C, David P, Elger CE (2001) Indications of nonlinear deterministic and finite-dimensional structures in time series of brain electrical activity: dependence on recording region and brain state. Phys Rev E 64(6):061907
14. Davies ME, James CJ (2007) Source separation using single channel ICA. Sig Process 87:1819–1832
15. Mijovic B, Vos MD, Gligorijevic I, Taelman J, Huffel SV (2010) Source separation from single-channel recordings by combining empirical-mode decomposition and independent component analysis. IEEE Trans Biomed Eng 57:2189
16. Chaumon M, Bishop DVM, Busch NA (2015) A practical guide to the selection of independent components of the electroencephalogram for artifact correction. J Neurosci Methods 250:47–63
17. Fu K, Qu J, Chai Y, Zoub T (2015) Hilbert marginal spectrum analysis for automatic seizure detection in EEG signals. Biomed Sig Process Control 18:179–185
18. Acharya UR, Sree SV, Swapna G, Martis RJ, Suri JS (2013) Automated EEG analysis of epilepsy: a review. Knowl-Based Syst 45:147–165
19. Faust O, Acharya UR, Adeli H, Adeli A (2015) Review: wavelet-based EEG processing for computer-aided seizure detection and epilepsy diagnosis. Seizure 26:56–64

Comparison of Text Forum Summarization Depending on Query Type for Text Forums

Vladislav Grozin, Kseniya Buraya and Natalia Gusarova

Abstract Various approaches are developed for evaluation of query-oriented text summarization. However, for text forums this procedure is not well-defined, and standard approaches are not suitable. Evaluation of query-oriented text summarization greatly depends on the query type. We compare two typical scenarios of search of professionally significant information on Internet forums. Our subject of interest is the similarities and differences between relevance-oriented queries and usefulness-oriented queries. To compare these query types we have collected dataset, extracted textual, structural features and social graph features, constructed different ranking models, used suitable quality measure (NDCG), and applied feature selection techniques to investigate causes of differences. We have found out that these query types are very different by their nature, have weak correlation. Distinct model types and features should be used in order to create an efficient information retrieval system for each query type.

1 Introduction

Nowadays the value of professionally important information is rising steadily. Specialized web-forums are a valuable source of knowledge of that kind. Forums contain experience of people who actually used the technology and its features. Moreover, forums contain both positive and negative experiences—something that is not available from official documentation at all. But usually the majority of posts

V. Grozin (✉) · K. Buraya · N. Gusarova
Mechanics and Optics, National Research University of Information Technologies, Saint-Petersburg 197101, Russia
e-mail: 161397@niuitmo.ru

K. Buraya
e-mail: ks.buraya@gmail.com

N. Gusarova
e-mail: natfed@list.ru

P.J. Soh et al. (eds.), *Advances in Machine Learning and Signal Processing*,
Lecture Notes in Electrical Engineering 387, DOI 10.1007/978-3-319-32213-1_24

at forums are useless and superfluous, containing a lot of hackneyed, repeated and irrelevant information. The obvious solution for this problem is using techniques of text summarization.

The text summarization is one of the tasks of information retrieval. It is about automatically extracting the main gist of the given documents to indicate the main aspects in them. This task is being actively investigated yielding a wide range of approaches, search mechanisms, results management and presentation (see, for example, [1]).

A crucial issue of the text summarization is evaluation problem, involving information retrieval effectiveness, or assessing consumers's satisfaction with the system [2]. Various approaches are developed for an assessment of text summarization. First of all, the approaches based on the 'bag of words' model are widely used. Typically the experimental queries are generated by extracting keywords from the list of terms frequently searched for within the field of interest (see, for example, [3]).

Besides, there is a set of search evaluation initiatives and competitions like TREC, DUC and MUC. They have created methodologies that can be conducted in a controlled lab-based setting. The most used is the Cranfield methodology [2] based on specialized test collection containing a set of predefined topics describing typical users' information needs.

Evaluation from a user-oriented perspective goes beyond the traditional Cranfield style experiments [2]. A common approach is to investigate users behavior in retrieval tasks in a controlled lab-based environment. Questions identified by the researcher are used here instead of predefined queries.

However, there is no track devoted to web-forums within the list of tracks managed by these evaluation initiatives [1].

Therefore, the paper deals with applying standardized evaluation approaches for text forums. Evaluation greatly depends on query type. TREC distinguishes two query types: usefulness and relevance-oriented. This paper is concerned with finding similarities and differences between these query types in order to find whether this approach is applicable for text forums.

2 Related Works

2.1 Terminology

There is a various terminology for information retrieval evaluation. Saracevic et al. [4] distinguish six levels of evaluation for information systems (including IR systems). The first three levels are referred to measuring system performance, the last three levels correspond to user-oriented evaluation. These may be assessed by different terms, including efficiency, utility, informativeness, usefulness, usability, satisfaction and the users search success [2].

The term relevance is vaguely used in literature. Some authors use it to refer to the degree of match between a document and a question; in contrast, other authors distinguished between relevance (similar to system relevance assessed by an external judge/expert) and pertinence (user-relevance assessed only by the real user with the information need represented in the question) (see the discussion in [2]).

Saracevic et al. [4] consider utility to be a more appropriate measure for evaluating information retrieval systems. A document has utility if it is pertinent (relevant as perceived by the user) and also contributes to the user knowledge in the context of the query (by providing information that was previously unknown). In our paper, we follow this opinion and adopt utility as a measure of usefulness and worth of the answers provided by the system to its users.

2.2 Methods of Forum Summarization

There are different approaches to the problem of text summarization. Main classifications are extraction-based and abstraction-based summarization as well as single-document and multi-document approaches. The majority of works in the area of forum summarization use extraction-based techniques and single-document approach [5]. Extractive forum summarization tasks are in turn divided into generic summarization (obtaining a generic summary or abstract of the whole thread) and query relevant summarization, sometimes called query-based summarization, which summarizes posts specific to a query [6].

The large variety of algorithms is used in both variants including naive Bayes classifier, statistical language models, topic modeling, graph-based algorithms etc. [3, 5–10]. In this paper we use algorithms of gradient boosting and linear regression which have already proved the efficiency for text forum summarization in our previous work [6, 7]. We also use for comparison a query-oriented algorithm based on LDA (see below for details).

2.3 Nearest Researches

We managed to find several researches with the aim close to our work in literature. Grozin et al. [7] consider reviews posted in web, assessing “Review Pertinence” as the correlation among review and its article. Tang et al. [8] consider the sentence relevance and redundancy within the summarized text. Their maximum coverage and minimum redundant (MCMR) text summarization system computes sentence relevance as its similarity to the document set vector. This idea is also used in [9] for cross-lingual multi-document summarization.

Some articles [10, 11] are devoted to comparing system effectiveness and user utility. Oufaida [10] compared traditional TREC procedure of batch evaluation and user searching on the same subject. Petrelli [11] confirmed that test collections and their associated evaluation measures do predict user preferences across multiple information retrieval systems. They found that NDCG metric most effectively modeled user preferences.

To sum up there are no articles dedicated in deep details to the problem discussed in our article.

3 Experiment

Our goal is to create models that efficiently retrieve posts for different query types from text forums that will satisfy users needs, and investigate differences and similarities between query types. In our work, we examine two query types (and thus, construct two ranking model types):

- Query which target is to retrieve objective and interesting information in the domain of subject of interest (informativeness). This query type focuses on extracting pieces of information that contribute towards user's knowledge.
- Query which target is to retrieve any information related to the query (relevance). Text forum can contain posts that are relevant (related) to the query; the goal of this scenario is to fetch these posts.

Therefore, we have to study informativeness-oriented queries and relevance-oriented queries, their similarities and differences. Note that these post informativeness and relevance maybe be independent: posts can be irrelevant, yet informative (detailed explanation of something that is related to the domain of the query, but not related to the user query itself), and posts can be relevant, yet non-informative (thread-starting questions).

3.1 Data Collection

To collect our data, we used following algorithm:

1. Select a forum and a narrow user query within. The query is defined as a set of keywords.
2. Select some threads within the forum which titles contain query keywords. This is done to reduce amount of obviously non-informative and irrelevant posts, and reduce amount of required expert time.
3. Copy information about all the posts from these threads: post text, author, and thread URL.
4. Mark down sentiment value, informativeness and relevance of each post.

Formal criteria for marking up informativeness, Relevance and Sentiment are listed in Table 1. The forums used in our work are listed at Table 2. Each thread collected from forum contains at least 400 posts.

Table 1 Formal markup criteria

Parameter	Value	Comment
Informativeness	0	Post contains no useful information
	1	Post gives some useful information, but most of it is not useful
	2	Post gives some useful information, but it is
	3	Post contains useful information, but explanations and arguments are missing
	4	Post contains useful information, but explanations and arguments are incomplete
	5	Post contains a lot of useful information with rich explanations and arguments
Relevance	0	Post is completely irrelevant to the query/topic
	1	Posts theme weakly intersects with query/topic
	2	Post contains mostly irrelevant information, but some parts of it are relevant
	3	Post contains mostly relevant information, but some parts of it are irrelevant
	4	Post is relevant to the query/topic, but contains some extending information
	5	Post is completely relevant to the query/topic
Sentiment value	−2	Post contains clearly expressed negative emotions
	−1	Post contains humble negative emotions or sarcasm
	0	Post has neutral sentiment value
	1	Post is overall cheerful and contains signs of joy or happiness
	2	Post contains clearly expressed positive emotions and exaltation

Table 2 The chosen Internet forums

Forum		Query	URL
1	iXBT (hardware forum)	Choosing of ADSL modem	http://forum.ixbt.com/
2	Fashion, style, health	Diets for overweight people	http://mail.figgery.com/
3	Kinopoisk (cinema forum)	"Sex at the city" series	http://forum.kinopoisk.ru/
4	Housebuilding forum	Building a house using 6×6 wooden planks	https://www.forumhouse.ru/

3.2 Models and Parameters

We have to construct set of models to estimate informativeness and relevance. Two models were used to estimate each target parameter:

- Linear model. It is interpretable, and it captures linear dependencies well. We used non-regularized linear model.

- Gradient boosting model. It is interpretable, and it can capture nonlinear dependencies. We used three CV folds to estimate the best amount of trees; number of trees were capped to 2000, and shrinkage factor was 0.001. Indirection level value (number of splits for each tree) was set to 3.
- LDA. This robust interpretable model splits available posts into subsets (topics) according to their texts using bag-of-words approach. Each topic can be interpreted as a set of keywords, and we used presence of these keywords to estimate target variables. It is expected that these subsets will have different properties (for example, "offtopic" and "on-topic"). For hyperparameters we have chosen 100 iterations and 3 topics.

Models for each target variable were constructed independently, but using the same technique, same train and test sets, and same set of features for linear and gradient boosting models.

Despite the fact that our target variables have six discrete grades, we treated them as quasi-continuous and used models in regression mode to avoid sparse class population because we have multiple strictly ordered classes.

To fit models we divided the data from each forum into train (70 % of each forum) and test (30 %) sets. To ensure model stability we used bootstrap-like method. The data was resampled with replacement, then it was split into test and train sets, after that, models were fit, and model qualities were estimated. This process is repeated 200 times, and model qualities are averaged and confidence interval is calculated.

3.3 Quality Estimation

Widely used recall/precision metrics are not useful in our context, because we have ordered multiple classes for each target variable. It is recommended to use cumulative gain metrics to evaluate retrieval system quality [2]. We used normalized cumulative gain. It is a cross-query comparable metric that lies between 0 and 1. It is calculated using formula:

$$NDCG_N = \frac{DCG_N}{IDCG_N} \tag{1}$$

$$DCG_N = rel_1 + \sum_{i=2}^{N} \frac{rel_i}{log_2(i)} \tag{2}$$

where N is the size of resulting set (how many documents to retrieve), rel_i is true value of target variable (relevance or usefulness) of ith post in the retrieved set, and $IDCG_N$ is maximum possible DCG_N for specified forum and N, i.e. DCG_N for ideal algorithm. The full procedure of model quality estimation for both query types is:

1. Fit models to train set of each forum for each target variable (usefulness,relevance) and apply them to test set of each forum. This gives $Usefulness_{est}$ and $Relevance_{est}$, some approximation of true usefulness and relevance values of test set.
2. Sort posts by decreasing target variable approximation ($Usefulness_{est}$ or $Relevance_{est}$) and take N top posts. This gives selection of N best posts according to the model.
3. Calculate NDCG metric for the selection using true usefulness and relevance values of this N best posts subset.

We variated N from 2 to 30 to investigate how models behave in case of different selection windows.

3.4 Features

We have to extract features for linear and gradient boosting models that will hint us on how useful or relevant is the specific post. There are a lot of possible features we can extract; we used the ones that are suitable for our case. Chosen features are listed at Table 3.

Table 3 Features

Type	Feature	What this feature means
Post's author graph features	Betweenness, non-sentiment graph inDegree, non-sentiment graph outDegree, non-sentiment graph	Author's social importance How many times author was quoted How many times author quoted someone
	Betweenness, sentiment graph	Author's social importance
	inDegree, sentiment graph	With which sentiment author was quoted
	outDegree, sentiment graph	Author's quotes sentiment
Post's author features	Number of threads author is participating in	Author activity
Thread-based post features	Position in thread	Chance of off-topic
	Times quoted	Post's impact on forum
Text features	Length	Number of arguments and length of explanations
	Links	Number of external sources/images
	Sentiment value (calculated using sentiment keywords)	Post's usefulness
	Number of query keywords	Topic conformity
	Most used topic keyword count	Topic conformity

Sentiment value was marked down by experts and is used as a feature. It is expected that posts with a positive attitude will be more useful.

Also, simple non-semantic text features were extracted: text length in characters, number of links and number of keywords within text. We used two algorithms of keyword extraction. First one splits the query into words, and treats them as keywords. A more extensive list of keywords would mean a search for synonyms and equivalents; it requires semantic analysis and is not available for every language and for every query domain. The second algorithm creates frequency table for each thread, and takes top 5 most popular words. In both algorithms, stopwords were stripped.

We represented social structure in the form of a social graph, where the nodes are the users, and edges indicate a link between two users. For the creation of the social graph we have used citation analysis: if person A quotes person B by explicitly mentioning his name in text, there is a guaranteed connection between A and B. We used two methods: a non-sentiment graph (edge weight is always 1) and a sentiment graph (edge weight is related to the post's sentiment value). After the creation of the graph parallel edges' weights were summed. Then, the weights of the edges were inverted.

Node centrality is often used to find people who are important members of society. We considered some proven [12] metric to evaluate node centrality: Betweenness centrality—the number of shortest paths between all pairs of nodes that pass through the node; inDegree—the total weight of incoming edges; outDegree—the total weight of the outgoing edges.

Position in thread is calculated as position of post in chronological order (first post has position in thread equal to one, next post has value of two etc.).

4 Results and Discussion

Correlation between usefulness and relevance on all forums is 0.36. This is an evidence of that these parameters are different, and query types expect IR system to do different things. Also, distribution of relevance is skewed towards 5 (see Fig. 1b), while distribution of usefulness has peak around 3 (see Fig. 1a). The skew of relevance is explained by the procedure of data collection: we choose posts from already relevant threads, so it is expected that most of marked posts have high relevance. Distribution of usefulness shows that great portion of posts has moderate (2–3) usefulness, and only a small portion of posts have marginally high or low usefulness.

Figure 2 shows result of application of the procedure described at Quality estimation section. Plotted lines have 99.5 % error bands.

As one can notice, linear model is better at selecting relevant posts, and gradient boosting model is better at estimating usefulness. This means that relevance can be better described as a linear combination of the features, and usefulness is best approximated as a non-linear construction over calculated features.

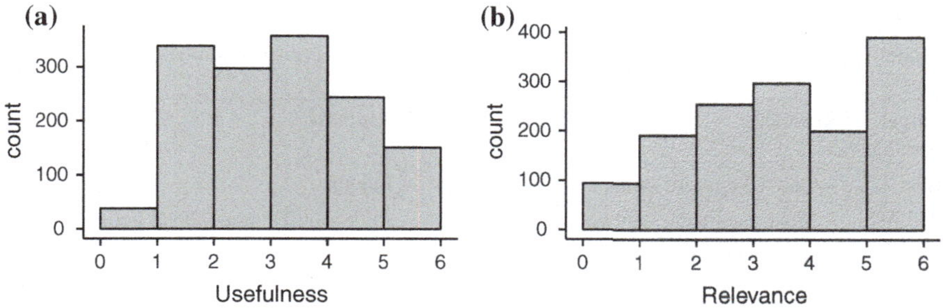

Fig. 1 Distribution of target variables. **a** Distribution of usefulness. **b** Distribution of relevance

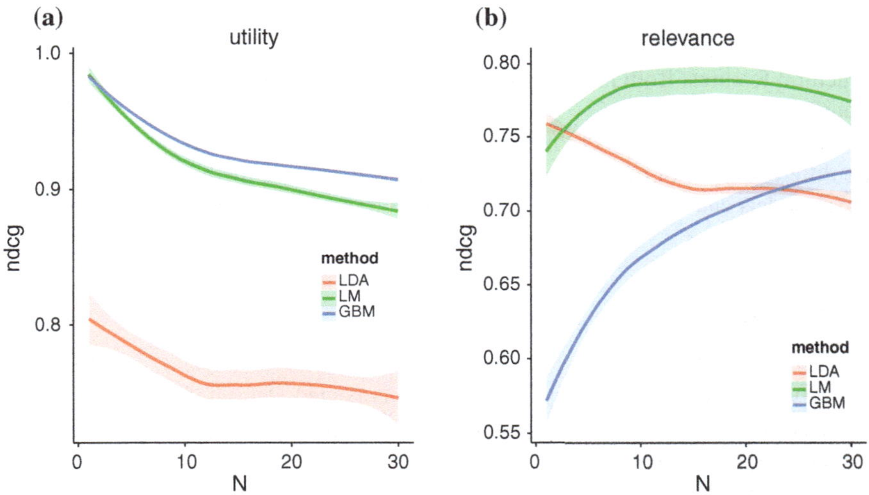

Fig. 2 Dependence of NDCG on target variable and model type. **a** Target variable is usefulness. **b** Target variable is relevance

For better comparison of query types, we have to investigate which features were best in each model. To do this we have chosen most important features (significance level of 0.001) from linear model constructed for relevance, and best features from gradient boosting model (the best models for each target variable). Feature selection from GBM was done by selecting top 4 features using relative influence metric [13]. The results are presented at Table 4.

Relevance is best estimated using keyword-related features, and usefulness is best estimated using post length and position in thread. Also, graph features appear in best feature list. This means that relevance-oriented are quite different from usefulness-oriented queries. Relevance-oriented queries can be handled by keyword-based features, and usefulness-oriented queries require simple textual and

Table 4 Best features

Model type	Linear regression	Gradient boosting model
Target variable	Relevance	Usefulness
Best features	– Query keyword count	– Length
	– Most used topic keyword count	– Author outDegree (social graph feature)
	– Author inDegree (social graph feature)	– Author outDegreeSent (social graph constructed using sentiment values feature)
	– Author inDegreeSent (social graph constructed using sentiment values feature)	– Post position in thread

structural features. Both model types can be improved by incorporating social graph features.

Note that despite the fact that relevance-oriented and usefulness-oriented queries are different types of queries that require different ranking methods, in real systems these models can be merged [14] in order to retrieve both relevant and useful posts.

5 Conclusion

We have defined query types to consider, collected dataset from four forums, constructed features and models, estimated model quality and interpreted the results to compare query types. The usefulness-oriented and relevance-oriented queries are different by nature, and have weak correlation of their target variables. Relevance-oriented queries are best handled using keywords-based features and linear model while usefulness-oriented queries are best handled using gradient boosting model and textual and structural features.

References

1. Nenkova A, McKeown K (2012) A survey of text summarization techniques. Min Text Data Springer 43–76
2. Elbedweihy Khadija M, Wrigley Stuart N, Clough Paul, Ciravegna Fabio (2015) An overview of semantic search evaluation initiatives. Web Semant Sci Serv Agents World Wide Web 30:82–105
3. Bhatia S, Mitra P (2010) Adopting inference networks for online thread retrieval. In: Proceedings of the twenty-fourth AAAI conference on artificial intelligence. Atlanta, Georgia, pp 1300–1305
4. Saracevic T, Kantor P, Chamis AY, Trivison D (1988) A study of information seeking and retrieving. I Backgr Methodol J Am Soc Inf Sci 39:161–176

5. Ren Zh, Ma J, Wang Sh, Liu Y (2011) Summarizing web forum threads based on a latent topic propagation process. In: CIKM11, October 2428. Glasgow, Scotland
6. Grozin VA, Gusarova NF, Dobrenko NV (2015) Feature selection for language-independent text forum summarization. In: International conference on knowledge engineering and semantic web, pp 63–71
7. Grozin VA, Dobrenko NV, Gusarova NF, Ning T (2015) The application of machine learning methods for analysis of text forums for creating learning objects. In: Computational linguistics and intellectual technologies, vol 1, pp 199–209
8. Tang J, Yao L, Chen D (2009) Multi-topic based query-oriented summarization. In: Society for industrial and applied mathematics—9th SIAM international conference on data mining 2009, Proceedings in applied mathematics 3, pp 1141–1152
9. Wang Jun-ze, Yan Zheng, Yang Laurence T (2015) Ben-xiong Huang An approach to rank reviews by fusing and mining opinions based on review pertinence. Inf Fusion 23:3–15
10. Oufaida Houda, Nouali Omar, Blache Philippe (2014) Minimum redundancy and maximum relevance for single and multi-document Arabic text summarization. J King Saud Univ Comput Inf Sci 26:450–461
11. Petrelli Daniela (2008) On the role of user-centred evaluation in the advancement of interactive information retrieval. Inf Process Manage 44:22–38
12. Borgatti Steve (2005) Centrality and network flow. Soc Netw 27(1):55–71
13. Friedman J (2001) Greedy boosting approximation: a gradient boosting machine. Ann Stat 29:1189–1232
14. Croft WB (2015) Combining approaches to information retrieval. In: Croft WB (ed) Advances in information retrieval. Kluwer Academic Publishers, Boston

Vibro-Acoustic Fault Analysis of Bearing Using FFT, EMD, EEMD and CEEMDAN and Their Implications

Satish Mohanty, Karunesh Kumar Gupta and Kota Solomon Raju

Abstract This paper analyses the vibro-acoustic characteristics of the bearing using FFT (Fast Fourier Transform), EMD (Empirical Mode Decomposition), EEMD (Ensemble EMD) and CEEMDAN (Complete EEMD with Adaptive Noise) algorithms. The main objective is to find out the best algorithm that avoids mode mixing problems while decomposing the signal and also enhance the feature extraction. It is observed that even though acoustic and vibration can be used for the fault detection in the bearing, duo follow differently interns of their statistical distributions. The feature of the bearing is acquired using acoustic and vibration sensors and analyzed using non-linear and non-stationary signal processing techniques. The statistical distribution of the data plays a major role in truly extracting the components using signal processing techniques. All the algorithms are data driven, as per the conditional events of the system, these algorithms efficiency increases or decreases. Here, the vibro-acoustic feature of the normally distributed acoustic and vibration signature are extracted effectively using CEEMDAN with least computational time and efficient signal extraction.

1 Introduction

Bearings are the vital element in almost all industries and daily life. It has wide application and the preventive measures need to be taken care to avoid any kind of disaster. Bearing fault generally occur due improper uses i.e., harsh environmental

S. Mohanty (✉) · K.K. Gupta
Birla Institute of Technology and Science, Pilani 333031, India
e-mail: satish.mohanty@pilani.bits-pilani.ac.in

K.K. Gupta
e-mail: kgupta@bits-pilani.ac.in

S. Mohanty · K.S. Raju
Central Electronics Engineering Research Institute (CEERI),
Council of Scientific and Industrial Research (CSIR), Pilani 333031, India
e-mail: solomon@ceeri.ernet.in

P.J. Soh et al. (eds.), *Advances in Machine Learning and Signal Processing*,
Lecture Notes in Electrical Engineering 387, DOI 10.1007/978-3-319-32213-1_25

condition and improper uses [1]. The condition of failure in bearing depends on various parameters such as bearing types, applications, environmental conditions, or any manufacturing defects [2, 3].

The condition in the bearing can be evaluated and analyzed using various sensing and signal processing techniques. As far as the fault frequencies identifications are ascertain, the feature can be acquired different types of sensors i.e., vibrations, acoustic/sound pressure monitoring, acoustic emission (AE) monitoring, ultrasonic emission, temperature monitoring, chemical analysis, laser monitoring, current monitoring and perception based monitoring etc. [4–6]. Each techniques having their significant contribution in detecting fault. AE are mostly used for qualitative analysis as compared to the quantitative analysis by ultrasonic emission, but both can be used for early fault detection. These sensing methods are costly as compared to microphone and accelerometer. Microphone with higher sensitivity can be used to detect the fault in the bearing, but it is prone to high external noise [8]. Accelerometer sensor can be used for fault analysis, but early fault cannot be detected using this technique. To infer the features of the bearing, combination of different sensing technologies can be used as an asset to discover the problem persists in the bearings.

The signal processing can be done through different techniques i.e., Fast Fourier Transform (FFT), Short Time Fourier Transform (STFT), Wavelet Transform [7] and Hilbert Huang Transform (HHT) [8, 9], EMD, EEMD, CEEMDAN [10, 11] and Variational Mode Decomposition (VMD) [12]. FFT has higher extraction efficiency than that of all other algorithms, but, it suffers from the condition of non-linearity and non-stationary. The global to local decomposition method adopted by STFT can be used to analyze the non-stationary signal, but it suffers from the non-linearity condition and lacks in multi resolution analysis. The window and the signal behavior must match statistically to extract the actual information present in the signal, which is least considered in signal analysis. Wavelet transform is better option than FFT, but the improper selection of the basis function can affect the analysis. Even though, it can be used for multi resolution signal analysis i.e., only for frequency modulated signals not for amplitude modulated. To better analyze amplitude and frequency modulated signals, EMD can be used as it behave like a dyadic filter. In literature, EMD has been used by many researchers, but EMD abide by noise and sampling rate issues. EMD is a dyadic filter, the reaction to noise and sampling inhibit its application in industrial noisy environment. The vibration signal obtained from the experimental setup is complex and are of multi tone signals. EMD fails to decompose close multi tone signals and it can be better performed using VMD [12]. VMD cannot be used for time frequency analysis and the selection of constraint bandwidth and the resulting modes cannot be decided adaptively. Here the analysis is carried out using EMD, EEMD CEEMDAN [10] techniques and their significance in fault detection of bearing to certain extent. The goal of this paper is to use non linear and non stationary signal processing technique in fault detection of bearing. EEMD uses the Gaussian noise to avoid the mode mixing problem occurred in case of EMD. The solution leads to the significant residual noise and the decomposition level increases. CEEMDAN follows the same trend of adding noise, but it has better spectral separation and the decomposition time also reduces. The mode mixing in the signal is too complex and can be carried out with statistical signal processing or wavelet packet transform followed by CEEMDAN to extract the feature of the faults [13, 14]. The performance of these signal processing techniques are tested to on vibro-acoustic signals to identify the fault in the bearing.

2 Mathematical Interpretation of Frequency

In general, the frequency of vibration of the ball bearing is estimated from the mathematical formulation and the same is compared with experimental signals to identify the nature of the fault. The defects frequencies calculated mathematically for outer race, inner race, ball spin and fundamental train frequencies are defined in (1–4).

$$f_{or} = f_s \frac{N_b}{2}\left(1 - \frac{B_d}{P_d}\cos\varphi\right) \tag{1}$$

$$f_{ir} = f_s \frac{N_b}{2}\left(1 + \frac{B_d}{P_d}\cos\varphi\right) \tag{2}$$

$$f_{bs} = f_s \frac{P_d}{2B_d}\left(1 - \frac{B_d^2}{P_d^2}\cos^2\varphi\right) \tag{3}$$

$$f_{ftf} = \frac{f_s}{2}\left(1 - \frac{B_d}{P_d}\cos\varphi\right) \tag{4}$$

Where f_{or} is outer race defect frequency, f_{ir} is the inner race defect frequency, f_{bs} is the ball spin frequency, f_{ftf} is the train frequency, f_s is the spin frequency of shaft, N_b is the number of balls in the bearing, B_d is the ball diameter, P_d is the pitch diameter and φ is the contact angle.

3 Empirical Mode Decomposition

The acoustic and vibration signal obtained from the experimental setup is complex and are of multi tone signals. EMD is used to decompose the signal into number of Intrinsic Mode Function (IMF's). The decomposition method extract from higher to lower frequencies till the residual monotonic signal is achieved. The decomposition of the real time data x(t) is as follows,

1. Sample the time domain signal $x(t)$, depending on the sample rate of acquisition device (DAQ card) and the required sample for the type of applications.
2. Identify all maxima and minima for the sampled data points $x(n)$.
3. Generate upper and lower envelope i.e., $e_{min}(n)$ and $e_{max}(n)$ using Cubic Spline interpolation.
4. Calculate the mean m (n) for upper and lower envelope.
5. $m(n) = (e_{min}(n) + e_{max}(n))/2$.
6. Extract the mean from the time series and define the difference of x(n) and m(n) as d(n).
7. $h(n) = x(n) - m(n)$;
8. Check the properties of h(n). If SD > 0.3, repeat steps 1–7 until the residual satisfies some stopping criterion. Standard deviation (SD) is calculated as;

$$SD = \sum \frac{(\text{prev}(h) - h)^2}{\text{prev}(h)^2}$$

9. In the end the signal x(n) can be represented as in (5).

$$X(n) = \sum_{i=1}^{n} c_i(n) + r_n(n) \tag{5}$$

Once the IMF's are obtained, FFT is applied to the IMF's to get the spectral components of the original decomposed signals as in (6).

$$X(k) = \sum_{n=0}^{N-1} \text{IMF}(n) e^{-j2\pi kn/N} \tag{6}$$

Where N is number of discrete sample points, and is 10,000 for this experiment.

4 Experimental Setup and Methodology

The experimentations are performed to identify the acoustic and vibration feature in global and local domain using different signal processing techniques. The accelerometer sensors are mounted onto the surface of the ball bearing using stud mounting and the data are acquired from the sensors using NI USB 4432. Vibration and acoustic signals are acquired at a sampling rate of 5120 samples/sec using two different types of sensors i.e., ±50 g, ±1 g accelerometer and GRAS array microphone as shown in Fig. 1. In case of acoustic signal, acquisition preamplifier is used to amplify the signal from the microphone to enhance the strength of the signal. For, practicalities of the paper only ±50 g and GRAS array microphones are used to analyze the extracted features.

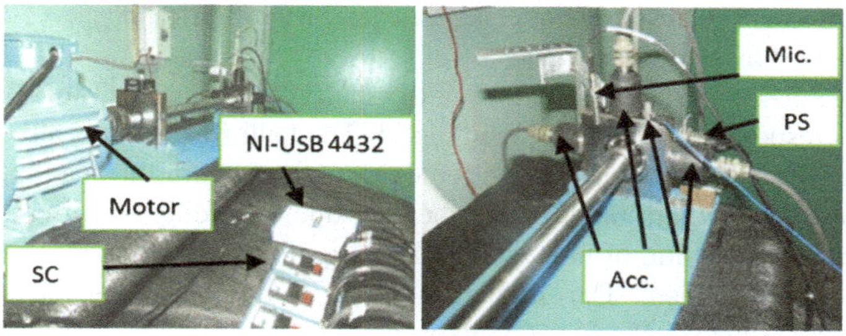

Fig. 1 Experimental setup of ball bearing simulator with array microphone, accelerometers, proximity sensor. **Acc.* (Accelerometer), *Mic.* (Microphone), *PS* (Proximity Sensor), *SC* (Signal Conditioner)

Table 1 Ball bearing configuration

Bearing type	Pitch dia (in)	Rolling element dia (in)	Number of rolling element
6205(DGBB)	1.537	0.3125	9

Table 2 Ball bearing frequencies

Shaft speed (RPM)	f_{ir} (Hz)	f_{or} (Hz)	f_{bs} (Hz)	f_{ftf} (Hz)
3000 (50 Hz)	270.747	179.253	117.877	19.917

For fault identification, SKF-6205, deep groove ball bearing (DGBB) is used for analysis. Before the diagnosis of the fault, the bearing is subjected to load for a period of 25 h in a Spectra Quest bearing prognostic simulator. The developed fault is further analyzed using the fabricated experimental setup as shown in Fig. 1. The bearing configuration and the fault frequencies are listed in Tables 1 and 2.

5 Results and Analysis

The vibro-acoustic features can be used simultaneously to detect diagnosis of fault. The behavioral patterns for the four different algorithms are investigated in the detection of the fault as well as the exact IMF (intrinsic mode functions) identification that exactly emulated the faulty state of the bearing. The mode mixing problem in EMD is investigated further using EEMD and CEEMDAN.

5.1 Fast Fourier Transform

FFT is the hidden basic building block of all the signal processing and decoding algorithms, even though the extraction method changes with bit variation in the basis functions. The result for the acoustic and vibration response of the faulty bearing is as shown in Figs. 2 and 3.

It can be observed that the rotation of the shaft of 50 Hz as in Table 2. is traced with its corresponding harmonics as in Fig. 3. The fault frequency is also identified as 270 Hz, which closely matches with the inner race fault of the bearing. It can be drawn that the maximum failure in the bearing due to loading is caused due to the inner race. The detection of inner race fault is significant as compared to the outer race under radial load. The limitation of FFT in analyzing non-linear and non-stationary signals calls for new algorithms.

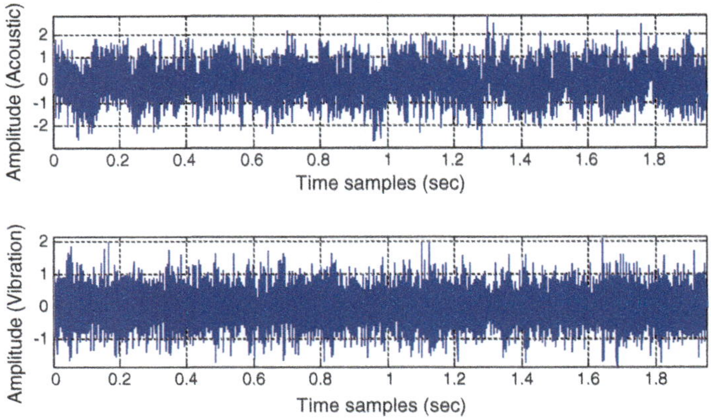

Fig. 2 Time response of acoustic (*top*) and vibration (*bottom*) data samples

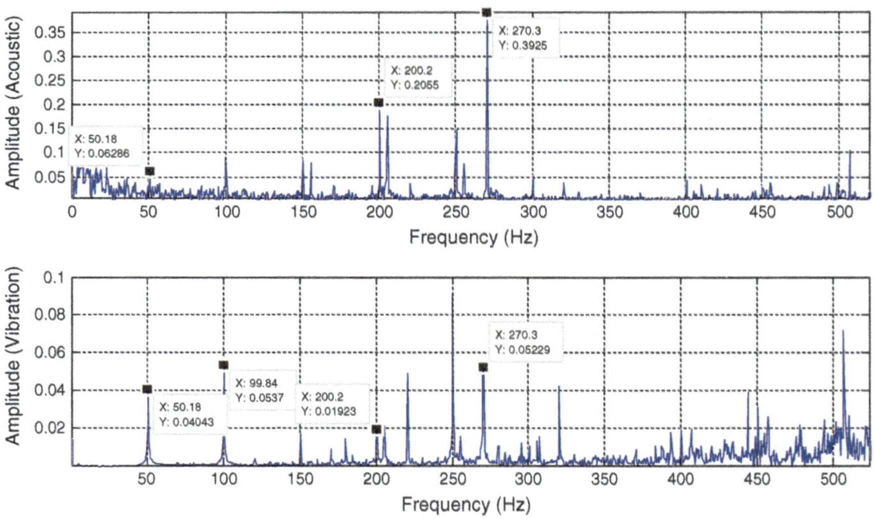

Fig. 3 Frequency response of acoustic (*top*) and vibration (*bottom*) data samples

These algorithms generally deal with noises that are Gaussian. The statistical distributions of the time domain signals for acoustic and vibration signatures are as in Figs. 4 and 5. The ranking of the distributions are based on Chi-squared test. The

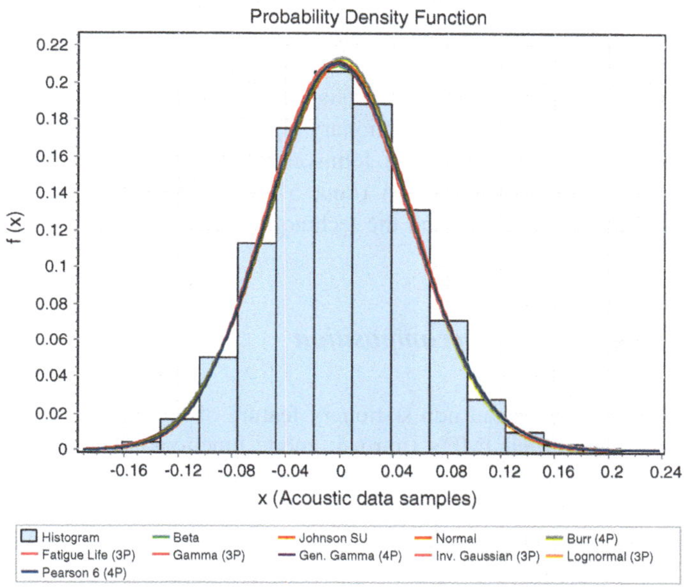

Fig. 4 Probability density functions for acoustic signal

Fig. 5 Probability density function for the vibration signal

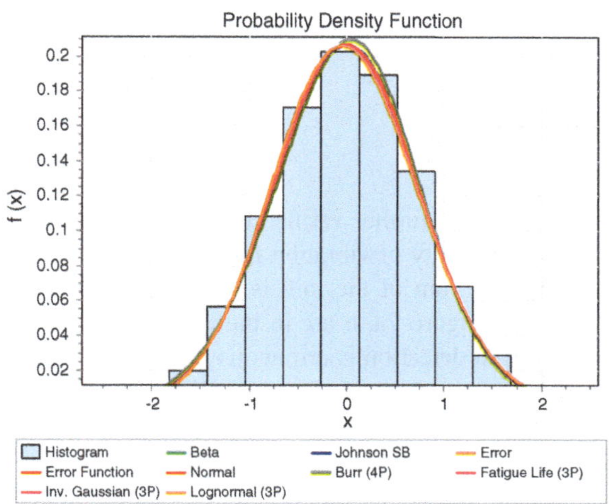

purpose is to check the exact distribution of acoustic and vibration and their probability density function.

It is observed from Fig. 4 that the acoustic data follows normal distribution (rank 3) and the fitness function of this distribution is higher compared all other distributions. This is true for our experimental data; it is not always true that the data matches to the normal distribution. If the data matches to the perfect normal distribution then Principal component analysis can be used for blind source separation as mode mixing are concerned. If the signal distribution are non-Gaussian then ICA (Independent component analysis) can be used just after is processed by the any of the non-linear and non-stationary algorithms considered in [15].

Figure 5 shows that the Beta and Johnson SB distributions are best suited as compared to the normal distribution (rank 5). The distribution has tremendous impact on the analysis process and the techniques used.

5.2 *Empirical Mode Decomposition*

To extract the non-linear and non stationary feature of the bearing, further signals are decomposed into their IMF's (intrinsic mode functions). The extracted results for the acoustic and vibrations are as shown in Fig. 6. It is observed from Fig. 6a that the IMF 3, 4 have the same significant peak at 270 Hz. Even though the algorithm could able to trace the fault signature and matches to Fig. 3, the selection of IMF is now difficult as the same frequency reflects at multiple decomposition levels. It can be observed from Fig. 6b that for vibration signal analysis the mode mix-up happens at the fifth order harmonics i.e., 249.4 Hz of the rpm rather at fault frequencies. It means the signal of acoustic and vibration even though looks for the same source i.e., bearing; they have different statistical distributions as observed in Figs. 4 and 5.

5.3 *EEMD*

The analysis is further verified using EEMD technique. It is observed from Fig. 7a that, the intensity of vibration falls to lower level as compared to the EMD, but the detection of rpm of the mill is stable for both acoustic and vibration signals i.e., 50.18 Hz. Figure 7a, b are in more congruence as compared to Fig. 6a, b.

The rpm detection is erroneous in case of EMD for acoustic signal. The EEMD algorithm is best suited to extract all the information independent of the data types

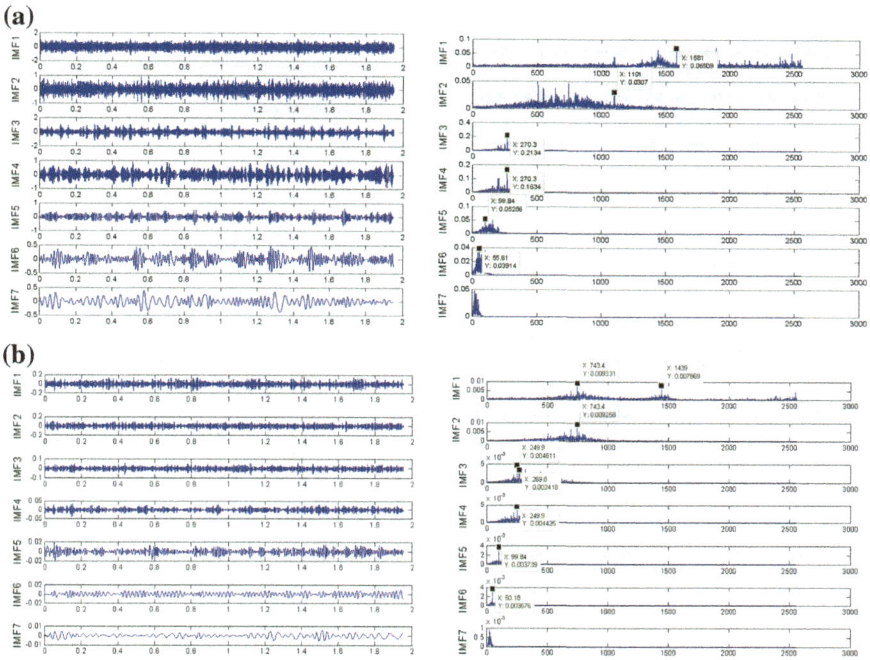

Fig. 6 Time (*left*) and frequency (*right*) response of **a** acoustic and **b** vibration signals using EMD

i.e., acoustic or vibration as compared to EMD. As the problem of mode mixing is concerned EEMD also has the same problem as that of EMD.

5.4 *CEEMDAN*

The detection and analysis is further validated using CEEMDAN. It can be observed from the acoustic patterns in Fig. 8a, that the acoustic pattern is well verse with the Fig. 7a. The same analysis for vibration results shows that the Figs. 7b and 8b are in congruence. There is no significant difference between EEMD and CEEMDAN. The computational extraction level increases as the decomposition level increases as in Table 3. The computational complexity of CEEMDAN is

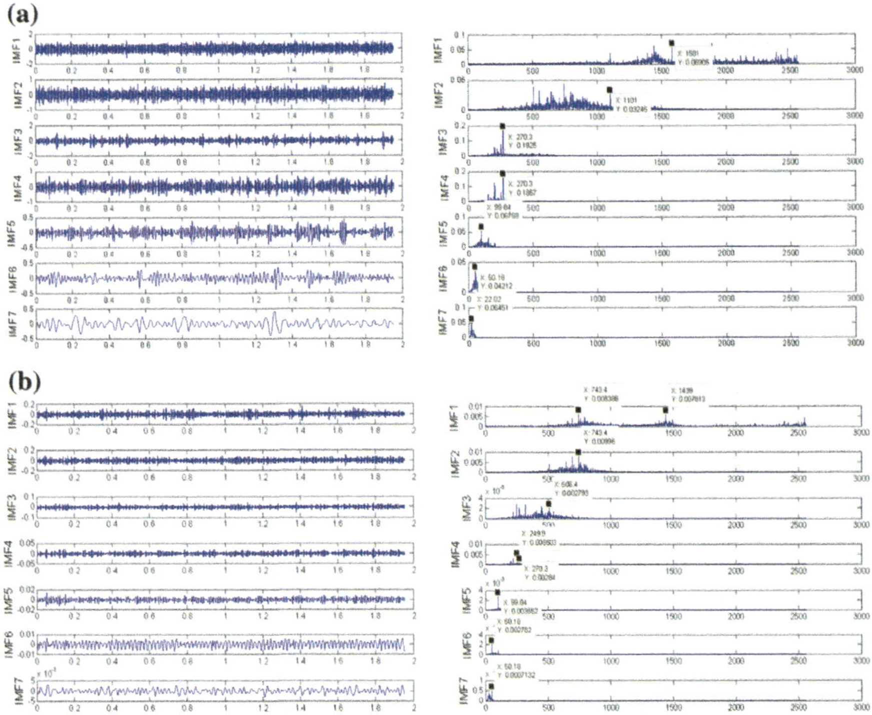

Fig. 7 Time (*left*) and frequency (*right*) response of **a** acoustic and **b** vibration signals using EEMD

lower than that of EEMD with effective signal extraction for both acoustic and vibration signals.

All the algorithms are very effective in extracting the inner race fault of the bearing effectively, if the mathematical Eqs. (1–4) are known. All these algorithms having short fall in avoiding mode mixing problem occurred in the bearing fault analysis. These techniques are adaptively decomposes into number of levels as compared to the VMD algorithms, but the mode mixing problem can be avoided using VMD, but VMD is not applicable for non stationary signal analysis. To avoid mode mixing, the data need to be statistically distributed using statistical tool and further the data need to adaptively un-correlate the correlated IMF components using ICA (independent component analysis) [15].

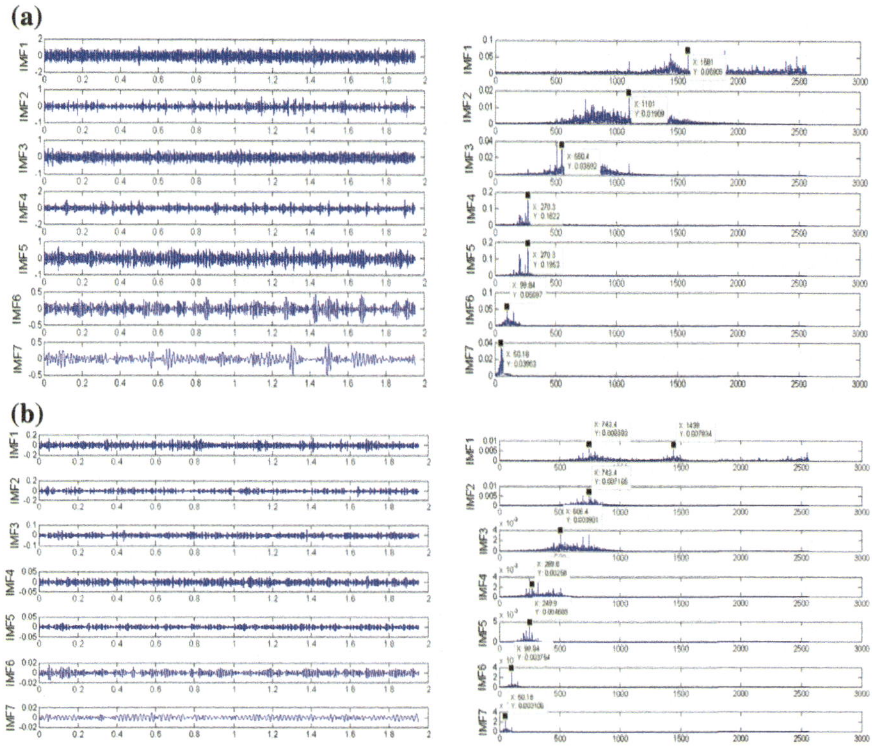

Fig. 8 Time (*left*) and frequency (*right*) response of **a** acoustic and **b** vibration signal using CEEMDAN

Table 3 Decomposition levels by algorithms

Algorithms	Number of modes produced for acoustic signal	Number of modes produced for vibration signals
EMD	13	13
EEMD	15	15
CEEMDAN	14	14

6 Conclusions

CEEMDAN perform better than all other algorithms in terms of signal detection and computational time, but lags to avoid the mode mixing problem. These algorithms can be used to detect amplitude and frequency modulated fault signals adaptively. All the algorithms except FFT can be used for nonlinear and non stationary signal analysis with effective identification of the inner race fault in the bearing. These algorithms are prone to mode mixing problems, even though they effectively extract the information content. These algorithms computational cost increases as the standard deviation is chosen to a lower value and the sample length selection is higher. In future, the extraction of the feature using these algorithms followed by statistical distribution analysis and Independent component analysis can be used to eliminate the mode mixing problems.

References

1. T. T. Company (2011) Timken bearing damage analysis with lubrication reference guide. Timken Co, pp 1–39
2. Upadhyay RK, Kumaraswamidhas LA, Azam MS (2013) Rolling element bearing failure analysis: A case study. Case Stud Eng Fail Anal 1(1):15–17
3. Sadeghi F, Jalalahmadi B, Slack TS, Raje N, Arakere NK (2009) A review of rolling contact fatigue. J Tribol 131(4):041403
4. Al-Ghamd AM, Mba D (2006) A comparative experimental study on the use of acoustic emission and vibration analysis for bearing defect identification and estimation of defect size. Mech Syst Sig Process 20(7):1537–1571
5. Tandon N, Nakra BC (1992) Comparison of vibration and acoustic measurement techniques for the condition monitoring of rolling element bearings. Tribol Int 25:205–212
6. Rezaei A, Dadouche A, Wickramasinghe V, Dmochowski W (2011) A comparison study between acoustic sensors for bearing fault detection under different speed and load using a variety of signal processing techniques. Tribol Trans 54:179–186
7. Kankar PK, Sharma SC, Harsha SP (2011) Rolling element bearing fault diagnosis using wavelet transform. Neurocomputing 74(10):1638–1645
8. Lei Y, Lin J, He Z, Zuo MJ (2013) A review on empirical mode decomposition in fault diagnosis of rotating machinery. Mech Syst Sig Process 35(1–2):108–126
9. Mohanty S, Gupta KK, Raju KS, Singh A, Snigdha S (2013) Vibro acoustic signal analysis in fault finding of bearing using empirical mode decomposition. Int Conf Adv Electron Syst 29–33
10. Torres ME, Colominas MA, Schlotthauer G, Flandrin P (2011) A complete ensemble empirical mode decomposition with adaptive noise. In: ICASSP, IEEE International Conference on Acoustics, Speech and Signal Processing, pp 4144–4147
11. Colominas MA, Schlotthauer G, Torres ME, Flandrin P (2012) Noise-assisted Emd methods in action. Adv Adapt Data Anal 4(4):1250025
12. Dragomiretskiy K, Zosso D (2014) Variational mode decomposition. IEEE Trans Sig Process 62(3):531–544
13. Li W, Zhu Z, Jiang F, Zhou G, Chen G (2015) Fault diagnosis of rotating machinery with a novel statistical feature extraction and evaluation method. Mech Syst Sig Process 50–51: 414–426
14. Mohanty S, Gupta KK, Raju KS (2015) Multi-channel vibro-acoustic fault analysis of ball bearing using wavelet based multi-scale principal component analysis. Twenty First Natl Conf Commun 1–6
15. Tang B, Dong S, Song T (2012) Method for eliminating mode mixing of empirical mode decomposition based on the revised blind source separation. Sig Process 92(1):248–258

An Evaluation of Convolutional Neural Nets for Medical Image Anatomy Classification

Sameer Ahmad Khan and Suet-Peng Yong

Abstract Classification of the anatomical structures is an important precondition for several computer aided detection and diagnosis systems. Attaining extraordinary precision for automatic classification is a stimulating job because of vast amount of variation in the anatomical structures. Current trend in object recognition is driven by "Deep learning" methods that are outperforming the contemporary methods in classification of images. Till now these "Deep learning" methods have been applied on natural images. In this study, we compare the performance of three main Deep learning architectures i.e. LeNet, AlexNet, GoogLeNet on medical imaging data containing five anatomical structures for anatomic specific classification.

1 Introduction

Classification of medical images is considered to be an important component of computer aided detection and diagnosis systems [1]. Automatic localization or identification is very useful in initializing organ specific processing such as detecting liver tumors [2]. It is a challenging task to achieve high accuracies for automated classification of anatomy, because of the variability's in the anatomical structures due to varying contrast, deformed shapes due to pathologies and occlusion. In image classification problems the descriptiveness and discriminative power of features extracted are important to achieve good classification results. The feature extraction techniques that have been used in medical imaging commonly include filter based features [3] and the very popular scale invariant feature transform (SIFT) [4].

S.A. Khan (✉) · S.-P. Yong
Department of Computer and Information Sciences,
Universiti Teknologi Petronas, Seri Iskandar, Malaysia
e-mail: sameer15khan@gmail.com

S.-P. Yong
e-mail: yongsuetpeng@petronas.com.my

P.J. Soh et al. (eds.), *Advances in Machine Learning and Signal Processing*,
Lecture Notes in Electrical Engineering 387, DOI 10.1007/978-3-319-32213-1_26

Neural Networks (NN) has been studied for many years to solve complex classification problems including image classification. The distinct advantage of neural network is that the algorithm could be generalized to solve different kinds of problems using similar designs. Convolutional Neural Network (CNN) is a successful example of attempts to model mammal visual cortex using NN. The reason for using convolutional neural nets (CNNs) for anatomy specific classification is that these CNNs outperformed the contemporary methods in natural image classification [5]. Also CNNs have made substantial advancements in biomedical applications [6]. In addition to this recent work has shown how the implementation of CNNs can significantly improve the performance of the state-of-the-art computer aided detection systems (CADe) [7–9]. In this study we are evaluating the comparative performance of three milestones in the development of Convolutional Neural Networks for anatomy specific classification, i.e. LeNet [10], AlexNet [5] and GoogLeNet [11].

2 Related Work

2.1 Convolutional Neural Nets

Convolutional Neural Networks (CNNs) are a special kind of deep prototypes that are in charge for numerous exhilarating recent results in computer vision. Initially proposed in the 1980's by K. Fukushima and after that developed by Y. LeCun and teammates as LeNet [10], CNNs picked up acclaim through the accomplishment of LeNet on the challenging task of handwritten digit recognition in 1989 It took a few decades for CNNs to create another leap forward in computer vision, commencing with AlexNet [5] in 2012, which won the overall ImageNet challenge.

In a CNN, the key calculation is the convolution of a feature detector with an input signal. Convolution with a pool of filters, like the learned filters in Fig. 1, augments the representation at the first layer of a CNN, the components go from individual pixels to straightforward primitives like even and vertical lines, circles, and fixes of shading. Rather than ordinary single-channel picture processing filters, these CNN filters are processed over all of the input channels. Convolutional filters are translation-invariant so they yield a high reaction wherever a feature element is identified.

2.1.1 LeNet

LeNet [10] comprises of five layers that contains trainable parameters as shown in Fig. 2. The input is a 28×28 pixel image.

Layer 1 represents convolutional layer that contains 20 feature maps with kernel size of 5, which depicts that each unit in each feature map is connected to 5×5 neighborhood in the input. Conv1 contains 1520 learned parameters. Layer 2

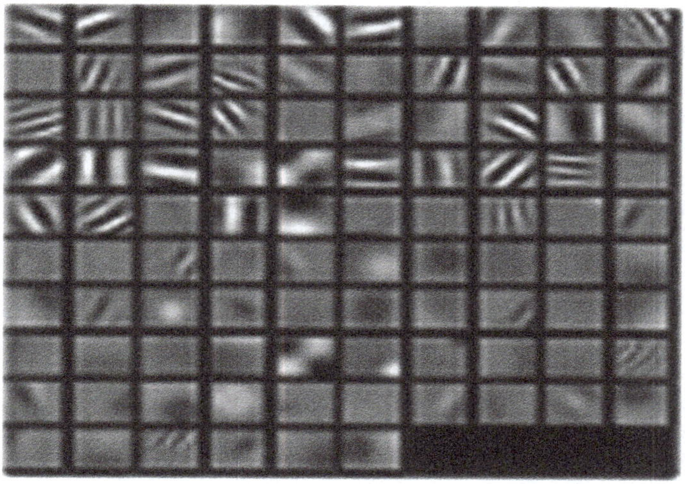

Fig. 1 Filters learned by convolutional layer

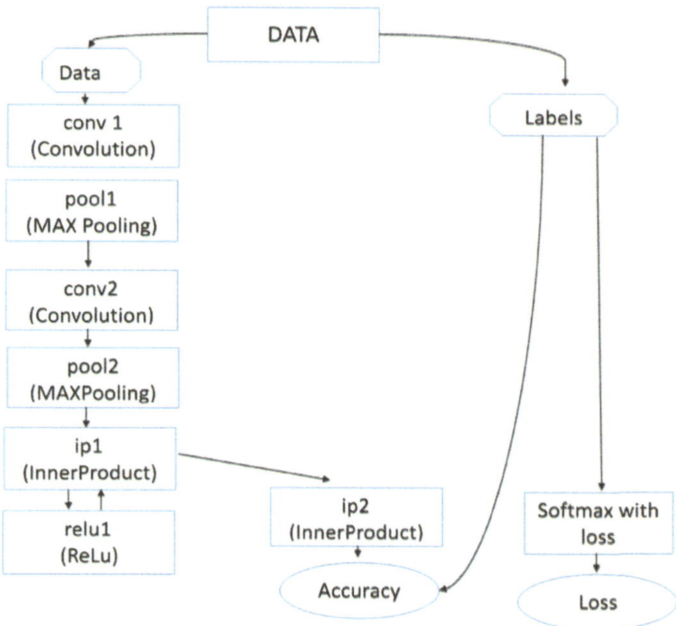

Fig. 2 LeNet architecture

i.e. pool1 is a pooling layer that aggregates the learned parameters to make the invariant to the transformations. Pool1 represents a layer with 20 feature maps of size 12 × 12. Layer 3 is again a convolutional layer conv2 that produces 25,050 learned parameters by convolving the pooled feature maps. Layer 4 i.e. pool2 aggregates the convolved features from layer 3 i.e. conv2. After convolutions and pooling, in layer 5, i.e. ip1, an inner product operation trailed by rectified linear unit activation (ReLU) function is applied, that resulted in 400,500 learned parameters. After this in layer 6 i.e. ip2 an inner product operation is again applied, that resulted in a reduced set of learned parameters, i.e. 2505. So a total of 429,575 parameters are learned which are then passed to a softmax classifier to determine the loss from the actual output.

2.1.2 AlexNet

AlexNet [5] proposed by Alex Krizhhevsky as shown in Fig. 3 is a convolutional neural net that revolutionized the image classification task by beating the state of the art image classification methods in 2012.

AlexNet comprises of 11 layers. i.e. conv1 added with relu1 and norm1, with kernel size 11 and stride of 4, which means after every four pixels perform the convolution. Which produces some learned parameters. The first layer i.e. conv1 layer is followed by pooling i.e. pool1 as explained above for the LeNet. The kernel size for the pooling is set to 3 with stride 2. Pool1 is followed by convolution conv2 with kernel size 5 and stride 2. On conv2 parameters relu2 is applied, that is followed by norm2. The conv2 parameters are again pooled in pool2 layers by applying

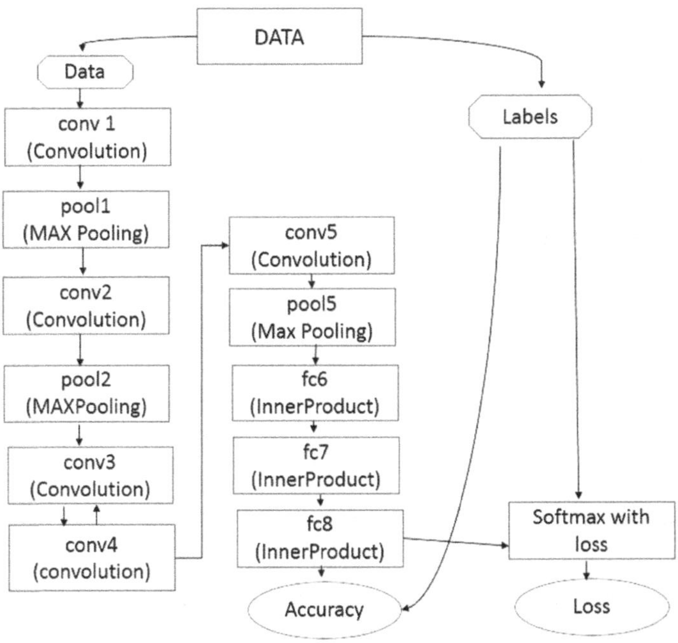

Fig. 3 AlexNet architecture comprising of 8 layers

maxpooling with kernel size 3 and stride 2. The pooled feature maps are again convolved in layer conv3, with parameter setting of kernel size equal to 3, stride of 1 and padding of 1. These convolved features are again convolved in layer conv4 with parameter setting same as in layer conv3, followed by relu4. The features from layer conv4 are again convolved in layer conv5, with the same parameter setting as in layer conv4 followed by relu5. The features from layer conv5 are pooled in layer pool5. Which is followed by fully connected layers, i.e. fc6, fc7 and fc8. In the layer fc6 two operations are applied, i.e. relu6 and drop6. Dropout operation prevents the deep nets from over fitting. The layer fc6 is followed by fc7, which is accompanied with relu7 and drop7. The features are finally fully connected through layer fc8 to the softmax classifier that determines the loss from the actual output.

2.1.3 GoogLeNet

GoogleNet [11] is a deep learning framework in which authors proposed an inception architecture that is based on how an optimal local sparse structure in a convolutional vision network can be approximated and covered by available components [11]. The architecture is based on the Hebbian principle, which states that neurons that fire together-wire together. According to this architecture and Hebbian principle, in images correlation tend to be local cover very local clusters by 1×1 convolutions. After that cover more spread out clusters by 3×3 convolutions as illustrated in Fig. 4.

After 3×3 convolution, the cluster that are more spread out cover those with 5×5 convolution, that will result in a heterogeneous set of convolutions. GoogLeNet comprises of 9 inception modules.

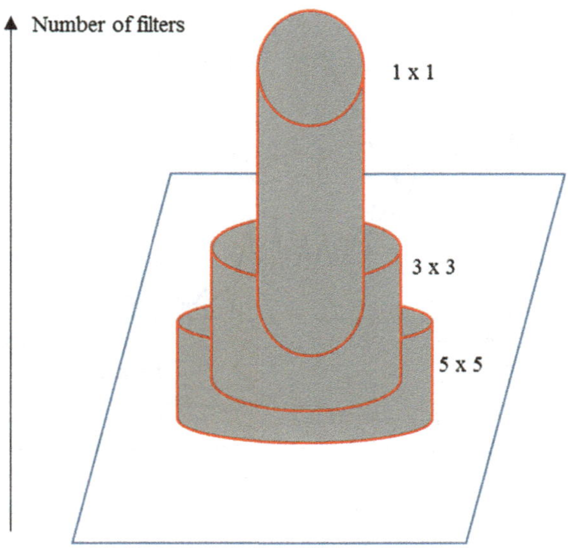

Fig. 4 Convolution for local and spread out clusters that are correlated

3 Experimental Evaluation of LeNet, AlexNet and GoogLeNet for Anatomy Specific Classification

We started our experimentation with the data set acquired from the U.S. National Library of medicine, national Institutes of Health, Department of Health and Human Services. This is an open access medical image database that contains thousands of anonymous medical imaging data, ranging from various modalities like CT, MRI, PET, XRAY etc. this database also contain images with various pathologies. For our experimental evaluation we downloaded 5500 images of various anatomies. The anatomies we considered for our experimentation are lung, liver, heart, kidney and lumbar spine. We downloaded the normal and pathological images, so that these frameworks should be generalized to classify any image of the same organ if it varies in shape or contrast. We supplied 1000 images per category for the training purpose, out of which 25 % were used validation. For the testing purpose we used the different test set also acquired from the same database. The test set contains 66 images of different anatomies as mentioned above. We used 3851 images for training and 1149 for validation.

3.1 Experimental Evaluation of LeNet

We started our experimentation with LeNet. Before training the net we resized the images to the size of 28 × 28 and preprocessed them by subtracting the mean

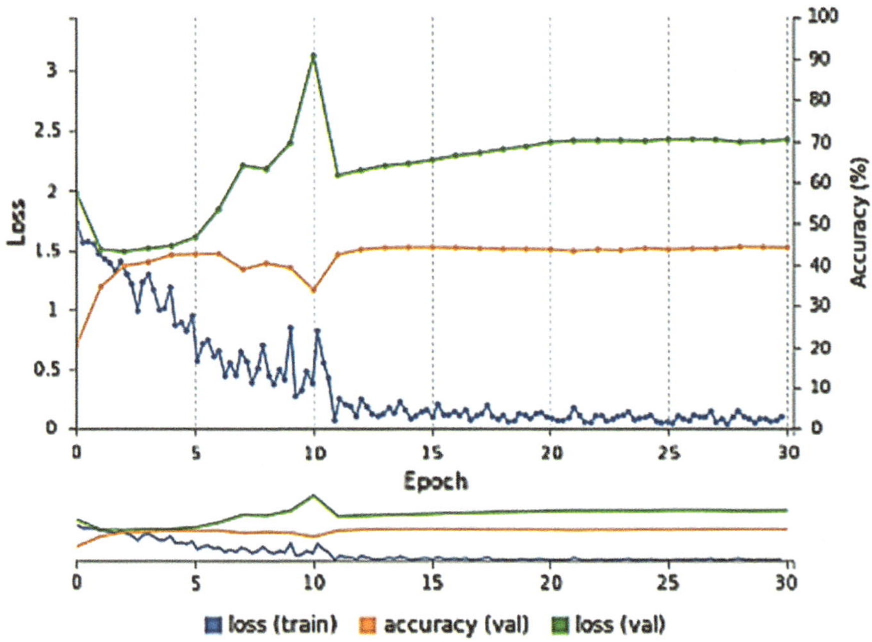

Fig. 5 Training loss and validation accuracy with each epoch

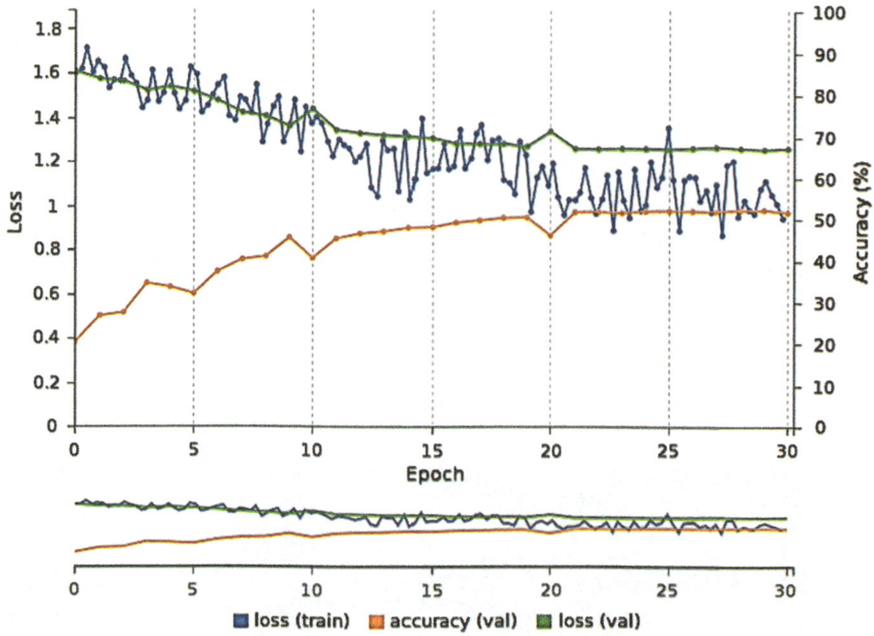

Fig. 6 Training loss and validation accuracy for AlexNet

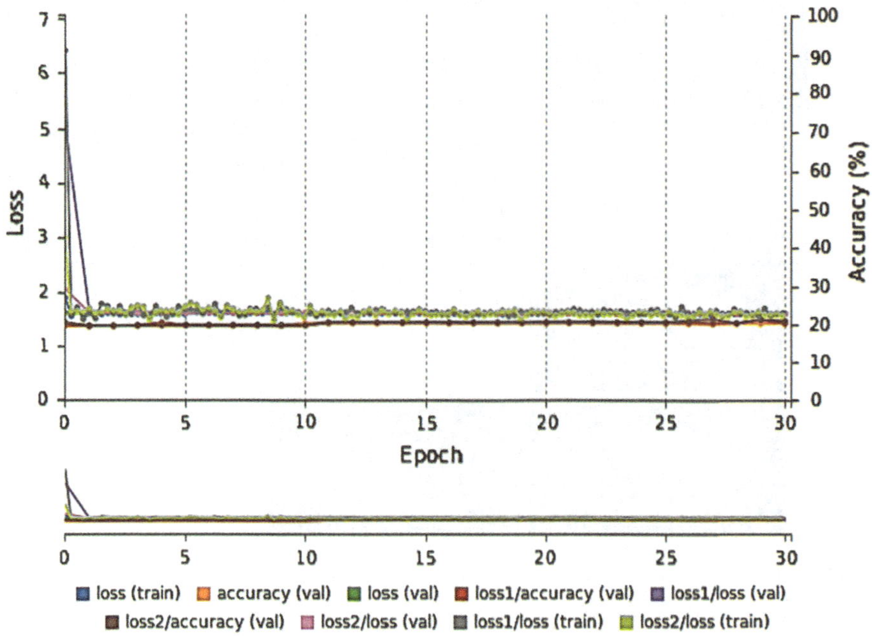

Fig. 7 Training loss and validation accuracy for GoogLeNet

image from each pixel. After that we trained the LeNet with the batch size of 50. Which means 50 images were supplied at a time for each epoch for training and we used stochastic gradient descent as a training algorithm with a learning rate of 0.01. The training of the LeNet is shown in Fig. 5, which depicts how the accuracy and training loss goes with each iteration.

This figure gives us the accuracy of 45 % on the validation data, whereas the training loss decreases and validation loss is greater than the validation accuracy with each iteration depicting that the model is over fitting. After that we tried to see how this network performs on the unknown data i.e. the test data. The test data is evaluated on AlexNet and the top nine predictions to classify the data into respective classes is shown in Figs. 8 and 9. The summarized results of training and validation is shown in Table 1.

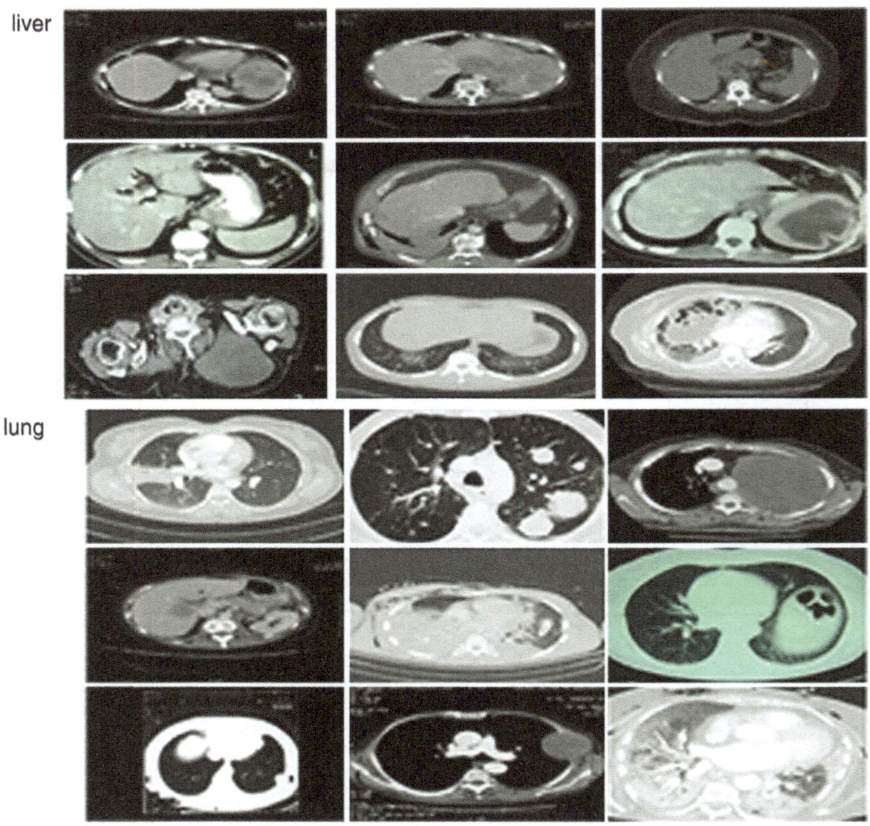

Fig. 8 Top 9 predictions of AlexNet for each anatomy

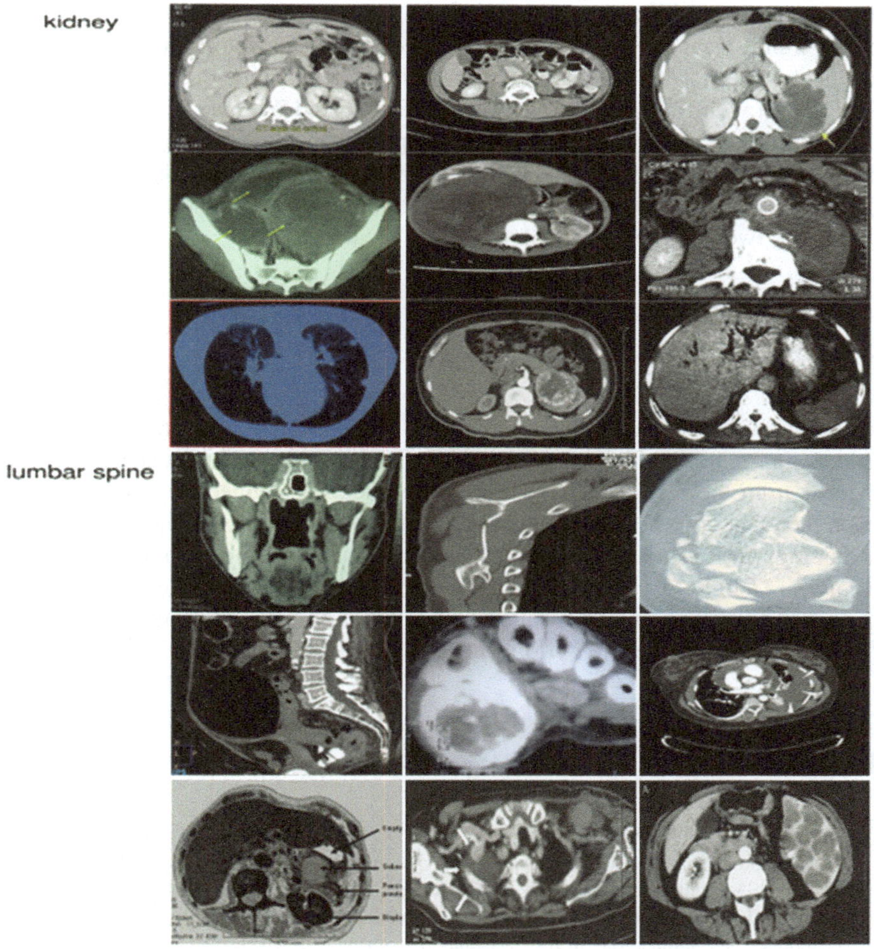

Fig. 9 Top 9 predictions of AlexNet for each anatomy

Table 1 Comparative results of LeNet, AlexNet and GoogLeNet

CNN	Training loss	Validation accuracy (%)
LeNet [10]	0.1	45
AlexNet [5]	0.9	51
GoogLeNet [11]	1.7	20

3.2 Experimental Evaluation of AlexNet and GoogLeNet

The parameter setting for AlexNet is different from LeNet. The image dimensions for AlexNet are set as 256 × 256. The images are mean subtracted also and network is trained with the same training algorithm i.e. stochastic gradient descent.

The batch size for AlexNet is 50 while as the default batch size is 100. But because of the limiting capability of our machine we choose the 50 batch size and same setting has been adopted for the GoogLeNet. The training of the AlexNet and GoogLeNet with each iteration is shown in Figs. 6 and 7 respectively. It is evident from the figures that GoogLeNet does not perform well on the medical imaging data, whereas AlexNet has much higher validation accuracy then LeNet and GoogLeNet. But its training error increases with each epoch but still performs better than other two CNNs in terms of validation accuracy.

4 Conclusion

In this study we compared three state-of-the-art convolutional neural networks for anatomy specific classification. We experimented with five different anatomies. It is evident from the results that CNN with the AlexNet architecture performs quite good then other two architectures. While as one of the good outcomes of this study is that it gave an insight into an important factor i.e. increasing the number of layers in case of GoogLeNet does not always increase the performance. So in order to get the better accuracies an optimization with solution to over fitting is needed in the future to train these nets to perform better on medical image data.

References

1. Roth HR, Lee CT, Shin H-C, Seff A, Kim L, Yao J et al (2015) Anatomy-specific classification of medical images using deep convolutional nets. arXiv:1504.04003
2. Criminisi A, Shotton J, Robertson D, Konukoglu E (2011) Regression forests for efficient anatomy detection and localization in CT studies. In: Medical computer vision. Recognition techniques and applications in medical imaging. Springer, pp 106–117
3. Song Y, Cai W, Zhou Y, Feng DD (2013) Feature-based image patch approximation for lung tissue classification. IEEE Trans Med Imaging 797–808
4. Zhang F, Song Y, Cai W, Lee M-Z, Zhou Y, Huang H et al (2014) Lung nodule classification with multilevel patch-based context analysis. IEEE Trans Biomed Eng 1155–1166
5. Krizhevsky A, Sutskever I, Hinton GE (2012) Imagenet classification with deep convolutional neural networks. In: Advances in neural information processing systems, pp 1097–1105
6. Cireşan DC, Giusti A, Gambardella LM, Schmidhuber J (2013) Mitosis detection in breast cancer histology images with deep neural networks. In: Medical image computing and computer-assisted intervention–MICCAI 2013. Springer, pp 411–418
7. Prasoon A, Petersen K, Igel C, Lauze F, Dam E, Nielsen M (2013) Deep feature learning for knee cartilage segmentation using a triplanar convolutional neural network. In: Medical image computing and computer-assisted intervention–MICCAI 2013. Springer, pp 246–253
8. Roth HR, Lu L, Seff A, Cherry KM, Hoffman J, Wang S et al (2014) A new 2.5 D representation for lymph node detection using random sets of deep convolutional neural network observations. In: Medical image computing and computer-assisted intervention–MICCAI 2014. Springer, pp 520–527

9. Li Q, Cai W, Wang X, Zhou Y, Feng DD, Chen M (2014) Medical image classification with convolutional neural network. In: 13th international conference on control automation robotics & vision (ICARCV), pp 844–848
10. LeCun Y, Bottou L, Bengio Y, Haffner P (1998) Gradient-based learning applied to document recognition. In: Proceedings of the IEEE, pp 2278–2324
11. Szegedy C, Liu W, Jia Y, Sermanet P, Reed S, Anguelov D et al (2014) Going deeper with convolutions. arXiv:1409.4842

Ant Colony Optimization and Feature Selection for Intrusion Detection

Tahir Mehmod and Helmi B. Md Rais

Abstract Network intrusion detection gained a lot of attention from the security expert. Intrusion detection system has been designed for the purpose detecting attack and comprises of detection method that can be anomaly based or it can be signature based. These detection method, however, highly depends on the quality of the input features. Supervised learning approach for the detection method finds the relationship between the feature and its class. Therefore, irrelevant, redundant, and noisy features must be eliminated before applying supervised algorithm. This can be done by feature selection method. In this paper ant colony optimization has been applied for feature selection on KDD99 dataset. The reduced dataset is validated using support vector machine. Results show that accuracy of the SVM is significantly improved with reduced feature set.

1 Introduction

As many organizations are facing the cyber-attack, confidentiality, integrity, and availability of the data is become a major issue. Intrusion detection system is designed to detect intrusion in a single host or in a network [1]. Formal type is called host based intrusion detection system while the later one is the second type of intrusion detection system and is called, network based intrusion detection system. Host based intrusion detection system use system log files and other logging mechanism to identify any attack. It resides on a single host system and that is why it highly depends on operating system architecture. Any shortcoming of operating system may compromise intrusion detection system as well. On the other hand, network based intrusion detection system is deployed on a network segment and it

T. Mehmod (✉) · H.B.M. Rais
Department of Computer and Information Sciences,
Universiti Teknologi Petronas, Seri Iskandar, Malaysia
e-mail: tahirmehmood.seecs@gmail.com

H.B.M. Rais
e-mail: helmim@petronas.com

P.J. Soh et al. (eds.), *Advances in Machine Learning and Signal Processing*,
Lecture Notes in Electrical Engineering 387, DOI 10.1007/978-3-319-32213-1_27

analyzes the network packet for any attack [2]. Network based intrusion detection system is independent of the operating system, in fact it is transparent to the operating system of the host as it do not reside on the host system. Both types of intrusion detection system use intrusion detection method for the detection of intrusion.

There are two types of intrusion detection method. One is called signature based intrusion detection method and the second type is called anomaly based intrusion detection method [3]. Signature based intrusion detection method also known as misuse based detection method looks for pattern or signature in a data that complies within a malware [4]. Signature based IDS has a database consists of signatures of the attacks. Signature based IDS uses the stored signatures of the malware for the detection. Therefore, this method has high true positive rate. The problem with the signature based technique is that it cannot detect novel attacks as no signature exist yet for the novel attack [5]. Contrarily anomaly based detection method can detect novel attack as it looks for abnormal behavior that do not comply with the normal operation of the system or network. Major drawback of this method is that it has high false positive rate since it is difficult to define normal behavior of the system or network [6].

Many machine learning algorithms have been used for the implementation of the detection methods. These machine learning algorithms highly depend on the input features. Irrelevant, redundant, and noisy features causes the machine learning algorithm to develop the detection model with low accuracy rate and with high false positive rate. Therefore, these feature must be eliminated at the preprocessing step.

Rest of the report is organized as follows; Sect. 2 gives introduction about feature selection. Section 3 contains introduction of ant colony optimization (ACO) which is followed by related work in Sect. 4. Proposed methodology has been discussed in Sect. 5 and results are given and discussed in Sect. 6. At the end conclusion is given in Sect. 7.

2 Feature Selection

For classification problem feature selection is used for the elimination of irrelevant, redundant, and noisy features to improve the accuracy of the classification algorithm. This is done at the preprocessing step before applying any machine learning algorithm. Feature selection process selects a subset of features that represents the whole feature set [7]. Which features should be included or excluded is being decided in this process. Relevancy and redundancy are the two decisive factors for feature selection process [8]. Relevant features are those that envisage the desired system response, on the other hand, redundant features have a high degree of correlation among themselves. Thus, removal of the redundant features is desired. Features that are highly correlated with each other give no additional information. While robust features have a high degree of correlation relevant to desired decision and uncorrelated with other features in feature set. By using feature selection at preprocessing step, the predictive accuracy of the machine learning algorithms can be increased. Robust feature set also reduces the training time of the classifier as robust features are invariant in nature and reduces the dimension in high dimensional data. Reduced dataset also decreases dataset which acquire less storage space.

Feature selection process has four steps as shown in Fig. 1. Subset generation, subset evaluation, stopping criteria, and result validation [9]. Subset generation generates a different subset of features, and each feature subset is evaluated in subset evaluation process. If the current feature subset is better than previous feature subset than it replaces the previous one. Subset generation is a searching process, which can be complete, heuristic or random search. Generated feature subset is then validated by some tests.

In network intrusion detection, features are extracted from protocols header at different layers of network architecture and contents of data packets. Due to this reason noise in channels propagate to extracted features, this leads to false intrusion alarm. There are two types of feature selection methods: Filter and wrapper. Filter method selects the subset of features without involving learning algorithm in evaluation phase and is mainly based on ranking of features, which represents the relevancy of the features [10]. In contrast, wrapper method evaluates a subset of features using learning algorithm [11]. This evaluating algorithm is called iteratively unless a robust subset of the feature is selected. Filter based approach is computationally fast compared to wrapper based as it doesn't involve any learning algorithm during ranking of features. However, wrapper based feature subset accomplishes good accuracy rate as it involves learning algorithm in the subset evaluation phase [12].

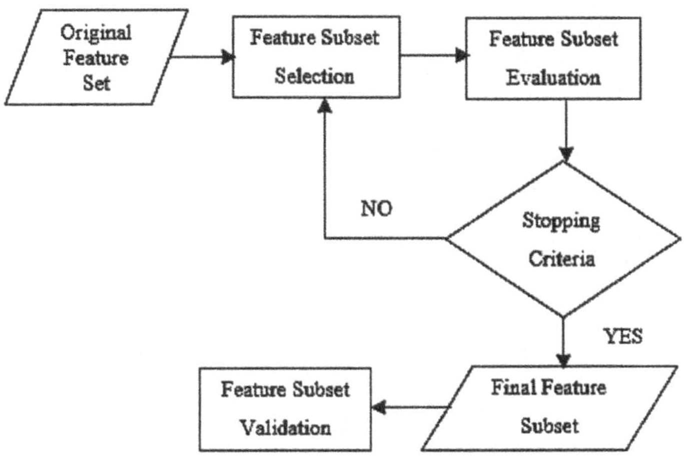

Fig. 1 Feature selection process

3 Ant Colony Optimization

Ants used chemical substance for the communication and is called, pheromone. Ants used it to remember the path from source of food to nest. More intensity of pheromone attracts more ants. Initially ant's foraging like behavior was used for traveling salesman problem (TSP) and the model was named ant system [13]. Ant colony optimization (ACO) has produced efficient results for NP-hard set problems. ACO has many variations. In this paper we used ant colony system (ACS), which uses two level pheromone update i.e. local pheromone and global pheromone update. In global pheromone update only those edges get pheromone update that belongs to best ant. A digital ant selects next node using some transition probability rule.

4 Related Work

Since feature selection is NP-complete problem that is why ACO has been widely adapted for feature selection. Below lists some of them.

George [14] utilized principle component analysis for feature reduction. Using principle component analysis 28 features were selected. The reduced feature set was validated using SVM. Tsang et al. [15] used independent component analysis and principle component analysis for his proposed model called, ant colony clustering model.

Gao et al. [16] proposed ant colony optimization method for KDD99 feature selection. The proposed method mapped the features into graph which were connected to each other, giving opportunity to each ant to select any feature. Selection of next feature by an ant was based on the heuristic information and pheromone value. Fisher discrimination rate was used as heuristic information. Edges contained the pheromone value and only the ant that resulted less squared error used global pheromone update on the edges visited during solution construction.

Nadia and Marcus [17] proposed a wrapper based feature selection based on ACO. The proposed method does not used traditional graph method, instead each feature is represented by 1 and 0 which indicated the selection of feature. An ant select next feature using a probability function which uses pheromone value and heuristic information. Each feature possessed pheromone value. While heuristic information described the desirability of feature which calculated by the number of ants visited that feature. Local pheromone update was used at each construction step while global pheromone was used by the best ant which update the pheromone value for all features that were selected by best ant.

Alwan and Mahamud [18] used mixed variable ant colony optimization for feature selection and at the mean time regulating C and γ parameters for SVM. SVM used RBF kernel function and the applied kernel function highly depends on C and γ value. During the feature selection mixed variable ant colony optimization method also searches for C and γ values that can improve the accuracy of SVM.

Heuristic information adopted fisher discrimination rate. Pheromone value lied on the edges and only the ant that produced high accuracy for SVM was allowed to update the pheromone value on the edges used during solution construction.

5 Proposed Methodology

Ant colony optimization for feature selection has been proposed in this paper. Features are represented in a completely connected graph problem thus choice of selecting next feature is given to each ant. An ant moves to next feature using given transition probability.

$$p_{i,j} = \max(\tau_j)^{\beta}.\eta_j \tag{1}$$

Pheromone value (τ) is related to each feature instead on edges. At the start of solution pheromone value to each feature is initialized by its entropy value to the prediction of the class. β controls the importance of the pheromone value during selection of next feature. If β is 0 than pheromone value for the feature is completely ignored. Number of time the feature visited is considered as heuristic information (η). Initially heuristic information is kept 1 so that no feature can get biased heuristic at the start of the constructing solution by the ants.

At each solution construction, local pheromone value is updated using given formula,

$$\tau_j = (1 - \rho).\tau_j + \Delta_j \cdot \sigma \tag{2}$$

where

$$\Delta_j = \begin{cases} \rho & if\, j\, \varepsilon\, S^+ \\ 0 & otherwise \end{cases}$$

S^+ is the set of features visited for that particular run. $\sigma \in (0,1]$ controls the value for Δ_j. After completing tour each ant passes its dataset to naïve bayes classifier. Ant's dataset that results high accuracy rate gets global pheromone update and uses following equation.

$$\tau_j = (1 - \rho).\tau_j + \emptyset_j \cdot \sigma \tag{3}$$

Where

$$\emptyset_j = \begin{cases} \rho & if\, j\, \varepsilon\, global\, best\, ant\, tour \\ 0 & otherwise \end{cases}$$

This whole method is repeated until stop criteria is met which is if none of the ants can improve accuracy for naive bayes classifier compare to the previous best ant result.

6 Results and Discusssions

In this paper KDD99 dataset has been used for evaluation of detection model. Feature subset is validated using LibSVM in Weka [19]. Binary classification has been used in the experiment. This dataset is widely adapted for the evaluation of detection model [20]. Dataset contains one normal class data and four attack classes' data namely, Denial of service (DoS), Probe, Remote-to-Local (R2L), and User-to-Root (U2R). Training and testing dataset exist for this dataset. Training dataset hold 494,021 network records while testing dataset comprises of 311,029 network records. Each instance is represented by 41 features. Both datasets contains redundant instances which were removed. Subset of training dataset is generated which contains 5823 instances for each two classes, normal class and attack class. Since we used binary SVM therefore all the four attack classes are merged into single attack class. The test dataset used contains 77,287 instances.

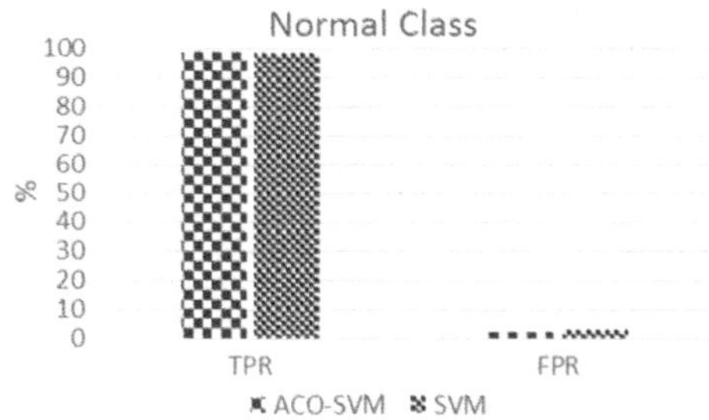

Fig. 2 Result comparison for normal class

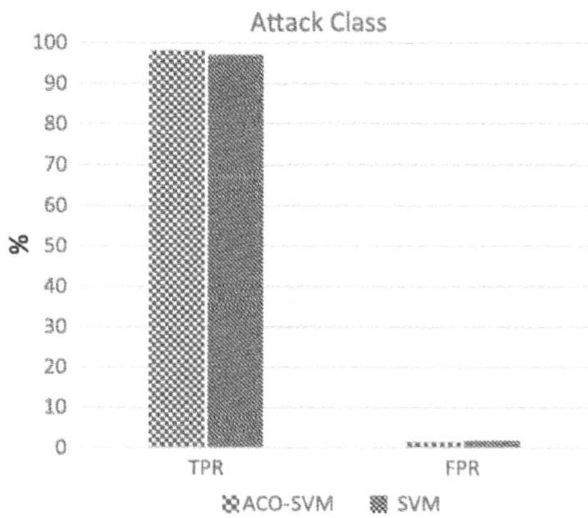

Fig. 3 Result comparison for attack class

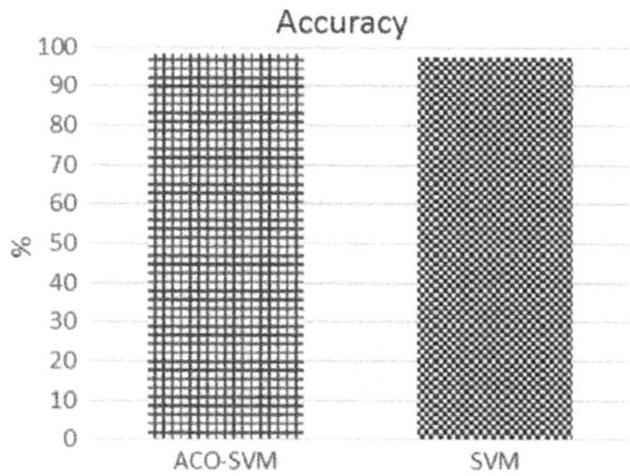

Fig. 4 Accuracy comparison for feature set

By using ACO 14 features were selected. Feature subset is than validated using SVM. Result is compared with full feature set result. Figure 1 shows the result for normal class. It can be seen that true positive rate (TPR) for reduced feature set is improve to 98.5 % from 98.2 % for whole feature set. Moreover, false positive rate (FPR) for reduced feature set is 2 % compared to full feature set i.e. 3 %. Figure 2 depicts the result for the attack class. From result it can be seen that reduced feature set gave TPR 98 % which is better than full feature set results 97 %. Accuracy for both feature set is shown in Fig. 3. Accuracy rate for SVM has been improved from 97.72 to 98.29 % when classified with the reduced feature set. (See Fig. 4)

7 Conclusion

High amount of data and irrelevant, redundant features make it difficult to build the prediction model for anomaly detection method. Features selection play vital role to build the prediction model in machine learning. In this paper feature selection method for anomaly detection has been presented. Ant colony optimization (ACO) has been proposed in the work due to its capability of utilizing previous information in the form of pheromones. SVM is used to build the anomaly detection model and the selected features are validated using this model. This work shows that robust features can be selected using ACO. Also fast and efficient detection method can be achieved using these robust features. This leads to real time detection of the intrusions in networks.

References

1. Kenkre PS, Pai A, Colaco L (2015) Real time intrusion detection and prevention system. In: Proceedings of the 3rd international conference on frontiers of intelligent computing: theory and applications (FICTA) 2014, pp 405–411

2. Bhuyan MH, Bhattacharyya DK, Kalita JK (2014) Network anomaly detection: methods, systems and tools. IEEE Commun Surv Tutorials 16(1):303–336
3. Othman ZA, Muda Z, Theng LM, Othman MR (2014) Record to record feature selection algorithm for network intrusion detection. Int J Adv Comput Technol 6(2):163
4. García-Teodoro P, Díaz-Verdejo J, Maciá-Fernández G, Vázquez E (2009) Anomaly-based network intrusion detection: techniques, systems and challenges. Comput Secur 28(1–2):18–28
5. Hämäläinen T (2014) Artificial immune system based intrusion detection: innate immunity using an unsupervised learning approach
6. Friedberg I, Skopik F, Fiedler R (2015) Cyber situational awareness through network anomaly detection: state of the art and new approaches. e i Elektrotechnik und Informationstechnik 132 (2):101–105
7. García S, Luengo J, Herrera F (2015) Feature selection. In: Data preprocessing in data mining SE—7, vol 72. Springer, pp 163–193
8. Düntsch I, Gediga G (2000) Rough set data analysis–a road to non-invasive knowledge discovery
9. Liu H, Yu L (2005) Toward integrating feature selection algorithms for classification and clustering. Knowl Data Eng IEEE Trans 17(4):491–502
10. Zhang F, Chan PPK, Biggio B, Yeung DS, Roli F (2015) Adversarial feature selection against evasion attacks
11. Pitt E, Nayak R (2007) The use of various data mining and feature selection methods in the analysis of a population survey dataset. In: Proceedings of the 2nd international workshop on Integrating artificial intelligence and data mining, vol 84, pp 83–93
12. Wang A, An N, Chen G, Li L, Alterovitz G (2015) Accelerating wrapper-based feature selection with K-nearest-neighbor. Knowl-Based Syst 83:81–91
13. Rais HM, Othman ZA, Hamdan AR (2007) Improved dynamic ant colony system (DACS) on symmetric traveling salesman problem (TSP). Int Conf Intell Adv Syst ICIAS 2007, pp 43–48
14. George A (2012) Anomaly detection based on machine learning: dimensionality reduction using PCA and classification using SVM. Int J Comput Appl Vol
15. Tsang C-H, Kwong S (2005) Multi-agent intrusion detection system in industrial network using ant colony clustering approach and unsupervised feature extraction. In: IEEE international conference on industrial technology ICIT 2005, pp 51–56
16. Gao H, Yang H, Wang X (2005) Ant colony optimization based network intrusion feature selection and detection, pp 18–21
17. Abd-Alsabour N, Randall M (2010) Feature selection for classification using an ant colony system. In: 2010 Sixth IEEE international conference on e-Science work, pp 86–91
18. Hiba Basim Alwan KRK-M (2013) Mixed variable ant colony optimization technique for feature subset selection and model selection, no 025, pp 24–31
19. Hall M, Frank E, Holmes G, Pfahringer B, Reutemann P, Witten IH (2009) The WEKA data mining software: an update. ACM SIGKDD Explor Newsl 11:10–18
20. Tavallaee M, Bagheri E, Lu W, Ghorbani A-A (2009) A detailed analysis of the KDD CUP 99 data set. In: Proceedings of the second IEEE symposium on computational intelligence for security and defence applications 2009

The manufacturer's authorised representative in the EU is Springer Nature Customer Service Centre GmbH, Europaplatz 3, 69115 Heidelberg, Germany. If you have any concerns regarding our products, please contact ProductSafety@springernature.com

Printed and bound by CPI Group (UK) Ltd, Croydon, CR0 4YY
30/09/2026
02225229-0009